W0091112

Cambridge Lower Secondary
Complete
Mathematics

Second Edition

Deborah Barton

Review team:
Bhavana Kotwal, Domnick Odipo,
Diano Pesidas, Preeti Verma

8

Great Clarendon Street, Oxford, OX2 6DP, United Kingdom

Oxford University Press is a department of the University of Oxford. It furthers the University's objective of excellence in research, scholarship, and education by publishing worldwide. Oxford is a registered trade mark of Oxford University Press in the UK and in certain other countries

British Library Cataloguing in Publication Data

Data available

978-1-38-201875-3

Digital edition: 978-1-38-201881-4

10 9 8 7 6 5

Paper used in the production of this book is a natural, recyclable product made from wood grown in sustainable forests. The manufacturing process conforms to the environmental regulations of the country of origin.

Printed in India by Manipal Technologies Limited

Acknowledgements

Artworks by Integra

p7: Gorodenkoff/Shutterstock; **p25:** Jorg Hackemann/Shutterstock; **p26:** Mark Weiss/Corbis; **p27:** Feng Yu/Shutterstock; **p37:** MalDix/Shutterstock; **p40 (T):** Rainer Lesniewski/Shutterstock; **p40 (M):** Ken Rinkel; **p40 (B):** archiZG/Shutterstock; **p41 (L):** Denys Po/Shutterstock; **p41 (R):** Gorodenkoff/Shutterstock; **p59:** insta_photos/Shutterstock; **p60:** Pixsooz/Shutterstock; **p65:** Donald Joski/Shutterstock; **p87:** Jochen Tack; **p99 (T):** insta_photos/Shutterstock; **p99 (B):** Oksana Kuzmina/Shutterstock; **p106:** Bankoo/Shutterstock; **p126:** Peter Parks/AFP IMAGES; **p138:** DAJ/amana images RF/Getty images; **p149:** Ugurhan BETIN; **p158 (T):** Pal Teravagimov/Shutterstock; **p158 (B):** Villiers Steyn/Shutterstock; **p173:** Science Photo Library; **p184:** Elena Larina/Shutterstock; **p193:** Preechar Bowonkitwanchai/Shutterstock; **p212:** Sally and Richard Greenhill/Alamy Stock Photo; **p230:** Dmitry Naumov/Shutterstock; **p249:** ID1974/Shutterstock;

Every effort has been made to contact copyright holders of material reproduced in this book. Any omissions will be rectified in subsequent printings if notice is given to the publisher.

This Student Book refers to the Cambridge Lower Secondary Mathematics (0862) Syllabus published by Cambridge Assessment International Education.

This work has been developed independently from and is not endorsed by or otherwise connected with Cambridge Assessment International Education.

The manufacturer's authorised representative in the EU for product safety is Oxford University Press España S.A. of el Parque Empresarial San Fernando de Henares, Avenida de Castilla, 2 – 28830 Madrid (www.oup.es/en).

Contents

Contents

About this book

This book follows the Cambridge Lower Secondary Mathematics curriculum in preparation for the assessments at this level, but also for study at IGCSE. It has been written by a highly experienced teacher, examiner and author.

This book is part of a series of nine books. There are three student textbooks covering stages 7, 8 and 9 and three homework books written to match the textbooks closely, as well as a teacher book for each stage.

The books are carefully balanced between all the content areas in the framework: number, algebra, geometry, measure and handling data. Some of the questions in the exercises and the investigations within the book are underpinned by thinking and working mathematically methods, providing a structure for the application of mathematical skills.

Features of the book include:

- **Objectives** – showing skills required for the Cambridge Secondary 1 framework.
- **What's the point?** – providing rationale for inclusion of topics in a real–world setting.
- **Before you start** – for each chapter to assess whether the student has the required prior knowledge.
- **Notes and worked examples** – in a clear style using accessible English and culturally appropriate material.
- **Exercises** – carefully designed to increase gradually in difficulty, providing plenty of practice of techniques.
- **Considerable variation in question style** – encouraging deeper thinking and learning, including open questions.
- **Comprehensive practice** – plenty of initial questions for practice followed by varied questions for stretch, challenge, crossover between topics and links to the real world with questions set in context.
- **Extension questions** – providing stretch and challenge for students:
  - questions with a box, e.g. **1**, provide challenge for the average student
  - questions with a filled box, e.g. **1**, provide extra challenge for high-attaining students.

- **Investigation and activity boxes** – providing extra fun, challenge and interest.
- **Full colour presentation with modern artwork** – pleasing to the eye, interesting to look at, drawing the attention of the reader.
- **Consolidation examples and exercises** – providing review material on each chapter.
- **Summary and Check out** – providing a quick review of each chapter's key points, aiding revision and enabling you to to assess progress.
- **Review exercises** – provided every five chapters with mixed questions covering all topics.
- **Bonus chapter** – the work from Chapter 16 is not in the Cambridge Secondary 1 Mathematics curriculum. It is in the Cambridge IGCSE® curriculum and is included to stretch and challenge high-attaining students.

A note from the author

If you don't already love maths as much as I do, I hope that after working through this book you will enjoy it more. Maths is more than just learning concepts and applying them. It isn't just about right and wrong answers. It is a wonderful subject full of challenges, puzzles and beautiful proofs. Studying maths develops your analysis and problem-solving skills and improves your logical thinking – all important skills in the workplace.

Be a responsible learner – if you don't understand something, ask or look it up. Be determined and courageous. Keep trying without giving up when things go wrong. No one needs to be 'bad at maths'. Anyone can improve with hard work and practice in just the same way athletes improve their skills through training. Look for challenges, then maths will never be boring.

Most of all, enjoy the book. Do the 'training', enjoy the challenges and have fun!

Deborah Barton

Integers, powers and roots

Objectives

In this chapter you will learn about:

- multiplying and dividing integers
- squares, square roots, cubes and cube roots
- laws of indices
- the order of operations
- lowest common multiple, highest common factor and prime factorization
- natural numbers and rational numbers.

What's the point?

Algorithms (methods) can be used to find how to write relatively small numbers as products of prime factors. Trying to do this for extremely large numbers would take a long time, even using the most powerful super computers. It is easy to multiply together two large prime numbers to get another number. Reversing the process is much harder. A process that is easy to do but hard to undo is very useful for encryption and cyber security.

Before you start

You should know ...

1 a Subtracting a negative number is the same as adding a positive number.
For example:
$5 - {}^{-}2 = 5 + 2 = 7$

b Adding a negative number is the same as subtracting a positive number.
For example:
${}^{-}3 + {}^{-}5 = {}^{-}3 - 5 = {}^{-}8$

Check in

1 Work out:
a ${}^{-}7 + 4$
b ${}^{-}3 + {}^{-}7$
c ${}^{-}3 - {}^{-}9$
d $15 + {}^{-}4$

2 The order of operations (BIDMAS).

For example:

$(2 \times 5) - 3 \times 2^2$ Brackets first

$= 10 - 3 \times 2^2$ then Indices

$= 10 - 3 \times 4$ then Multiplication

$= 10 - 12$ then Subtraction.

$= {}^-2$

3 How to find highest common factors (HCFs) and lowest common multiples (LCMs).

For example:
Factors of 20 are: **1**, 2, 4, **5**, 10, 20
Factors of 15 are: **1**, 3, **5**, 15
Common factors of 20 and 15 are: **1** and **5**, the HCF is 5.

Multiples of 20 are: 20, 40, **60**, 80, 100, **120**, …
Multiples of 15 are: 15, 30, 45, **60**, 75, 90, 105, **120**, …
Common multiples of 20 and 15 are: **60**, **120**, … the LCM is 60.

2 Work out:

a $4 \times 5^2 - (6 \times 7)$
b $6 \times 5 - 12 \div (5 - 2)$
c $(3^3 + 4) - 2^2 \times 5$
d $\frac{4^2 + 5}{7}$

3 Find the HCF and LCM of:

a   10, 15
b   18, 24
c   12, 30
d   15, 40

1.1 Multiplying and dividing negative integers

Multiplying negative integers

You can show your times tables on a graph.
The diagram shows a graph of the 2-times table.
Notice all the points lie in a straight line.

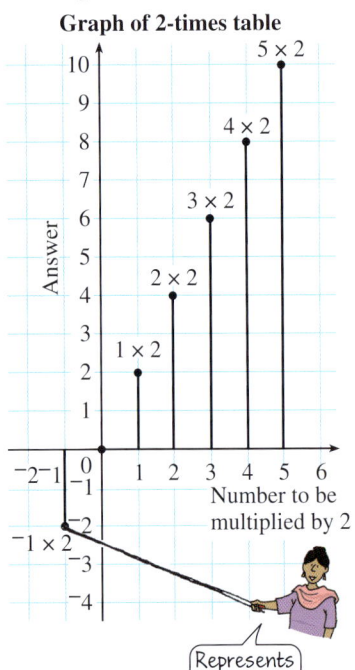

Graph of 2-times table

Represents
$^-1 \times 2$

Exercise 1A

1 a Draw a graph of the 3-times table for multiplying numbers from $^-3$ to 4. The answer axis will need to show numbers from $^-9$ to 12.
   b Do all the answers lie on a straight line?
   c If you extended the line, what would be the answers to $^-4 \times 3$ and $^-5 \times 3$?

2 a Draw a graph of the 4-times table for multiplying numbers from $^-4$ to 3.
   b If you extended the line, what answers would you find for $^-5 \times 4$ and $^-7 \times 4$?

3 a Write down the answers to:
     i $3 \times {}^-2$ **ii** $4 \times {}^-2$ **iii** $5 \times {}^-2$

b Copy the graph. Complete it with your answers to part **a**.

Graph of ⁻2-times table

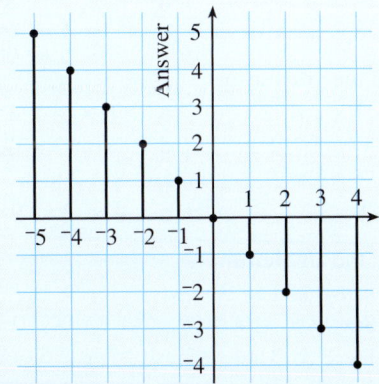

Number to be multiplied by ⁻2

c Lay a ruler along the points. What answer does the graph give for ⁻1 × ⁻2?

d Extend the graph to find the answer to ⁻2 × ⁻2.

4 What does this graph show?

5 Use the graph in question **4** to write down the answer to:

a 4 × ⁻1

b ⁻4 × ⁻1

c ⁻2 × ⁻1

6 Look at the graph of the ⁻3-times table.

a What is 0 × ⁻3?

b What answers does the graph suggest for ⁻1 × ⁻3, ⁻2 × ⁻3, ⁻3 × ⁻3?

Graph of ⁻3-times table

⁻2 × ⁻3 = 6?

Number to be multiplied by ⁻3

7 Copy this table. Match up the multiplications and answers using arrows.

⁻3 × ⁻2	6
3 × ⁻2	6
⁻3 × 2	6
3 × 2	⁻6

8 Copy and complete the multiplication table. Use your graphs of the ⁻1-times, ⁻2-times and ⁻3-times tables to help you. Follow the number patterns up the columns to complete the upper left-hand corner of the table, where negative numbers are multiplied together.

×	⁻5	⁻4	⁻3	⁻2	⁻1	0	1	2	3	4	5
⁻5											
⁻4											
⁻3											
⁻2											
⁻1											
0											
1											
2					⁻2					8	
3								6			
4											
5											

3 × 2 = 6

Remember these rules.
- When you multiply a positive number by a negative number the answer is negative (mix means minus).
- When you multiply two negative numbers together the answer is positive (two minuses make a plus).

Work out:
a $3 \times {}^-5$ **b** $^-3 \times {}^-5$

a positive × negative = negative:
$3 \times {}^-5 = {}^-15$

b negative × negative = positive:
$^-3 \times {}^-5 = 15$

The same rules hold when you multiply more than two numbers.

Find $^-3 \times 4 \times {}^-2$

$$^-3 \times 4 \times {}^-2 = ({}^-3 \times 4) \times {}^-2$$
$$= {}^-12 \times {}^-2$$
$$= 24$$
$$\text{or } {}^-3 \times 4 \times {}^-2 = {}^-3 \times (4 \times {}^-2)$$
$$= {}^-3 \times {}^-8$$
$$= 24$$

Exercise 1B

1 Write down the answers.
     a 4×5 **b** $^-4 \times 5$ **c** $4 \times {}^-5$
     d $^-4 \times {}^-5$ **e** $2 \times {}^-5$ **f** $^-4 \times {}^-3$
     g 2×4 **h** $^-2 \times {}^-4$ **i** $^-2 \times 4$

2 Write down the answers.
     a $^-6 \times {}^-8$ **b** $6 \times {}^-8$
     c $^-2 \times 5$ **d** $^-7 \times {}^-5$
     e 14×5 **f** $^-5 \times {}^-14$
     g $100 \times {}^-1$ **h** $37 \times {}^-2$
     i $^-1 \times {}^-1$ **j** $^-3 \times {}^-3$
     k $^-20 \times {}^-20$ **l** $20 \times {}^-20$

3 Use the multiplication table in question **8** of Exercise 1A to help find the missing numbers.
     a $3 \times \square = 12$ **b** $3 \times \square = {}^-12$
     c $^-3 \times \square = {}^-12$ **d** $^-3 \times \square = 12$
     e $5 \times \square = {}^-10$ **f** $\square \times {}^-2 = 10$
     g $\square \times {}^-4 = 8$ **h** $\square \times {}^-4 = {}^-8$
     i $^-5 \times \square = 15$ **j** $\square \times 4 = {}^-16$

4 Find the answer to:
     a $^-1 \times 3 \times {}^-2$ **b** $^-6 \times {}^-4 \times 2$
     c $4 \times {}^-1 \times 2$ **d** $^-8 \times {}^-3 \times {}^-2$
     e $^-5 \times 2 \times 2$ **f** $8 \times 2 \times {}^-1$
     g $^-8 \times {}^-1 \times {}^-3$ **h** $4 \times {}^-3 \times {}^-9$

5 Find the answer to:
     a $6 \times 2 \times {}^-1 \times {}^-3$
     b $8 \times {}^-3 \times {}^-2 \times 2$
     c $1 \times {}^-2 \times {}^-3 \times 4$
     d $^-1 \times {}^-2 \times {}^-4 \times {}^-8$
     e $5 \times {}^-4 \times {}^-2 \times {}^-3$
     f $^-1 \times {}^-4 \times 4 \times 5$

6 Find the missing number.

   a $3 \times \square \times {}^-2 = 12$

   b $4 \times \square \times 4 = 64$

   c ${}^-3 \times \square \times {}^-3 = {}^-27$

   d ${}^-1 \times \square \times 8 = {}^-16$

   e $2 \times \square \times {}^-6 = 72$

   f ${}^-3 \times 4 \times \square = 72$

   g $\square \times {}^-3 \times {}^-4 = 72$

   h ${}^-9 \times \square \times {}^-8 = {}^-72$

7 Find the answer to:

   a ${}^-1 \times {}^-1$

   b ${}^-1 \times {}^-1 \times {}^-1$

   c ${}^-1 \times {}^-1 \times {}^-1 \times {}^-1$

   d ${}^-1 \times {}^-1 \times {}^-1 \times {}^-1 \times {}^-1$

8 Is the answer positive or negative, when you multiply together:

   a three negative numbers

   b four negative numbers

   c five negative numbers?

9 Find a rule that tells you whether the answer will be positive or negative when you multiply several negative numbers together.

Dividing negative integers

Multiplication and division of positive numbers are connected like this:

$5 \times 4 = 20$ so $20 \div 4 = 5$

For negative numbers:

${}^-5 \times {}^-4 = 20$ so $20 \div {}^-4 = {}^-5$

Exercise 1C

1 Work out:

   a ${}^-4 \times {}^-3$

   b ${}^-5 \times {}^-2$

   c ${}^-10 \times {}^-6$

   d $7 \times {}^-8$

   e ${}^-9 \times 5$

2 Use your answers from question **1** to work out:

   a $12 \div {}^-4$

   b $10 \div {}^-2$

   c $60 \div {}^-6$

   d ${}^-56 \div {}^-7$

   e $\dfrac{-45}{-9}$

3 How can you tell whether the answer to a division will be a positive or a negative number?

4 Copy this table. Match each division with its answer using an arrow.

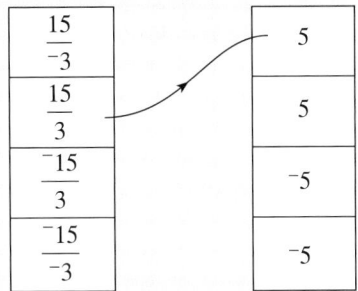

5 Multiplication and division of positive and negative numbers follow a pattern. Copy and complete these tables.

	Second number	
×	Positive	Negative
Positive	Positive	
Negative		

First number (row label for left column)

	Second number	
÷	Positive	Negative
Positive		
Negative		

First number (row label for left column)

6 Find the value of:

   a $\dfrac{20}{-2}$ **b** $\dfrac{8}{-4}$ **c** $\dfrac{-6}{3}$

   d $\dfrac{-100}{50}$ **e** $\dfrac{-21}{3}$ **f** $\dfrac{-21}{7}$

   g $\dfrac{-16}{-4}$ **h** $\dfrac{-4}{-4}$ **I** $\dfrac{-1}{1}$

7 Write these calculations in two lists: one list with the heading 'correct', one list with the heading 'not correct'. Do not use a calculator for any of the questions.

${}^-6 \times {}^-7 = {}^-42$

${}^-6 \div {}^-3 = 2$

$({}^-5)^2 = {}^-25$

${}^-5^2 = {}^-25$

${}^-6 \times {}^-2 \div {}^-3 = {}^-4$

${}^-142 \times {}^-35 = {}^-4970$

8 Work out:
  a $(^-6)^2$
  b $^-4^2$
  c $^-3 \times ^-8 \div ^-6$

9 Find the answer to:
  a $\dfrac{^-4 \times ^-9}{^-3}$
  b $\dfrac{^-21 \times ^-2}{^-7}$
  c $\dfrac{^-70}{10} + \dfrac{^-35}{^-7}$
  d $\dfrac{6}{^-2} - \dfrac{^-16}{4}$

10 a What is the square of $^-3$?
  b Give two possible values of $\sqrt{9}$.
  c Can you find a number that has the square $^-9$? Explain your answer.

1.2 Squares, cubes, roots and indices

Squares and square roots

Square numbers come from squaring integers (multiplying a whole number by itself). 1, 4 and 9 are the first three square numbers, which come from 1×1, 2×2 and 3×3. There is a short way to write these using **indices**. For example, the fourth square number is $4 \times 4 = 16$ and this can be written as $4^2 = 16$.

Any number can be squared. Sometimes you may want to use your calculator.

> **Example 3**
>
> Find the square of 2.7
>
> ....
>
> $2.7 \times 2.7 = 2.7^2$
> Key into your calculator $\boxed{2}\ \boxed{\cdot}\ \boxed{7}\ \boxed{x^2}\ \boxed{=}$
> $2.7^2 = 7.29$

Note that 7.29 is not a square number because 2.7 is not an integer.

Finding the **square root** of a number is the inverse of squaring a number.

$\sqrt{49}$ means $? \times ? = 49$ (where both numbers ? are the same)

$7 \times 7 = 49$ so $\sqrt{49} = 7$

Cubes and cube roots

The first three cube numbers are 1, 8 and 27, which come from $1 \times 1 \times 1$, $2 \times 2 \times 2$ and $3 \times 3 \times 3$.

The fifth cube number is $5^3 = 5 \times 5 \times 5 = 125$. Finding the cube root of a number is the inverse of cubing the number. We use a symbol similar to the square root symbol with a small raised 3. So, $\sqrt[3]{64}$ means $? \times ? \times ? = 64$ (where all the numbers ? are the same). $4 \times 4 \times 4 = 64$ so $\sqrt[3]{64} = 4$.

Many calculators work differently. Check that you know how to use the square, cube, square root and cube root functions on your calculator by using it to work out these expressions.

$$0.3^2 \qquad 1.8^3 \qquad \sqrt{1.44} \qquad \sqrt[3]{0.343}$$

You should have found:

$$0.3^2 = 0.09 \qquad\qquad 1.8^3 = 5.832$$
$$\sqrt{1.44} = 1.2 \qquad\qquad \sqrt[3]{0.343} = 0.7$$

> **Example 4**
>
> a Find the area of a square with side length 15 cm.
>
>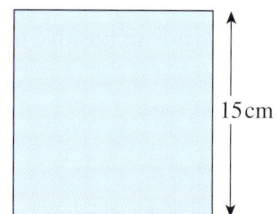
>
> 15 cm
>
> b Find the side length of a cube with volume $1.728\ \text{m}^3$.
>
>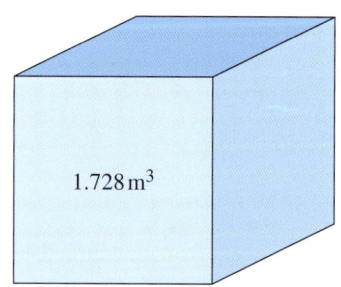
>
> $1.728\ \text{m}^3$
>
> ....
>
> a Area of square is $15^2 = 225\ \text{cm}^2$
> b Side length of the cube is $\sqrt[3]{1.728} = 1.2\ \text{m}$

From your work on negative numbers you know that $^-3 \times ^-3 = 9$ as well as $3 \times 3 = 9$. So if you want to work out the square root of 9 there are two possible answers: $\pm\sqrt{9} = 3$ or $^-3$. If you are working with side lengths of squares it does not make sense to write down the negative square root as you cannot have a negative side length. In some cases, though, the negative square root is very important – you will learn about this in your Cambridge IGCSE® course. If a positive and negative square root are needed, $\pm\sqrt{\ }$ is used.

From your work on negative numbers you know that $^-2 \times ^-2 \times ^-2 = ^-8$. So if you want to work out the cube root of $^-8$, the answer is $^-2$. Note that you cannot find the square root of a negative number on your calculator – you will get an error message if you try.

Exercise 1D

1 Work out these. (Where there are two possible answers, write both.)

a 9^2 **b** $\pm\sqrt{121}$

c 14^2 **d** $\pm\sqrt{0}$

e 13^2 **f** $\pm\sqrt{64}$

2 Work out these. (Where there are two possible answers, write both.)

a 3.6^2 **b** $\pm\sqrt{16.81}$

c 0.32^2 **d** $\pm\sqrt{0.16}$

e 11.4^2 **f** $\pm\sqrt{0.64}$

3 Emma says that the answer to $(^-10)^2$ and $^-10^2$ are not the same. Explain how you know she is correct.

4 Work out:

a $(^-9)^2$ **b** $^-3^2$

c $(^-0.5)^2$ **d** $^-1.5^2$

5 Work out:

a 2^3 **b** $(^-3)^3$ **c** $\sqrt[3]{1000}$

d $(^-4)^3$ **e** $\sqrt[3]{^-1}$ **f** $\sqrt[3]{^-125}$

6 Work out:

a 1.5^3 **b** $(^-3.2)^3$

c $\sqrt[3]{^-4.913}$ **d** $\sqrt[3]{^-0.001}$

7 Find the square or square root. Round the answers to 1 decimal place. In parts **b** and **d**, give positive and negative roots.

a 1.13^2 **b** $\pm\sqrt{6.4}$

c 3.1^2 **d** $\pm\sqrt{2.5}$

8 What is the side length of a square with area
a $7.84\,\text{cm}^2$ **b** $60.84\,\text{m}^2$ **c** $156.25\,\text{mm}^2$?

9 What is the area of a square with side length
a $3.2\,\text{m}$ **b** $6.9\,\text{mm}$ **c** $1.7\,\text{cm}$?

10 Find the side length of a cube with volume $2.744\,\text{m}^3$.

11 Magnus says $\sqrt[3]{^-729}$ is 9. Without using a calculator, explain how you know this is incorrect.

12 Complete this table. The first row has been done for you.

Question	Estimate without using a calculator	Answer correct to 2 d.p.
$-\sqrt{14.2}$	The answer is between the integers $^-3$ and $^-4$ (because $-\sqrt{9} = ^-3$ and $-\sqrt{16} = ^-4$ and 14.2 is between 9 and 16)	$^-3.77$
$-\sqrt{26.1}$	The answer is between the integers … and …	
$\sqrt[3]{^-26}$	The answer is between the integers … and …	
$\sqrt[3]{^-12.3}$	The answer is between the integers … and …	

13 To work out the square root of a fraction you find the square roots of the top and bottom separately:

$$\sqrt{\frac{9}{25}} = \frac{\sqrt{9}}{\sqrt{25}} = \frac{3}{5}$$

Using this method, work out:

a $\sqrt{\dfrac{16}{49}}$ **b** $\sqrt{\dfrac{81}{100}}$ **c** $\sqrt{\dfrac{25}{64}}$

14 To work out $\sqrt{0.25}$ without a calculator you can write the decimal as a fraction. (The numbers in the fraction should be square numbers.)

So $\sqrt{0.25} = \sqrt{\dfrac{25}{100}}$, then find the square roots of the top and bottom separately:

$$\sqrt{\frac{25}{100}} = \frac{\sqrt{25}}{\sqrt{100}} = \frac{5}{10} = 0.5$$

Using this method, work out:

a $\sqrt{0.36}$ **b** $\sqrt{0.81}$ **c** $\sqrt{1.44}$

d $\sqrt[3]{0.064}$ **e** $\sqrt[3]{^-0.125}$

15 Can you think of a way of working out $\sqrt{\dfrac{32}{50}}$ without a calculator?

Investigation

A $3 \times 3 \times 3$ cube made up of 27 smaller cubes is painted green on its outside.

How many of the smaller cubes are unpainted?

How many of the smaller cubes have 1 face painted? How many have 2 faces painted? How many have 3 faces painted?

What about a $4 \times 4 \times 4$ cube made up of 64 smaller cubes?

Copy and complete the table.

Cube	Number of faces painted on smaller cubes			
	None	**1**	**2**	**3**
$2 \times 2 \times 2$				
$3 \times 3 \times 3$				
$4 \times 4 \times 4$				
$5 \times 5 \times 5$				

Can you spot any rules?

What about a $13 \times 13 \times 13$ cube?

1.3 Index notation

You have learned that five cubed or $5 \times 5 \times 5$ can be written as 5^3 for short. This is called **index notation**. 3 is the **power** (also called the **index**). 5 is the **base** number. We can do this with numbers other than squared or cubed numbers.

$2 \times 2 \times 2 \times 2 = 2^4$ — We say 'two to the power four'.

$8 \times 8 \times 8 \times 8 \times 8 \times 8 \times 8 = 8^7$ — We say 'eight to the power seven'.

2^4 can be worked out in your head or with a calculator using the button $\boxed{x^y}$, $\boxed{x^\square}$ or $\boxed{y^x}$ (some calculators require you to press the button $\boxed{\wedge}$ to find the power).

Find out what you need to press on your calculator. Check that your calculator gives $2^4 = 16$.

Exercise 1E

1 Write each of these using index notation.
   a $3 \times 3 \times 3 \times 3 \times 3$
   b $7 \times 7 \times 7 \times 7 \times 7 \times 7 \times 7$
   c $5 \times 5 \times 5 \times 5 \times 5 \times 5 \times 5 \times 5$
   d $^-2 \times ^-2 \times ^-2 \times ^-2$

2 Copy and complete (without a calculator if you can):
   a $2^5 = 2 \times 2 \times \ldots = 32$
   b $2^6 = 2 \times 2 \times \ldots = \square$
   c $4^4 = 4 \times 4 \times \ldots = \square$
   d $5^4 = 5 \times 5 \times \ldots = \square$
   e $1024 = 2 \times 2 \times \ldots = 2^\square$
   f $100\,000 = 10 \times 10 \times \ldots = 10^\square$

3 Using the power key on your calculator, work out:
   a 3.2^4 b 1.3^5
   c 2.2^6 d 1.1^5
   e $(^-7)^5$

4 Work out:
   a $(^-1)^2$ b $(^-1)^3$
   c $(^-1)^4$ d $(^-1)^5$
   e When the base is $^-1$, explain how you know whether the answer will be positive or negative for the power you use.
   f Use this rule to predict the answers to $(^-1)^{20}$ and $(^-1)^{37}$. Check these on a calculator. Were you correct?

5 Using your calculator, try working out 99^{99}. Do you get an error message? Do you get an error message for 99^{33}? (See your teacher if you don't understand the answer your calculator gives.) What is the highest power you can work out on your calculator for the base number 99?

6 What is the highest power you can work out on your calculator for the base number 9999?

Notice that

$4^3 \times 4^2 = 4 \times 4 \times 4 \times 4 \times 4 = 4^5$

That is, $4^3 \times 4^2 = 4^{3+2} = 4^5$

● As long as the two numbers are powers of the same number you can multiply them by adding their indices. Using symbols,

$a^m \times a^n = a^{m+n}$

Example 5

Work out:
a $2^3 \times 2^4$ **b** $2^2 \times 2^5 \times 2^7$
c 7×7^4

....

a $2^3 \times 2^4 = 2^{3+4} = 2^7$
b $2^2 \times 2^5 \times 2^7 = 2^{2+5+7} = 2^{14}$
c $7 \times 7^4 = 7^5$

Notice also that
$$4^3 \div 4^2 = \frac{4 \times \cancel{4} \times \cancel{4}}{\cancel{4} \times \cancel{4}} = 4^1$$
That is, $4^3 \div 4^2 = 4^{3-2} = 4^1$

- When two numbers are powers of the same number, you can divide them by subtracting their indices. Using symbols,
  $$a^m \div a^n = a^{m-n}$$

Example 6

Work out:
a $5^7 \div 5^3$ **b** $8^{12} \div 8^{10}$

....

a $5^7 \div 5^3 = 5^{7-3} = 5^4$
b $8^{12} \div 8^{10} = 8^{12-10} = 8^2$

What does 5^0 mean?
$5^2 \div 5^2 = 25 \div 25 = 1$, but
$5^2 \div 5^2 = 5^{2-2} = 5^0$
So $5^0 = 1$

Using symbols,
$$a^0 = 1$$

Exercise 1F

1 Copy and complete:
a $3^4 = 3 \times 3 \times \ldots = 81$
b $5^4 = 5 \times 5 \times \ldots = \square$
c $2^7 = 2 \times 2 \times \ldots = \square$
d $25 = 5^\square = 5 \times \ldots$
e $49 = 7^\square = 7 \times \ldots$
f $8 = 2^\square = 2 \times \ldots$
g $81 = 3^\square = 3 \times \ldots$
h $10\,000 = 10 \times 10 \times \ldots = 10^\square$
i $1\,000\,000 = 10 \times 10 \times \ldots = 10^\square$

2 Copy and complete:
a $3^2 \times 3^4 = 3 \times 3 \times \ldots = 3^\square$
b $2^3 \times 2^5 = 2 \times 2 \times \ldots = 2^\square$
c $7^5 \times 7 = 7 \times 7 \times \ldots = 7^\square$
d $5^6 \times 5^4 = 5 \times 5 \times \ldots = 5^\square$

3 Stephan thinks $2^3 \times 2^4 = 4^7$. Is Stephan correct? Explain how you know.

4 Simplify:
a $6^2 \times 6^3 \times 6^5$ **b** $7 \times 7^{10} \times 7^{12}$
c $3^2 \times 3^{10} \times 3^5$ **d** $10 \times 10 \times 10^3$

5 Simplify these, if possible, leaving the answer in index form. If it is not possible, explain why.
a $2^6 \times 2^7$ **b** $2^3 \times 3^2$
c $3^2 \times 4^2 \times 5^2$ **d** $4^8 \times 4^2 \times 4$

6 Copy and complete:
a $\dfrac{3^4}{3^2} = \dfrac{3 \times 3 \times \ldots}{3 \times \ldots} = 3^\square$
b $\dfrac{5^7}{5^3} = \dfrac{5 \times 5 \times \ldots}{5 \times \ldots} = 5^\square$
c $\dfrac{7^{10}}{7^4} = \dfrac{7 \times 7 \times \ldots}{7 \times \ldots} = 7^\square$

7 Rashmi thinks $2^{10} \div 2^2 = 2^5$. Explain why Rashmi is wrong.

8 Simplify:
a $\dfrac{2^6}{2^3}$ **b** $\dfrac{3^7}{3^4}$ **c** $\dfrac{4^8}{4^3}$
d $\dfrac{7^{10}}{7^5}$ **e** $\dfrac{9^7}{9}$ **f** $\dfrac{5^{12}}{5^8}$

9 Copy and complete:
a $2^3 \div 2^3 = 2^\square = 1$
b $3^5 \div 3^5 = 3^\square = \square$
c $7^7 \div 7^7 = 7^\square = \square$
d $9^{10} \div 9^{10} = 9^\square = \square$

10 Write down the value of:
a 5^0 **b** 7^0 **c** $(^-2)^0$
d 8^0 **e** $\left(\dfrac{1}{2}\right)^0$ **f** m^0

11 Simplify, leaving your answer in index form:
a $6^3 \div 6^2$ **b** $5^7 \div 5^4$
c $12^6 \div 12^3$ **d** $7^5 \div 7$
e $20^9 \div 20^4$ **f** $q^2 \div q$
g $b^6 \div b^5$ **h** $y^4 \div y^m$
i $8p^7 \div 2p^3$ **j** $6x^6 \div 2x^2$
k $10m^{10} \div 2m^2$

1.4 Prime factorization to find the HCF and LCM

Prime numbers

- A number with exactly two different factors is called a **prime number**.

Example 7

Find two prime numbers.

....

23 is a prime number.
It has exactly two factors: 1 and 23.
2 is a prime number.
It has exactly two factors: 1 and 2.

- A number that has more than two different factors is called a **composite number**.

Example 8

Find a composite number.

....

12 is a composite number.
It has six factors: 1, 2, 3, 4, 6, 12.

- The number 1 is special.

1 is not a prime number (as it doesn't have two factors).
1 is not a composite number (as it only has one factor).

Exercise 1G

1 a i What are the factors of 35?
   ii Is 35 a prime number?
  b i What are the factors of 37?
   ii Is 37 a prime number?

2 a Which numbers between 20 and 50 have just two factors?
  b Write down all the prime numbers from 1 to 50.

3 Copy and complete these sentences.
  a A number that has only two different factors, itself and 1, is called a . . . number.
  b A number that has three or more different factors is called a . . . number.

4 The following grid shows all the numbers from 1 to 100. Make a larger copy of the grid on squared paper (or ask your teacher for a copy of this grid from the Teacher Handbook).

1	2	3	4	5	6	7	8	9	10
11	12	13	14	15	16	17	18	19	20
21	22	23	24	25	26	27	28	29	30
31	32	33	34	35	36	37	38	39	40
41	42	43	44	45	46	47	48	49	50
51	52	53	54	55	56	57	58	59	60
61	62	63	64	65	66	67	68	69	70
71	72	73	74	75	76	77	78	79	80
81	82	83	84	85	86	87	88	89	90
91	92	93	94	95	96	97	98	99	100

The **sieve of Eratosthenes** is a simple, ancient method for finding all prime numbers up to a specified integer. Instructions for how to use it are below.

  a The first number in the grid is 1, cross it out as you know it is not prime.
  b The next number in the grid is 2, circle it as you know it is prime.
  c Cross out all of the rest of the multiples of 2 as these are composite numbers.
  d The next number in the grid is 3, circle it as you know it is prime.
  e Cross out the rest of the multiples of 3 as these are composite numbers.
  f The next number (not already crossed out) is 5, circle it as you know it is prime.
  g Cross out the rest of the multiples of 5 as these are composite numbers.
  h The next number (not already crossed out) is 7, circle it as you know it is prime.
  i Cross out the rest of the multiples of 7 as these are composite numbers.
  j You should find that the rest of the numbers in your grid, that are not crossed out, can be circled. They are prime. (You could continue this method by circling 11 and crossing out all multiples of 11, etc. and extending your grid beyond 100 if you want.)

5 List all the prime numbers between 1 and 100.

6 a Look at your grid for question **4**. How many times does a pair of primes occur together? Write down these primes.

  b How many times do three primes occur together? Can you explain why?

  c Apart from 2, what digits do all other primes end in?

7 3 and 5 differ by 2 and are both prime numbers. A pair of prime numbers with a difference of 2 like this are called twin primes.

  a What is the next pair of twin primes?

  b How many twin primes can you find between 0 and 100?

8 a Is 613 a prime number? How did you find out?

  b Find out whether 4999 is a prime number.

  c What about 30 031?

9 What is the biggest prime number currently known? (You may use the internet to find out.)

Investigation

The number 8 can be written as the sum of two prime numbers, 3 and 5:

$3 + 5 = 8$

The number 9 is the sum of the primes 2 and 7:

$2 + 7 = 9$

a Can all the numbers between 5 and 20 be written as the sum of two primes?

b Copy and complete this table where possible.

Number	Two primes equal to number
5	2 + 3
6	3 + 3
7	
8	

c Can any number be written as the sum of two primes?
Which ones can? Which ones cannot? Are there any rules?

Prime factorization

Prime numbers can be found using Eratosthenes' sieve (see Exercise 1G, question **4**). The prime numbers under 30 are 2, 3, 5, 7, 11, 13, 17, 19, 23 and 29.

Every number can be written as a product of its prime factors. The two most common ways to do this are by using factor trees and division by primes.

Example 9

Write 126 as a product of its prime factors using:
a the factor tree method
b the division by primes method.

a

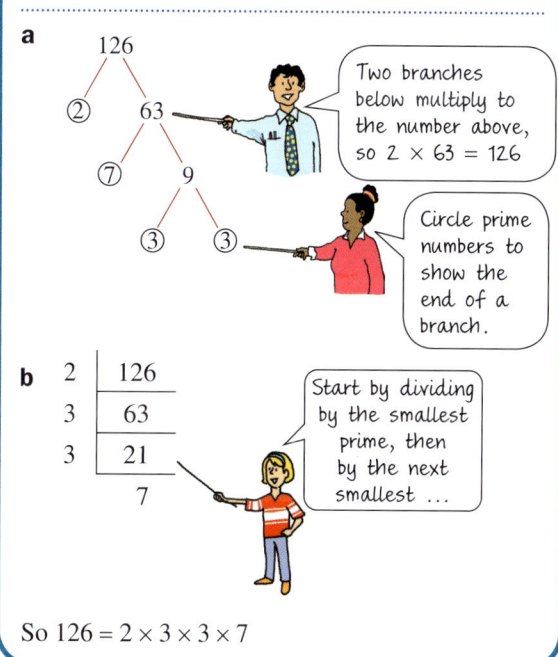

Two branches below multiply to the number above, so 2 × 63 = 126

Circle prime numbers to show the end of a branch.

b

2	126
3	63
3	21
	7

Start by dividing by the smallest prime, then by the next smallest ...

So $126 = 2 \times 3 \times 3 \times 7$

This can also be written in index form. This is useful if there are a lot of repeated prime factors.

$126 = 2 \times 3 \times 3 \times 7$; in index form this is $2 \times 3^2 \times 7$

Writing a number as a product of its primes is useful for working out harder HCF and LCM problems than those you did in Student Book 7, where you used listing to work them out.

Finding the HCF and LCM

Example 10

Find: **a** the HCF (highest common factor)
     b the LCM (lowest common multiple) of
     1260 and 600

....

a Using factor trees write 1260 and 600 as a
    product of their primes.
    $1260 = 2 \times 2 \times 3 \times 3 \times 5 \times 7$
    $600 = 2 \times 2 \times 2 \times 3 \times 5 \times 5$

    $\text{HCF} = 2 \times 2 \times 3 \times 5 = 60$

Use the common factors.

b Using factor trees write 1260 and 600 as a
    product of their primes using index notation.

    $1260 = 2^2 \times 3^2 \times 5 \times 7$
    $600 = 2^3 \times 3 \times 5^2$

    $\text{LCM} = 2^3 \times 3^2 \times 5^2 \times 7 = 12\,600$

Use the highest power of each common factor.

Example 10 can be done in a different way using repeated division. The working is shown in this table:

	1260	600
2	630	300
2	315	150
3	105	50
5	21	10

$1260 \div 2 = 630$ and $600 \div 2 = 300$

21 and 10 have no common prime factors so stop here

Divide both numbers by a prime number they have as a common factor then the next common prime factor (shown by the blue numbers). Repeat this until the numbers in the two right hand columns (the black numbers) have no more prime factors in common.

    $\text{HCF} = 2 \times 2 \times 3 \times 5 = 60$

To find the LCM use these numbers in blue from the table:

	1260	600
2	630	300
2	315	150
3	105	50
5	21	10

$\text{LCM} = 2 \times 2 \times 3 \times 5 \times 21 \times 10 = 12\,600$

Exercise 1H

1 Find the HCF and LCM of 25 and 30.

2 Explain why each of these statements is **not** correct.
  a 7 and 11 are both prime numbers so they don't have an HCF.
  b The HCF of 60 and 90 is 10.
  c The LCM of 6 and 12 is 6.

3 Find the HCF of these pairs.
  a 468, 324   **b** 540, 108
  c 450, 990   **d** 330, 910

4 Find the LCM of these pairs.
  a 108, 360   **b** 34, 39
  c 450, 180   **d** 150, 490

5 Find the HCF and LCM of:
  a 18, 20 and 30
  b 9, 16 and 12
  c 8, 18 and 50

6 Three strings of different lengths, 240 cm, 318 cm and 426 cm, are to be cut into equal lengths. What is the greatest possible length of each piece?

7 Two lighthouses flash their lights every 20 seconds and 30 seconds respectively. They flashed together at 7pm. When will they next flash together?

8 A man has a garden measuring 84 m by 56 m. He divides it into the minimum number of square plots. What is the length of each square plot?

9 Neha said she had a different way of working out LCMs. Here is Neha's method for finding the LCM of 90 and 300.

Step 1: Find the HCF first using the product of primes.
$90 = 2 \times 3 \times 3 \times 5$
$300 = 2 \times 2 \times 3 \times 5 \times 5$

Common factors are one each of 2, 3 and 5.
So HCF $= 2 \times 3 \times 5 = 30$

Step 2: LCM = HCF × numbers left over.
LCM $= 30 \times 2 \times 3 \times 5 = 900$

a Check Neha's method by using it to work out the answers to question **4** again.

b Rani said Neha's method wouldn't work if you had to find the LCM of three numbers. Is she correct? Try using Neha's method for question **5**.

c Are there any occasions when Neha's method will work with three numbers?

10 Use the digits 0, 1, 5, 6, 7 and 8 to make two 3-digit numbers with an HCF of 45 and an LCM under 3000.

1.5 Order of operations

You have already learned that the order in which you do operations is BIDMAS.

When following the rules of BIDMAS and calculating mentally, it is useful to jot down each step in the calculation rather than trying to do it all in one go.

When you study IGCSE® you will learn that the square root of a number and the cube root of a number can be written using indices. For example, $\sqrt{144} = 144^{\frac{1}{2}}$. So square roots and cube roots should be worked out at the same time as indices.

The order of operations, BIDMAS, is therefore:

Brackets first
then Indices
(including square roots and cube roots)
then Division and Multiplication (working left to right)
then Addition and Subtraction (working left to right).

Example 11

Work out: $\qquad 3^2 \times 8 \div (5-3)^2$

$\begin{aligned}
&3^2 \times 8 \div (5-3)^2 && \text{Brackets}\\
&= 3^2 \times 8 \div 2^2 && \text{Indices}\\
&= 9 \times 8 \div 4 && \text{Multiplication}\\
&= 72 \div 4 && \text{Division}\\
&= 18
\end{aligned}$

Example 12

Work out: $\qquad 5 \times 2^2 - 15 \div \sqrt{25}$

$\begin{aligned}
&5 \times 2^2 - 15 \div \sqrt{25} && \text{Indices and square root}\\
&= 5 \times 4 - 15 \div 5 && \text{Multiplication}\\
&= 20 - 15 \div 5 && \text{Division}\\
&= 20 - 3 && \text{Subtraction}\\
&= 17
\end{aligned}$

Exercise 1I

1 Jess wrote this working in her homework.

$\begin{aligned}
&6^2 - 5 \times 2 + \sqrt{9} && \text{Use BIDMAS}\\
& && \text{Indices first}\\
&= 36 - 5 \times 2 + 3 && \text{then Multiply}\\
&= 36 - 10 + 3 && \text{then Add}\\
&= 36 - 13 && \text{finally, Subtract.}\\
&= 23
\end{aligned}$

What mistake has Jess made? What is the correct answer?

2 Work out:

a $5^2 - 3 \times \sqrt{36}$ **b** $8 - 1.5 \times 2 + \sqrt[3]{64}$

c $2^3 \times \sqrt{64} \div (3+1)^2$ **d** $20 - \frac{15}{3} \times 27$

e $\sqrt{100} - 2 \div 2^2 + 7$

3 Work out:

a $(^-8 - ^-3) \times (^-3 + ^-2)$

b $(^-27 - 33) \div (^-25 + 19)$

c $5 + (^-7 - ^-4) \times 2^2$

4 Prahsant and Katy each try to work out the answer to $\sqrt{3^2 + 4^2}$. Prashant writes this working.

$\begin{aligned}
&\sqrt{3^2 + 4^2}\\
&= \sqrt{3^2} + \sqrt{4^2}\\
&= \sqrt{9} + \sqrt{16}\\
&= 3 + 4\\
&= 7
\end{aligned}$

Katy writes this working.

$\begin{aligned}
&\sqrt{3^2 + 4^2}\\
&= \sqrt{9 + 16}\\
&= \sqrt{25}\\
&= 5
\end{aligned}$

Who is correct?

5 Work out:

a $\sqrt{6^2 + 8^2}$

b $\sqrt{5^2 + 12^2}$

6 Work out:

a $\left(^-5 - ^-2\right) \times 4^2 + 50$

b $2.8 + (6 + ^-4) + 1.2^2$

c $\dfrac{4^2 + 2.6}{0.3}$

7 Write brackets to make these calculations correct.

a $\sqrt{9} + 4^2 \times 10 = 190$

b $5 + 14 \div 2 - ^-3 = 15$

c $10^2 - \sqrt[3]{1000} + 2 \times 8 - 3 = 80$

1.6 Natural numbers and rational numbers

Here are some definitions you need to know. **Natural numbers** are the numbers you use to count (they are positive whole numbers not including 0).

1, 2, 3, 4, . . .

Whole numbers are natural numbers including 0.

0, 1, 2, 3, . . .

Integers are positive and negative whole numbers and include 0.

. . . , $^-3$, $^-2$, $^-1$, 0, 1, 2, 3, . . .

A **rational number** can be made by dividing two integers. This is most (but not all) of the other numbers you will learn about. You will learn about irrational numbers in Student Book 9.

Rational numbers include:

- all fractions, e.g. $\dfrac{1}{4}$, $-\dfrac{735}{6189}$

- terminating and recurring decimals (as they can be made by dividing two integers), e.g. $2.5 = \dfrac{5}{2}$, $0.\dot{3} = \dfrac{1}{3}$

  (you will learn more about recurring decimals in Chapter 6).

You can represent this on a Venn diagram.

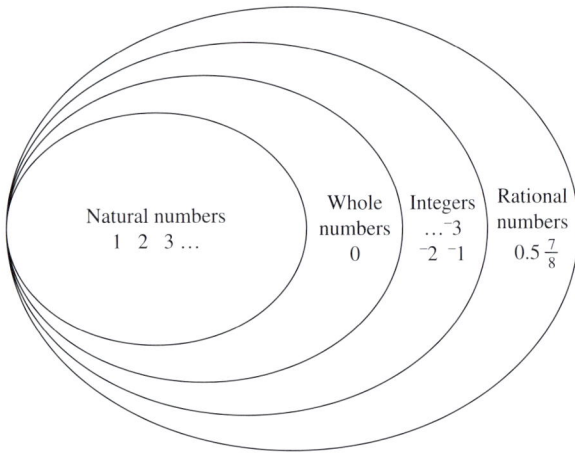

Exercise 1J

1 Alex says that all of the natural numbers are also rational numbers because 8 is the result of dividing two integers $\dfrac{8}{1}$. Is Alex correct?

2 Decide whether each of these statements is true or false.

a 72 is an integer.

b 72 is a whole number.

c 72 is a rational number.

d 72 is a natural number.

e $^-72$ is an integer.

f $^-72$ is a whole number.

g $^-72$ is a rational number.

h $^-72$ is a natural number.

3 Decide whether each of these statements is true or false.

a All natural numbers are integers.

b All integers are whole numbers.

c All integers are rational numbers.

d All rational numbers are natural numbers.

4 105, $^-3$, 0.2, 16, $\dfrac{5}{16}$, $0.\dot{6}$

From this list of numbers, write down all of the:

a natural numbers

b integers

c rational numbers

5 Copy this Venn diagram

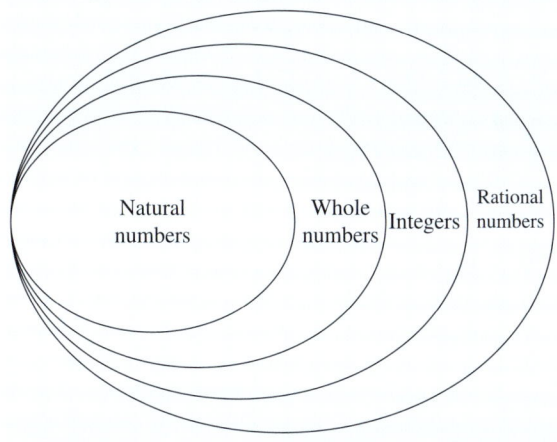

Write these numbers in the correct place on the Venn diagram.

$^-50, 5, \frac{3}{5}, ^-1, 62.3, 20, 0.1$

6 Wasim's teacher says the number π can be approximated by $\frac{22}{7}$. Wasim says π must be a rational number because it can be written as a fraction. Is he correct? Explain your answer.

Consolidation

Example 1

Work out:

a $^-6 \times ^-5$ **b** $60 \div ^-6$ **c** $^-35 \div ^-7$

....

a $^-6 \times ^-5 = 30$
b $60 \div ^-6 = ^-10$
c $^-35 \div ^-7 = 5$

Example 2

Simplify:

a $2^7 \times 2^3$ **b** $\frac{4^3 \times 4^4}{4^6}$

....

a $2^7 \times 2^3 = 2^{7+3} = 2^{10}$

b $\frac{4^3 \times 4^4}{4^7} = \frac{4^7}{4^7} = 4^{7-7} = 4^0 = 1$

Example 3

Work out the HCF and LCM of 300 and 180.

....

Rewrite 300 and 180 as products of primes using the factor tree or repeated division method.

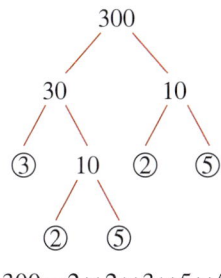

$$300 = 2 \times 2 \times 3 \times 5 \times 5$$

2	180
2	90
3	45
3	15
	5

$$180 = 2 \times 2 \times 3 \times 3 \times 5$$

$300 = 2 \times 2 \times 3 \times 5 \times 5$
$180 = 2 \times 2 \times 2 \times 3 \times 5$
$HCF = 2 \times 2 \times 3 \times 5 = 60$

> Use the common factors.

$300 = 2^2 \times 3 \times 5^2$
$180 = 2^2 \times 3^2 \times 5$
$LCM = 2^2 \times 3^2 \times 5^2 = 900$

> Use the highest power of each prime factor – this is easier to see if the primes are in index form.

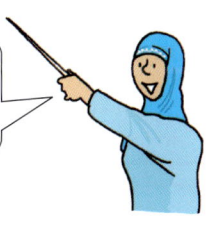

Example 4

Work out:
$36 \div \sqrt[3]{27} - 0.5 \times 16$

....

$36 \div \sqrt[3]{27} - 0.5 \times 16$ Use BIDMAS
$= 36 \div 3 - 0.5 \times 16$ Indices and cube root first
$= 12 - 0.5 \times 16$ then Division
$= 12 - 8$ then Multiplication
$= 4$ then Subtraction.

Exercise 1

1 Work out:

  a $^-3 \times {}^-4$ **b** $^-4 \times {}^-6$

  c $^-3 \times {}^-7$ **d** $^-13 \times {}^-14$

2 Work out:

  a $^-15 \div {}^-3$ **b** $^-16 \div {}^-4$

  c $^-144 \div {}^-9$ **d** $^-182 \div {}^-13$

3 Find the value of:

  a $\dfrac{^-3 \times {}^-9}{^-27}$ **b** $\dfrac{^-25}{^-5} - \dfrac{24}{^-4}$

  c $^-3 \times {}^-2 \times {}^-5$ **d** $7 \times {}^-4 \times {}^-2$

4 Work out:

  a $(^-3)^3$ **b** $\sqrt[3]{^-8}$

  c $(^-1)^2$ **d** $\pm\sqrt{121}$

5 Work out:

  a 2^7 **b** 4^4 **c** 5^3 **d** 3^4

6 Jade says that the square root of $^-25$ is $^-5$.
Katy says you cannot work out the square root of a negative number. Who is correct?

7 Write these numbers as products of their primes.

  a 15   **b** 18   **c** 30

  d 45   **e** 27   **f** 36

8 Work out:

  a $2^3 + \sqrt{25} - 16 \div \sqrt[3]{8}$

  b $\sqrt{10^2 - 8^2}$

  c $\sqrt[3]{125} \times 2 + 6^2 \div 3$

  d $\sqrt{150 - 29} - 2^2 \times 3$

9 Find the HCF and LCM of these numbers.

  a 56 and 140

  b 30 and 75

  c 35 and 42

  d 180 and 225

10 Simplify:

  a $2^4 \times 2^2$ **b** $5^8 \div 5^2$

  c $4^8 \div 4^4$ **d** $4^7 \times 4^3 \times 4^2$

  e $4^4 \times 4^3 \div 4^2$ **f** $\dfrac{9^{20}}{9^2 \times 9^{18}}$

11 The LCM of two numbers is 130. The HCF of the same two numbers is 13. Both numbers are less than 100. Find the two numbers.

12 What are the numbers p, q, r and s if:

  a $2^p = 4^q = 16^2 = 256$

  b $3^r = 9^s = 81^2 = 6561$

13 Yafeu says that if he starts with a number and squares it he ends up with the same number he started with. What could his number be? Is there more than one answer?

14 Find the highest common factor of 72, 120 and 42.

15 Two pieces of wood of different lengths, 150 cm and 210 cm, are to be cut into equal lengths. What is the greatest possible length of each piece?

Summary

You should know ...

1 a When you multiply two negative numbers together, the answer is positive.
For example: $^-5 \times {}^-3 = 15$ and $^-6 \times {}^-4 = 24$

Two minuses make a plus.

b When you multiply a positive number and a negative number together, the answer is negative.
For example: $^-3 \times 5 = {}^-15$ and $10 \times {}^-6 = {}^-60$

Mix means minus.

Check out

1 Calculate:

  a $3 \times {}^-40$

  b $^-7 \times 4$

  c $^-3 \times {}^-4$

  d $^-5 \times 11$

  e $^-4 \times {}^-6$

  f 4×5

2 a When you divide a negative number by a negative number, the answer is positive.

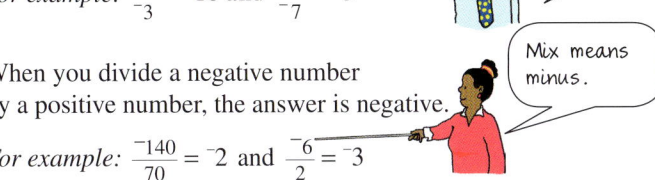

Two minuses make a plus.

For example: $\frac{^-6}{^-2} = 3$ and $\frac{^-30}{^-15} = 2$

b When you divide a positive number by a negative number, the answer is negative.

Mix means minus.

For example: $\frac{30}{^-3} = ^-10$ and $\frac{21}{^-7} = ^-3$

c When you divide a negative number by a positive number, the answer is negative.

Mix means minus.

For example: $\frac{^-140}{70} = ^-2$ and $\frac{^-6}{2} = ^-3$

3 Any positive number has two square roots; one is positive and one is negative.
For example: The square roots of 16 are 4 and $^-4$.

4 The cube root of a number means the number which, when multiplied by itself and then multiplied by itself again, makes the original number.
For example: $\sqrt[3]{512} = 8$, since $8 \times 8 \times 8 = 512$

5 You can use indices or powers to write a sum more simply.
For example: $3 \times 3 \times 3 \times 3 \times 3 = 3^5 = 243$

In 3^5, 3 is the base number, 5 is the power or index.

6 How to multiply and divide with indices, and about the zero index.
For example:
$3^2 \times 3^4 = 3^{(2+4)} = 3^6$
$6^5 \div 6^2 = 6^{(5-2)} = 6^3$
$4^0 = 1$

7 A number can be written as a product of its prime factors using a factor tree or repeated division by primes.
For example:

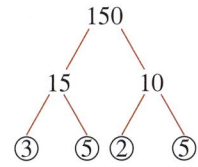

150

15   10

$150 = 2 \times 3 \times 5 \times 5$ ③   ⑤ ②   ⑤

In index form this is
$150 = 2 \times 3 \times 5^2$

---

2 Work out:
   a $15 \div ^-5$
   b $^-15 \div ^-3$
   c $^-24 \div 6$
   d $^-36 \div ^-9$

3 Work out:
   a $(^-3)^2$ **b** $^-9^2$
   c $\pm\sqrt{36}$ **d** $\pm\sqrt{49}$

4 Work out:
   a $\sqrt[3]{125}$ **b** $\sqrt[3]{^-8}$
   c $\sqrt[3]{^-1}$ **d** $\sqrt[3]{^-27}$

5 Work out:
   a 4^3 **b** 2^8
   c 5^4 **d** 3^4

6 Work these out. Write your answers in index form.
   a $2^2 \times 2^3$ **b** $5^4 \times 5^7$
   c $7^8 \times 7^3$ **d** $9^5 \times 9^2$
   e $4^5 \div 4^2$ **f** $3^3 \div 3^3$
   g $2^8 \div 2^4$ **h** 7^0

7 Write these numbers as products of their prime factors.
   a 45   **b** 72
   c 60   **d** 75

8 The highest common factor of two (or more) numbers is the highest factor common to those numbers. Use the common prime factors of each number to calculate it.
For example:
$90 = 2 \times 3 \times 3 \times 5$
$150 = 2 \times 3 \times 5 \times 5$
$HCF = 2 \times 3 \times 5 = 30$

The highest common factor of 90 and 150 is 30.

The lowest common multiple of two (or more) numbers is the lowest number that is a multiple of those numbers. Use the highest power of prime factors to calculate it.
For example:
$90 = 2 \times 3^2 \times 5$
$150 = 2 \times 3 \times 5^2$
$LCM = 2 \times 3^2 \times 5^2 = 450$

The lowest common multiple of 90 and 150 is 450.

For example:
$10 = 2 \times 5$
$15 = 3 \times 5$
$LCM = 2 \times 3 \times 5 = 30$
The lowest common multiple of 10 and 15 is 30.

8 Work out:
i the HCF and
ii the LCM of:
a 45 and 72
b 60 and 75

9 The order of operations is known as BIDMAS.
Brackets first
then **I**ndices (including square roots and cube roots)
then **D**ivision and **M**ultiplication (working left to right)
then **A**ddition and **S**ubtraction (working left to right).

For example:

$3 \times 2^3 - 18 \div \sqrt{36}$	Indices and square root
$= 3 \times 8 - 18 \div 6$	Multiplication
$= 24 - 18 \div 6$	Division
$= 24 - 3$	Subtraction
$= 21$	

9 Work out:
a $2^2 \times \sqrt{25} - 30 \div 5$
b $\sqrt{5^2 - 3^2}$
c $\sqrt[3]{125} + 3^3 \div 6$
d $\sqrt{140 + 4} + 2^4 \times 3$

10 a Natural numbers are the numbers you use to count (they are positive whole numbers, not including 0).
For example: 1, 2, 3, 4, …

b Whole numbers are natural numbers and 0.
For example: 0, 1, 2, 3, …

c Integers are positive and negative whole numbers and include 0.
For example: ⁻3, ⁻2, ⁻1, 0, 1, 2, 3, …

d A **rational number** can be made by dividing two integers.
For example: ⁻0.35 is a rational number because it can be made from dividing the two integers ⁻35 and 100.

10 Decide whether these are rational numbers.
a ⁻0.3
b $\frac{1}{3}$
c ⁻2
d 17

Expressions

Objectives

In this chapter you will learn about:
- constructing expressions
- expanding brackets
- factorizing expressions
- substituting into expressions.

What's the point?

Basic formulas are part of your life. When you want to find out how long a journey will take you will be using a formula:

journey time = distance ÷ speed

Before you start

You should know ...

1 How to work with negative numbers.
For example:
$7 - {}^-5 = 7 + 5 = 12$ (two minuses make a plus)
${}^-3 + {}^-4 = {}^-3 - 4 = {}^-7$ (mix means minus)

2 How to simplify basic algebra.
For example:
$a + a = 2a$
$3 \times b = 3b$ for short (no need to write the multiplication symbol)
$t \times 5 = 5t$ for short (write the number first, then the letter)

Check in

1 Work out:
  a ${}^-2 \quad {}^-6$
  b ${}^-4 + {}^-10$
  c ${}^-8 - {}^-3$
  d $20 + {}^-10$

2 Write these expressions in a shorter way.
  a $m + m + m$
  b $6 \times y$
  c $r \times 10$
  d $c + c + c + c + c$

3 This is the area of a rectangle, A.

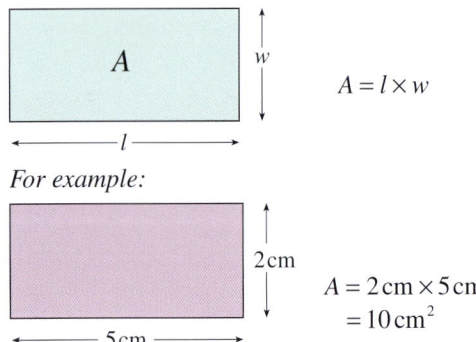

$A = l \times w$

For example:

$A = 2\,cm \times 5\,cm$
$= 10\,cm^2$

3 Work out the area of the rectangle with:
  a length 2 cm, width 3 cm
  b length 6 cm, width 6.5 cm
  c length 18 m, width 3.25 m

2.1 Expressions, formulae and equations

Here is some key language you need to know.

- An **unknown** is part of an **equation** that you don't know the value of so, in the equation $5x - 8 = 12$ x is the unknown. The equation can be solved to find the value of the unknown. There is only one possible value of x in a **linear** equation such as this one.

- $3x + 10$ is an **expression** and $3x$ and 10 are **terms** of the expression. 3 is the **coefficient** of $3x$, and 10 is the **constant** term. You cannot solve an expression as there is no equals sign.

- $s = d \div t$ is a **formula**. The letters are **variables** that are related by the formula and s is the **subject** of the formula. You cannot solve a formula as the formula just tells you the relationship between the variables.

- $y = 2x + 1$ is a **formula** and also a **linear function** in two variables (x and y). If you know a value of x you can work out the related value of y.

Constructing expressions

We can write algebraic expressions to help us derive formulae (you will learn more about formulae in Chapter 7). Algebraic expressions are a shorter way of writing something.

In Student Book 7 you learned how to construct simple expressions like the following.

If I have 3 pens costing $4 each the cost of these 3 pens is $3 \times 4 = \$12$

If I have 5 pens costing $4 each the cost of these 5 pens is $5 \times 4 = \$20$

If I have x pens costing $4 each the cost of these x pens is $x \times 4 = \$4x$

$4x$ is an expression that tells us the cost of an unknown number of pens each costing $4.

If you are not sure how to construct the expression, think what sum you would need to do if numbers were involved instead of letters. For instance, in Example 1 you could estimate the number of paper clips in the pot, think what sum to do using that number, then replace that number with a letter.

Example 1

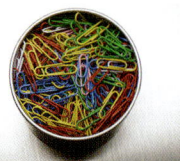

Write an expression for the number of paper clips if there are:
a 2 paper clips taken out of the pot
b 5 more paper clips in the pot
c 3 pots of paper clips exactly the same as this one
d 4 people sharing the pot of paper clips equally between them

There is an unknown number of paper clips in this pot, let's say n paper clips.
a $n - 2$
b $n + 5$
c $3n$
d $n \div 4$ or $\dfrac{n}{4}$

It doesn't matter what letter you use.

Example 2

Write an expression for the total time needed to roast a turkey with a mass of k kilograms, if it takes 20 minutes plus 35 minutes for each kilogram.

....

Time, in minutes, is $20 + 35k$

Exercise 2A

1 Decide whether each of the following is an equation, an expression or a formula.

   a $A = wl$

   b $3y + 2x$

   c $3p - 1 = 29$

   d $4t + 7$

   e $V = IR$

   f $7n + 3 = 2n + 23$

2

There are S sugar cubes in a bowl.
Write an expression for the number of sugar cubes if I:

   a take out 6 sugar cubes

   b put in 10 sugar cubes

   c have 5 bowls exactly the same

   d take out half of the sugar cubes

3 Write an expression for the perimeter (the distance around the edge) of these shapes. Simplify your expression where possible.

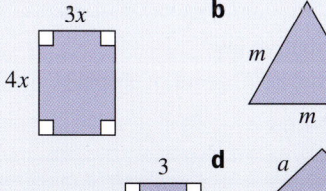
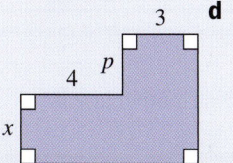
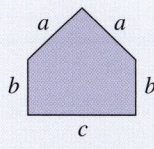

4 a Write an expression for the total cost of k pens at \$3 per pen and p pencils at \$1 per pencil.

   b Write an expression for the total cost of k pens at \$$m$ per pen and p pencils at \$$t$ per pencil.

5 Write an expression for the cost of hiring a taxi to travel K kilometres if there is a fixed cost of \$2 plus \$0.50 per kilometre.

6 To change a temperature in degrees Celsius to degrees Fahrenheit, multiply by 1.8 and then add 32. Write an expression to show the temperature in degrees Fahrenheit of something with a temperature of C degrees Celsius.

7 An exchange rate shows that you get 2 New Zealand dollars for every UK pound. Write an expression to show the number of New Zealand dollars you would get for P UK pounds.

8 a A book costs V dollars. A CD costs W dollars.
   Match each description with the correct expression. The first one is done for you.

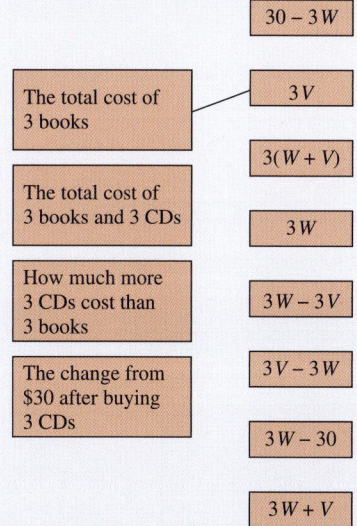

   b Choose any two expressions from **a** that are not paired with a description and write the meaning of those expressions in words.

9 Copy and complete this table about the unknown number n.

Expression	Meaning
$\frac{1}{2}n$	Half the number
$3(n-2)$	Start with the number, subtract 2 from it, then multiply that answer by 3

Expression	Meaning
$\frac{1}{2}n+5$	
	Add 5 to the number then halve it
$20-2n$	
	Start with the number, add 6 to it, then double it

10 Debbie has x pens; Kulwinder has y pens. What do each of the following mean?

   a $2x+y$

   b $3(x+y)$

   c $\frac{1}{2}x-y$

11 Write using algebra:

   a I think of a number v, add 7 then multiply by 5.

   b I think of a number w, subtract 2 then halve it.

   c I think of a number x and a number y and I find half the product of these two numbers.

   d I think of a number z, and multiply it by itself.

12 Find an expression for the area of these rectangles.

   a

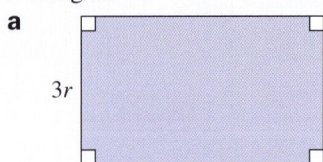

     $3r$ $3.5t$

   b

     $\frac{1}{2}n$ $5m$

13 A quadrilateral has two equal sides of length x cm. One of the other sides is four times as long as the shortest side and the final side is 15 cm longer than the shortest side. Write an expression for the perimeter of the quadrilateral. Give your answer in its simplest form.

14 Amy has x sweets, Gemma has 3 more sweets than Amy and Sho has twice as many sweets as Gemma.

   a Find an expression for the total number of sweets Amy, Gemma and Sho have. Give your answer in its simplest form.

   b If Amy has 4 sweets how many sweets does Sho have?

   c If Gemma has 9 sweets how many sweets do they have altogether?

   d If Sho has 24 sweets how many sweets do Amy and Gemma have altogether?

15 The monthly cost of local calls on a mobile phone is $8 plus 9 cents per call. Write an expression for the total cost, in dollars, of r local calls in x months.

16 Write an expression for the number of sugar cubes left in the bowl in question **2** if I take out 10% of the sugar cubes.

2.2 Expanding brackets

You can work out the multiplication:

   6×74

using the distributive law

$$6\times74 = 6\times(70+4)$$
$$= 6\times70+6\times4$$
$$= 420+24$$
$$= 444$$

In algebraic terms the distributive law is

$$a\times(b+c) = a\times b+a\times c$$

That is, everything inside the brackets is multiplied by what is outside.

This is called multiplying out brackets or **expanding brackets**.

Example 3

Expand the brackets.

a $3(x + 2y)$ **b** $x(x + 1)$

....

a $3(x + 2y) = 3 \times x + 3 \times 2y$
$$= 3x + 6y$$

b $x(x + 1) = x \times x + x \times 1$
$$= x^2 + x$$

In Example 3 you expanded $3(x + 2y)$. This is like working out the area of this rectangle.

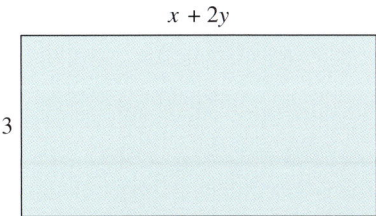

$x + 2y$

3

It can be divided into two rectangles like this.

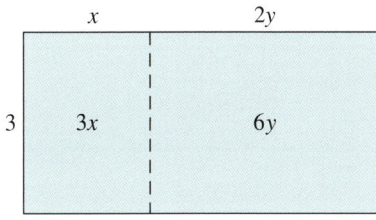

x $2y$

3 $3x$ $6y$

$$3(x + 2y) = 3x + 6y$$

You should know about the order of operations, BIDMAS, and how it applies to numbers.

Brackets first
then Indices
then Division and Multiplication
then Addition and Subtraction.

> BIDMAS tells you the order you should do operations in.

The same rules apply to algebra. In the next example, brackets are expanded (or multiplied out) before doing the addition and subtraction, to simplify.

Example 4

Simplify:

a $3(a - 2b) + a(4 - 2b)$
b $2a(3 - 2b) - a(4b - 2)$
c $7m + 5(6n + 3m) - 4n$

....

a $3(a - 2b) + a(4 - 2b)$
$$= 3 \times a - 3 \times 2b + a \times 4 - a \times 2b$$
$$= 3a - 6b + 4a - 2ab$$
$$= 3a + 4a - 6b - 2ab$$
$$= 7a - 6b - 2ab$$

b $2a(3 - 2b) - a(4b - 2)$
$$= 2a \times 3 - 2a \times 2b - a \times 4b + a \times 2$$
$$= 6a - 4ab - 4ab + 2a$$
$$= 6a + 2a - 4ab - 4ab$$
$$= 8a - 8ab$$

c $7m + 5(6n + 3m) - 4n$
$$= 7m + 30n + 15m - 4n$$
$$= 22m + 26n$$

> This is positive because multiplying two negatives makes a positive.

Exercise 2B

1 Expand the brackets.
   a $3(x + 2)$
   b $4(2x - 6)$
   c $5(3x - 3)$
   d $6(4 - 3x)$
   e $x(x + 5)$
   f $3x(x + 4)$
   g $2m(3m + 7)$
   h $5p(7 - 2p)$
   i $5m^2(m + 3c)$
   j $2x^2(5x - 7y)$

2 Work out the areas of the following rectangles.

   a

$x + 7$

$2x$

b

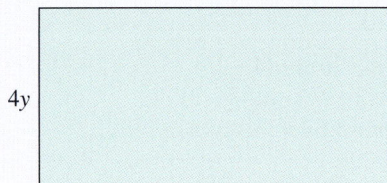

$7 - 3y$

$4y$

c

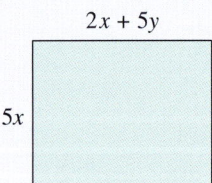

$2x + 5y$

$5x$

d

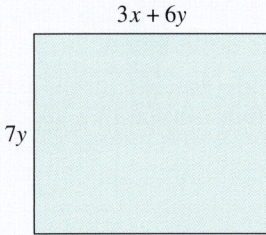

$3x + 6y$

$7y$

3 **a** Draw an area that shows the expression $4(x + 3)$.

    b Write a different expression that gives the same area.

4 **a** Draw an area that shows the expression $(4p)^2$.

    b Write a different expression that gives the same area.

5 **a** Draw an area that shows the expression $(t + 4)^2$.

    b Write a different expression that gives the same area.

6 Expand the brackets and simplify:

    a $3(4x - 2y) + 3(2x - 3y)$

    b $2(x - 4y) - 2(x + y)$

    c $3x(x + 4) + 5x(2x - 3)$

    d $4y(2 - 3y) - y(7 - 15y)$

7 Sunil, Ricardo, Diego and Helga each work out the answer to $4p^2(3p - 7m)$. Here are their answers.

Sunil: $12p^3 - 28p^2m$

Ricardo: $7p^3 - 11p^2m$

Diego: $12p^3 - 28pm^2$

Helga: $12p^3 - 28mp^2$

a Which **two** students got the correct answer? Is one answer better than the other? Why?

b Explain why these two answers mean the same even though they look different.

c Explain the mistakes made by the other two students.

8 Work out the areas of the two shaded regions below. Simplify your expressions.

a

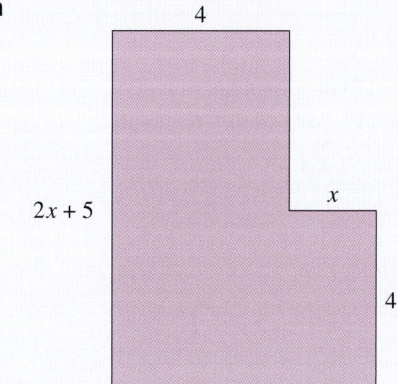

4

$2x + 5$

x

4

b

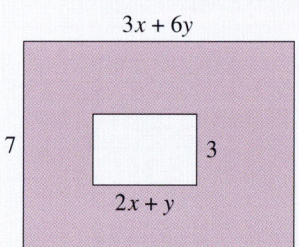

$3x + 6y$

7

3

$2x + y$

9 Pair up equivalent expressions to find the odd one out.

$^-6x - 5$

$6x - 11$

$5 - 2(3x + 5)$

$6 + 15x$

$5(3x + 2) - 9x - 7$

$4(2x - 1) - (2x - 7)$

$5 - 3(4 - 5x) + 13$

10 Expand the brackets.
  a $4x(3y + x - 2p)$
  b $8y^2(7x - 8t + 9y)$
  c $^-m^2(m - 8 - 9t + y)$

11 Work out the missing numbers or terms.
  a $2(x - \square) = 2x - 16$
  b $\square(3x + 4) = 18x + 24$
  c $\square(8g - 12) = 104g - 156$
  d $\square(6 + 13y) = 42 + \square$

12 Simplify:
  a $5y^2 - y(1 + 2y)$
  b $x(3 + 2x) + x^2(1 + 2x)$
  c $3x(1 - 2x) + x(x - 1)$
  d $5y(1 - y) - y(y + 3)$
  e $3y(1 + y - y^2) - y^2(2 - 3y)$
  f $7 - (c - 3) - 2c + 3(4 - c)$
  g $3m + 6 \times 2m - 20m$
  h $f - (h - 3f) + 5 \times 4h - 8f$

13 Find the missing numbers or expressions.
  a $7t(\square y + 8x) = 14ty - \square tx$
  b $3x(\square y - x) = 15xy - \square x^2$
  c $\square p(6y - x) = \square - 4px$

14 If the area of a rectangle is $8x + 12$ and the width is 4, what is the length of the rectangle?

15 Remove the brackets and simplify:
  a $3(x - 2y) + 2(x + y) - 3(x + 2y)$
  b $4x(1 + y) + 3y(2 - x) - 2(xy + 3y)$
  c $x(x^2 - 3) + x^2(1 - x) - 3x(3 + x)$

2.3 Factorizing expressions

The opposite of expanding brackets is to put an expression into brackets. This is called **factorizing**. You always put the highest common factor of all of the terms outside the brackets.

For example:

3 is the highest common factor of the terms $3l$, $6m$ and $9n$.
So $3l + 6m + 9n = 3(l + 2m + 3n)$

x is a common factor of the terms ax and bx.
So $ax + bx = x(a + b)$

x is a common factor of the terms x^2 and $3x$.
So $x^2 + 3x = x(x + 3)$
x and $(x + 3)$ are the factors of $x^2 + 3x$.

$2m$ is one factor of $2lm + 6mn + 10mp$.
$(l + 3n + 5p)$ is the other factor.

Exercise 2C

1 Copy and complete:
  a $3x + 3y = 3(\quad)$
  b $5a - 5b = 5(\quad)$
  c $4x + 4y + 4z = 4(\quad)$
  d $6a - 6b + 6c = 6(\quad)$
  e $2x + 6y = 2(\quad)$
  f $8a - 4b = 4(\quad)$
  g $3x + 6y + 9z = 3(\quad)$
  h $25a - 10b - 5c = 5(\quad)$

2 Copy and complete:
  a $ax + ay = a(\quad)$
  b $pa - pb = p(\quad)$
  c $px + py + pz = p(\quad)$
  d $ra - rb + rc = r(\quad)$
  e $qx + 3qy = q(\quad)$
  f $5sa - sb = s(\quad)$
  g $2tx + 5ty + tz = t(\quad)$
  h $7la - 4ib - lc = l(\quad)$

3 Copy and complete:
  a $3m + 5mn + m^2 = m(\quad)$
  b $2p + 3pr + p^2 = p(\quad)$
  c $6l + 2lm + 2l^2 = 2l(\quad)$
  d $5rs + 50rs^2 + 15r^2s = 5rs(\quad)$

4 Noah is fully factorizing the expression
$36x + 12y$
He writes $6(6x + 2y)$
Explain why this answer is incorrect.

5 Write down three expressions that cannot be factorized. Explain why they cannot be factorized.

6 Factorize:
  a $2a + 2b$ **b** $3a - 3b$
  c $4x + 12y$ **d** $9p - 6q$
  e $px + py$ **f** $ra - rb$
  g $7sx + 4sy$ **h** $2ta - 7tb$
  i $xa + xb + xc$ **j** $la - lb - lc$
  k $4rx + 5ry + rz$ **l** $pa - 6pb + 8pc$

7 The area of a rectangle is $2x^2 + 12x$. Find an expression for each of the length and width of the rectangle.

8 Factorize:
  a $lx + mx$ **b** $an - bn$
  c $7py + 2qy$ **d** $rt - 5st$
  e $pt + qt + rt$ **f** $an + bn - cn$
  g $5lx + mx + 2nx$ **h** $4kg - 21lg - mg$

9 Write down three different expressions that factorize to the form $n(3x+1)$ where n is an integer.

10 Factorize:
  a $4p+2pr+6pz$
  b $5m+15mp+25mg$
  c $9sr+3s+6s^2$
  d $4lm+2mn+8pmn$
  e $ab^2+6ab^3+2a^2b^2$
  f $2x^2y^2+3xy^2+x^2y^2$ (**Hint**: simplify first.)

11 The area of a rectangle is $8t^2+12t$. Find as many possible different side lengths for the rectangle as you can.

12 Factorize:
  a x^2+3x **b** y^2-5y
  c $2z^2+3z$ **d** $4m^2-m$
  e x^3+2xy **f** $4y^2z-y^3$
  g ab^2+a^2b **h** $x^2yz^2-xyz^3$
  i $\pi r^2+2\pi rh$ **j** $2lm^2+8l^2m$
  k $x^4+x^3+x^2$ **l** $32y+16y^3+8y^5$

13 The area of this rectangle is $24x^2+18x$

$4x+3$

Find an expression for the perimeter of the rectangle in its simplest form.

14 Factorize:
  a $abc^2+ab^2+a^2b$
  b $p^3q^2r+p^2q^2r^2+pq^2r^3$
  c $7axy+14bxy+21cxy$
  d $8x^6+16x^4+48x^3$
  e $2lmp-lm+5lm^2$
  f $f^4g^2-6f^2g^3+2fg^4$
  g $5abcd+35bcde$
  h $24k^2lm^2n-32k^2m^2n^3$
  i $16abcx-28bcdx-20cdex$

15 Expand the brackets, simplify, and factorize:
  a $3(x-5)+4(x-2)-5$
  b $x(y+3)+2x(4-y)-10xy$
  c $a(b+c)+b(c-a)+3c$
  d $lm(5l+m)+2l^2(m-3)+6l\left(m^2+l\right)$

16 The area of a rectangle is $36x^2+6x$. The perimeter of this rectangle is $30x+4$. Find an expression for each of the side lengths of the rectangle.

17 Factorize each expression.
  a $x^2y+3xy+4xy^2$
  b $2rs+18rst+8r^2s^2+10rst^2$
  c $14mn^2+2mn+8m^2n+8m^2n^2$
  d $\frac{1}{3}g^2h+\frac{5}{3}g^3h+2g^3h^3$
  e $x^2y+2xy^2+x^2y^3+x^4y^4$

2.4 Substitution into expressions

You have already learned about expressions such as $3x+4$, $2a+b$ and $7ty$. Notice there is no equals sign in an expression. When you know the values of the letters, you can find the value of the expression. The laws of arithmetic and order of operations also apply here.

Example 5

If $x=\,^-3$, find the value of:

a x^2 **b** $5x^2$
....
a $x^2=(^-3)^2=\,^-3\times\,^-3=9$
b $5x^2=5\times(^-3)^2=5\times9=45$

> In BIDMAS indices are calculated before multiplication.

You can also work out more complex expressions.

Example 6

If $x=\,^-2$, find the value of $2x^3-x^2+10x+6$
....
$=2(^-2)^3-(^-2)^2+10(^-2)+6$

$=2(^-8)-(4)+10(^-2)+6$

$=\,^-16-4+\,^-20+6$

$=\,^-40+6$

$=\,^-34$

Example 7

If $n = 5$, find the value of $\dfrac{5 + 3n}{5}$

$$\frac{5 + 3 \times 5}{5}$$

$$= \frac{5 + 15}{5}$$

$$= \frac{20}{5}$$

$$= 4$$

The numerator is found first as the fraction line works similarly to brackets.

Exercise 2D

1 If $x = 3$, $y = 4$ and $z = 5$, find the value of:
- **a** $xy + z$
- **b** $y^2 - 2z$
- **c** $(x + y)^2$
- **d** $\dfrac{8x - y}{z}$
- **e** $y(z + x)$
- **f** $\dfrac{z + 25x}{y}$
- **g** $2x^3 - yz$

2 If $m = 2$, $n = {}^-3$, $p = {}^-5$ and $t = 10$, find the value of:
- **a** $m + n - t$
- **b** $2t - m + 3p$
- **c** $n(m + t)$
- **d** $2p + 10t - 3n$
- **e** $6t(p + n)$
- **f** $3mp + 30$
- **g** $mp - tn$
- **h** $5(m + n + p)$
- **i** $p - n - m - t$
- **j** $4np - 6t$
- **k** $\dfrac{mnp}{t}$
- **l** $\dfrac{7m + 2n}{t + p + 3}$

3 Find the value of each of the following. The first one has been worked out for you.
- **a** $3x^2$ when $x = {}^-2$
  $$3x^2 = 3 \times ({}^-2)^2$$
  $$= 3 \times ({}^-2) \times ({}^-2)$$
  $$= 12$$
- **b** x^3 when $x = {}^-1$
- **c** $4x^3$ when $x = 2$
- **d** $3x^2$ when $x = 5$
- **e** $4t^2$ when $t = {}^-3$
- **f** $2p^3$ when $p = {}^-5$

4 Find the value of:
- **a** $x^2 + 4$ when $x = {}^-3$
- **b** $2x^2 + 5$ when $x = 2$
- **c** $3 + t^3$ when $t = {}^-1$
- **d** $4y^2 + y - 45$ when $y = {}^-5$
- **e** $24 + 3p^3$ when $p = {}^-2$
- **f** $100 - 2x^2$ when $x = 7$
- **g** $1 - m^3$ when $m = {}^-3$
- **h** $2w^2 + 3v^3 - r$ when $w = {}^-5$, $v = {}^-2$ and $r = {}^-8$

5 Find the value of each expression when:
    i $x = 2$ **ii** $x = {}^-2$
- **a** $3x^2 - x$
- **b** $2x^3 + x - 10$
- **c** $x^4 + 2x^2 + 7$
- **d** $4x^3 + 3x^2 + 2x + 1$
- **e** $x^3 - 5x^2 - 4x - 15$

6 Find the value of each expression.
- **a** $x^2 - x$ when $x = {}^-3$
- **b** $x + x^2 - x^3$ when $x = {}^-4$
- **c** $2x^2 - x - 1$ when $x = {}^-5$
- **d** $x^3 + 4x^2 - 16x + 19$ when $x = {}^-1$
- **e** $x^3 - 3x$ when $x = {}^-3$

7 Copy these expressions.

$$\boxed{2x + y^2} \quad \boxed{4xy} \quad \boxed{x^2 + y} \quad \boxed{4y^2 - x^2 + x}$$

Tick ($\checkmark$) which expressions have the same value when $x = 3$ and $y = {}^-2$.

8 Emil starts this substitution tree.

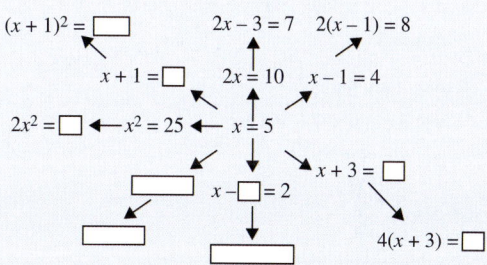

Complete the tree with possible values and expressions.

9 Choose your own values of n to substitute into each of these expressions. Try several values. What do you notice?

- **a** $\dfrac{2n + 14}{n + 7}$
- **b** $\dfrac{n^2 - 9}{n - 3} - n$

10 What values of m and p make these expressions equal?
$$2m + 5 \qquad 3(p + 1) \qquad 7m - 5p$$

11 **a** Work out the mean of these five expressions.
$$n + 2 \quad 2n + 6 \quad 3n - 5 \quad 5n + 4 \quad 4n + 3$$
   b Check by substituting $n = 6$ into the expressions and into your answer.

Consolidation

Example 1

Simplify:

$4(3x+5)-3(6-2x)$

$4(3x+5)-3(6-2x)$
$=12x+20-18+6x$
$=18x+2$

You get the plus sign here from multiplying together two negatives.

Example 2

Write an expression for half the distance travelled by a man walking at U kilometres per hour for t hours.

Distance = speed × time so distance is Ut

Half the distance is $\dfrac{Ut}{2}$

Example 3

Factorize:

a $4x - 12y$
b $3y^2 - 4y + 5y^3$

a $4x - 12y$
$= 4 \times x - 4 \times 3y$
$= 4(x - 3y)$

b $3y^2 - 4y + 5y^3$
$= y \times 3y - y \times 4 + y \times 5y^2$
$= y(3y - 4 + 5y^2)$

Example 4

If $d = {}^-3$ find the value of $4d^2 + 7d$

$4d^2 + 7d$ BIDMAS
$= 4 \times ({}^-3)^2 + 7 \times {}^-3$ Indices
$= 4 \times 9 + 7 \times {}^-3$ Multiplication
$= 36 + {}^-21$ Addition
$= 36 - 21$
$= 15$

Exercise 2

1 Decide whether each of the following is an equation, an expression or a formula.
$F = ma$
$7y - 2x$
$2p = 27 - p$
$y + 9$
$V = Al$
$6y + 1 = 3y + 7$

2 To change kilometres to miles you divide by 8 then multiply by 5. Write an expression for the number of miles in H kilometres.

3 Write using algebra:
  a I think of a number g, add 2 then multiply by 4.
  b I think of a number h, subtract 7 then double it.
  c I think of a number j and a number k and I find double the product of these two numbers.
  d I think of a number m, multiply it by itself, then halve it.

4 Find an expression for the area of these rectangles.

  a

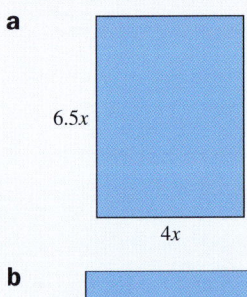

6.5x

4x

  b

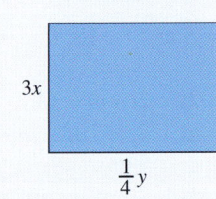

3x

$\frac{1}{4}y$

5 Expand the brackets.
  a $5(x-3)$ **b** $9(2x+8)$
  c $3(4+7y)$ **d** $12(7m-12p)$

6 Write down four expressions equivalent to $5m - 3t - m - 3t$ without using the numbers 5, 3 or 1. You may use brackets.

7 Expand the brackets and simplify:
  a $3(x+2)+4x$
  b $5(2+4y)+3(2y+6)$
  c $2+4(5x-7)$
  d $3(2t-1)-5(4t-3)$

8 Write down an expression for the areas of these rectangles.

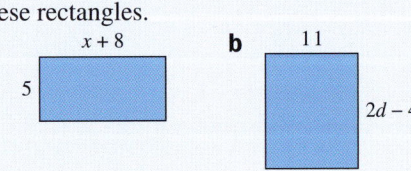

  a $x + 8$ **b**   11

5

$2d - 4$

9 Work out the missing numbers or terms.
   a $3(h - \square) = 3h - 15$
   b $\square(4x + 3) = 16x + 12$
   c $\square(4x - 3) = 40x - 30$
   d $\square(3 + 12m) = 15 + \square$

10 If the area of a rectangle is $4x + 6$ and the width is 2, what is the length?

11 Write down an expression for the shaded area in this shape.

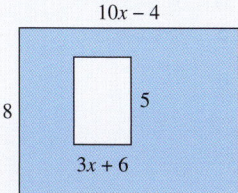

12 Factorize:
   a $3x + 9y$ **b** $2a - 4b$
   c $6x - 12y$ **d** $14x - 7y$

   e $15x + 18y$ **f** $6x + 72y$
   g $6x - 24y$ **h** $ax + ay$

13 Factorize:
   a $ax + 3ay$ **b** $an - 3am$
   c $6rx + 2ry$ **d** $3ax - 18ay$
   e $5mn - 5mp$ **f** $4pq - 3pr$
   g $6r + 4pr - 2qr$ **h** $3ab - 3a + ac$

14 If $t = 3$, $v = {}^-2$ and $r = {}^-5$, find the value of:
   a $v + 2 - r$ **b** $v(r + t)$
   c $2r - v + 3t$ **d** $3rv + 5t$

15 Find the value of:
   a $x^2 + 3$ when $x = {}^-3$
   b $3x^2 + x + 1$ when $x = 2$
   c $3v + v^3$ when $v = {}^-2$
   d $2y^2 + 3y - 4$ when $y = {}^-4$
   e $125 + m^3$ when $m = {}^-5$
   f $100 - 5x^2$ when $x = {}^-4$

Summary

You should know ...

1 You can construct expressions.
   For example:

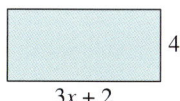

   $3x + 2$

   The area is width $\times$ length $= 4 \times (3x + 2)$
   $= 4(3x + 2)$
   $= 12x + 8$
   The perimeter is the distance around the outside
   $= 4 + 3x + 2 + 4 + 3x + 2$
   $= 6x + 12$

2 How to expand and simplify.
   For example:
   $6(3x - 2y) + 2(x - y)$
   $= 6 \times 3x - 6 \times 2y + 2 \times x - 2 \times y$
   $= 18x - 12y + 2x - 2y$
   $= 20x - 14y$

Check out

1 For these rectangles, construct expressions for:
   i the area
   ii the perimeter

   a
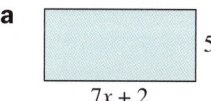

   b
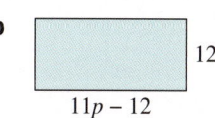

2 Expand the brackets and simplify:
   a $4(2x - 3)$
   b $6(2x + 7y)$
   c $3x(2 - 4y)$
   d $6x^2(3 - 4x)$
   e $3x(1 + 2y)$
      $-2y(1 - 2x)$
   f $4y(1 - y)$
      $+3y(2 - y)$

3 How to factorize an expression.
For example:
$3ax + 4ay = a(3x - 4y)$

4 How to substitute into expressions:
For example:
If $x = 2$, $y = {}^-3$ and $z = 1.5$, find the value of
$2y^2 - 4(x + z)$

                                       BIDMAS

$= 2 \times ({}^-3)^2 - 4(2 + 1.5)$ Brackets
$= 2 \times ({}^-3)^2 - 4 \times 3.5$ Indices
$= 2 \times 9 - 4 \times 3.5$ Multiplication
$= 18 - 14$
$= 4$

3 Factorize:
  a $20x + 15y$
  b $3x - x^2$
  c $4xy - x^2$
  d $6x + 72$

4 If $x = 2$, $y = {}^-3$ and $z = 1.5$, find the value of:
  a $xy + z$
  b $x^2 - 2y$
  c $5(x + y)^2$
  d $\dfrac{4z - y}{x}$

Shapes and mathematical drawings

Objectives

In this chapter you will learn about:

- properties of quadrilaterals
- Euler's formula
- plans, elevations and maps
- bearings
- constructions.

What's the point?

Architects, draughtsmen, engineers and carpenters are professionals who rely on accurate mathematical drawings for their work. There are many other professions that also use mathematical drawings, without which their work would be extremely difficult.

Before you start

You should know ...

1 How to use a protractor to draw and measure angles.

2 How to interpret a scale on a map.
For example: a distance of 10 m in real life is represented by 5 cm on a map. The scale of the map is 5 cm = 10 m. You can work out the value in real life for something that is 7 cm on the map using a scaling method like this:

$$
\begin{array}{ll}
\div 5 \Big\{ & 5\,\text{cm} \to 10\,\text{m} \\
& 1\,\text{cm} \to 2\,\text{m} \\
\times 7 \Big\{ & 7\,\text{cm} \to 14\,\text{m}
\end{array} \Big\} \div 5 \\ \Big\} \times 7
$$

Check in

1 Use your protractor to draw angles of:
   a 30° **b** 45°
   c 72° **d** 143°

2 A distance of 4 cm in a map represents 12 km in real life.
   a A distance of 11 cm on the map represents what length in real life?
   b A distance of 15 km in real life represents what length on the map?

3.1 Quadrilaterals

Classifying shapes

You can sort shapes in many different ways, for example by:

- the number of corners
- the number of curved edges
- the length of edges
- the size of angles
- the number of lines of symmetry.

'Bisect' means 'divide into two equal parts'.

Quadrilaterals

Quadrilaterals can be classified by their sides, angles or diagonals.

Quadrilateral		Sides	Angles	Diagonals
Trapezium		1 pair parallel	—	—
Isosceles trapezium		1 pair parallel, other pair equal	2 pairs of equal angles	Equal in length
Kite		2 pairs of adjacent sides equal	1 pair of opposite angles equal	One diagonal bisects the other at right angles
Parallelogram		Opposite sides parallel and equal	Opposite angles equal	Bisect each other
Rhombus		All sides equal, opposite sides parallel	Opposite angles equal	Bisect each other at right angles Diagonals bisect angles
Rectangle		Opposite sides parallel and equal	All angles 90°	Bisect each other and equal in length
Square		All sides equal, opposite sides parallel and equal	All angles 90°	Bisect each other at right angles and equal in length Diagonals bisect angles

Some shapes possess all the properties of another shape. For example, a square is also a rectangle because the square satisfies all the conditions required of a rectangle.

Exercise 3A

1 Copy and complete these sentences about quadrilaterals using words from the cards below. Cards may be used more than once.

| parallelogram | rhombus | trapezium |
| square | kite | rectangle |

a   A . . . has 4 right angles and all the sides are equal in length.

b   A . . . has diagonals equal in length that bisect each other at right angles.

c   A . . . has diagonals that are unequal in length but one diagonal bisects the other at right angles.

d   A square, a rhombus and a rectangle are also . . .s.

e   A . . . has only one pair of parallel sides.

f   A . . . and a . . . have diagonals that bisect their angles.

2 Are these statements true or false?
a All parallelograms are quadrilaterals.
b All rectangles are parallelograms.
c All squares are rhombuses.
d All rectangles are squares.

3 A pair of identical isosceles triangles are joined along an equal side. Draw diagrams to show that the resulting quadrilateral can be:
a a parallelogram
b a rhombus
c a kite.

4 ABC is a right-angled isosceles triangle with B = 90°. ACX is an isosceles triangle drawn on the side AC. Describe the triangle ACX if:

> In your description explain which are the pair of equal sides in triangle ACX and explain if you think ACX is congruent or not congruent to ABC.

a ABCX is a kite
b ACX = 105°
c ABCX is a square

5 AC and BD are diagonals of a quadrilateral intersecting at O. Name the quadrilateral if:
a OA = OC, OB > OD and AOB = 90°
b OA = OC, OB = OD and AOB > 90°
c OA = OB = OC = OD

3.2 3D shapes

Euler's formula

A **polyhedron** (**polyhedra** is the plural) is a three-dimensional shape with flat faces that are all polygons, straight edges and sharp corners or vertices. For example:

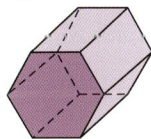

A hexagonal prism is a polyhedron, all its faces are polygons

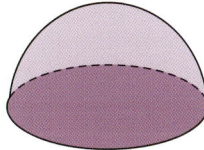

A hemisphere is not a polyhedron because it has a curved face and curved edges.

Leonard Euler was a Swiss mathematician born in the 18th century. He found a formula for connecting the number of vertices, V, faces, F, and edges, E, in polyhedra.

Euler's formula is $V + F - E = 2$

For the hexagonal prism shown on this page, there are 12 vertices, 8 faces and 18 edges.

$V + F - E = 2$

$12 + 8 - 18 = 2$

Example 1

A dodecahedron has 20 vertices and 30 edges. How many faces does a dodecahedron have?

....

$V + F - E = 2$

$20 + F - 30 = 2$

$F = 2 + 30 - 20$

$F = 12$

A dodecahedron has 12 faces.

Exercise 3B

1 What is the value of $F + V - E$ for a pentagonal prism?

2 A tetrahedron has 4 faces and 4 vertices. How many edges does it have?

3 A polyhedron has 7 vertices and 12 edges. How many faces does this polyhedron have?

4 A polyhedron has 5 faces and 9 edges. How many vertices does this polyhedron have?

5 A 3D shape has 3 faces, 0 vertices and 2 edges. Is it a polyhedron? Justify your answer.

6 Copy and complete this table about polyhedra.

Number of vertices	Number of faces	Number of edges
6	8	
8		12
	4	6
20		30
V	F	
	F	E

> In the last two rows you will be writing expressions in terms of V and F (5th row) or F and E (6th row).

7 A regular polyhedron has 8 vertices and 12 edges. What is the name of this regular polyhedron?

8 Explain how you know a pyramid with 5 faces has 8 edges.

Investigation

Part 1

Find out about the platonic solids.

Part 2

Euler's formula only works for polyhedra that follow certain rules. For example, you cannot stick two cubes together by one vertex. Find out what other rules there are.

Plans, elevations and maps

In Level 7 you learned about plans and elevations to represent 3D objects. These are 2D representations of 3D objects.

Maps are examples of plan drawings representing something from real life. These can be accurately drawn to scale, as in the map of the USA (above right), or not drawn to scale at all, as in the map of the zoo.

Map of the zoo

Another example of a plan drawing, that has not been drawn accurately to scale, can often be found on websites selling property, such as the following diagram of an apartment as viewed from above.

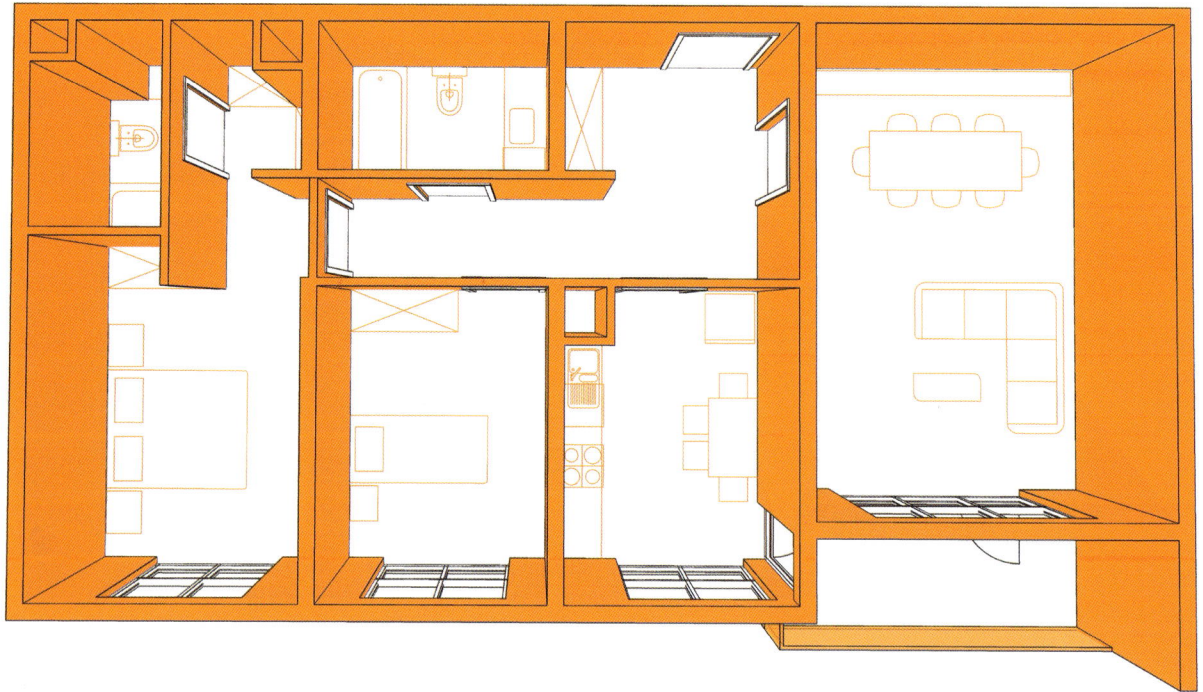

This side elevation of a car is an example of a technical scale drawing, which is drawn to scale and has essential measurements marked on it.

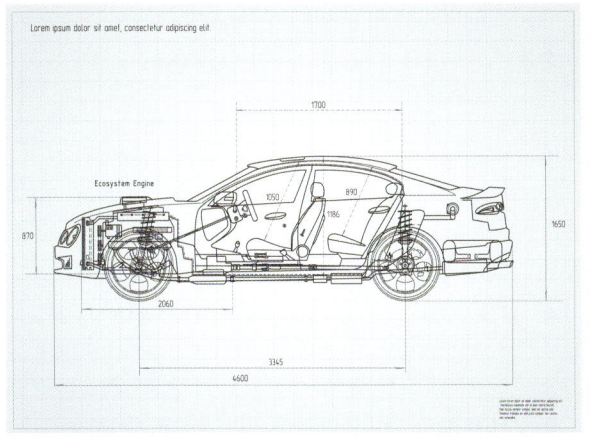

Accurate scale drawings, such as the car drawing, are needed to manufacture the object. Usually, if the drawing needs to be accurate, it is also drawn to scale. All lengths of 3 m in real life would all be the same length on the drawing, for example 3 cm.

Technical drawings used in manufacturing are usually now done on computers using CAD (Computer-Aided Design) software.

You can have 2D or 3D images in most CAD software. The images can be rotated to different angles and you can zoom in and out if you need to take large or small measurements from the images.

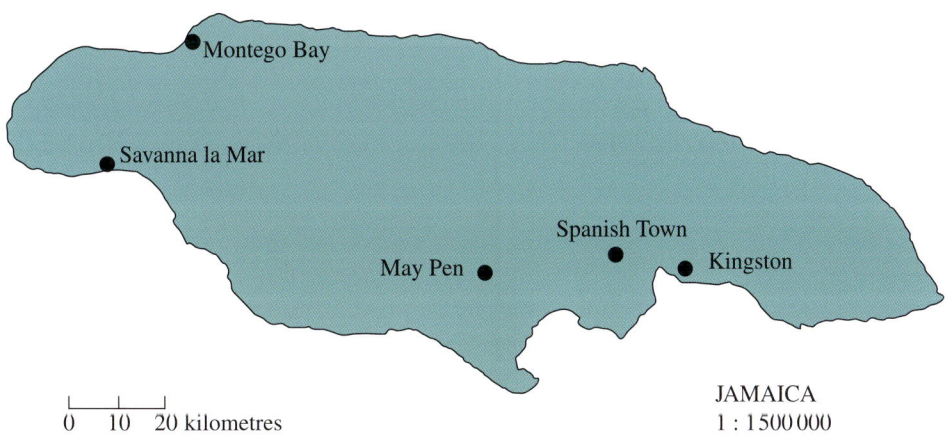

JAMAICA
1 : 1 500 000

Maps are usually drawn to a **scale**. The scale of the sketch map of Jamaica shown here is 1 : 1 500 000 This is the ratio of a length on the map to the actual distance in kilometres.

If you measure the distance on the map between May Pen and Spanish Town it is 1.8 cm.

Actual distance $= 1.8\,cm \times 1\,500\,000$

$= 2\,700\,000\,cm$

$= 27\,000\,m$

$= 27\,km$

Of course, this is the direct distance. The road distance may well be longer.

Example 2

A road map is drawn to a scale 1 : 50 000
Two towns are 5 km apart.
How far apart are they on the map?

....

$5\,km = 5 \times 1000\,m = 5000\,m$
$5000\,m = 5000 \times 100\,cm = 500\,000\,cm$
The scale is 1 : 50 000
$500\,000\,cm \div 50\,000 = 10\,cm$
The towns are 10 cm apart on the map.

Example 3

Draw the plan, front elevation and side elevation of this object.

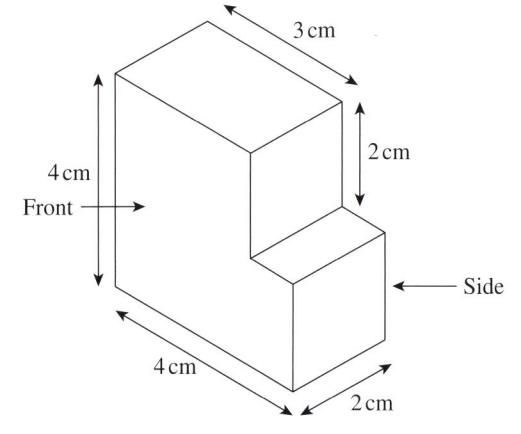

- -

Plan

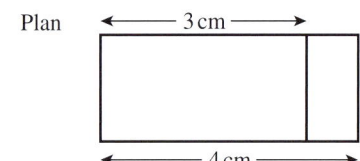

Notice that the plan includes the 'step' in the solid, shown by the vertical line inside the rectangle. This view doesn't tell you how deep the step is.

Side

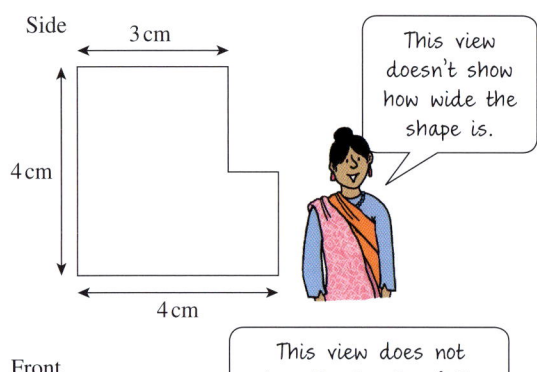

This view doesn't show how wide the shape is.

Front

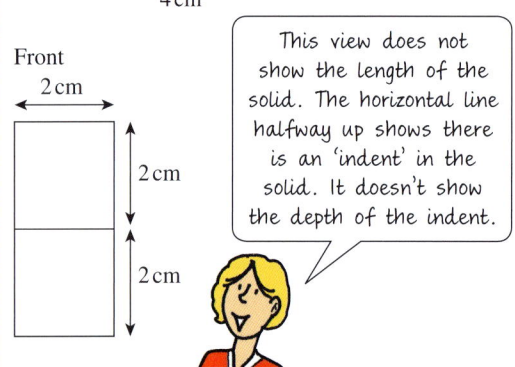

This view does not show the length of the solid. The horizontal line halfway up shows there is an 'indent' in the solid. It doesn't show the depth of the indent.

Exercise 3C

1. The distance on the sketch map on page 41 between Kingston and Spanish Town is 1 cm. What is the distance in kilometres?

2. Use a ruler to measure, in centimetres, the distance from Kingston to:
   a Montego Bay **b** May Pen
   What are the actual distances, in kilometres?

3. Use the map on page 41 to find, to the nearest kilometre, the greatest length of the island of Jamaica.

4. The distance from Kingston to Savanna la Mar is approximately 120 km. What is the distance on a map with a scale of 1 : 500 000?

5. This diagram shows a scale drawing of the plan and elevations of a cuboid.

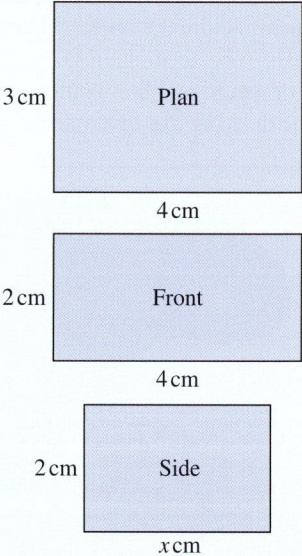

   a What is the height of the cuboid?
   b What is the value of x in the side elevation?
   c Amir uses a different scale. He draws the plan to these dimensions.

   What would be the sizes of the rectangles he draws for the front elevation and the side elevation?

6 This diagram shows the plan and elevations of a 3D shape. They are drawn to scale.

Plan

Front

Side

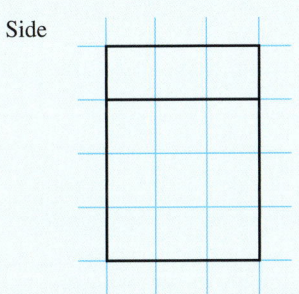

Here is a 3D diagram of the shape. Find the value of the missing side lengths.

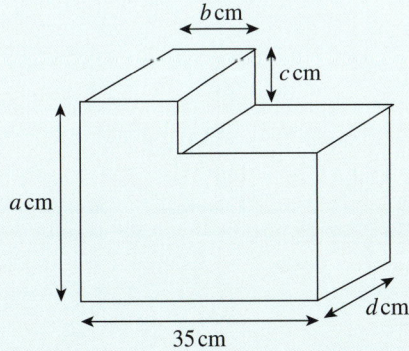

7 Here is a 3D diagram of a shape.

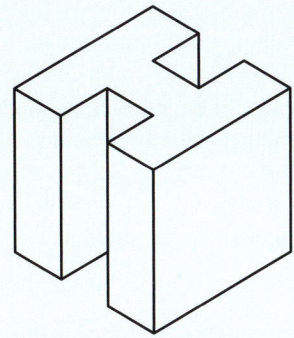

The front elevation is drawn as a scale drawing. It is a square with side length 6 cm. Draw the plan and side elevation using the same scale.

8 A map has a scale of 1 : 50 000
Find the actual distance between two places if the distance on the map is:
   a 1 cm   **b** 5 cm   **c** $\frac{1}{2}$ cm

Find the map distance if the actual distance is:
   d 5 km   **e** 10 km   **f** 1 km

9 On a map the scale is written 5 cm : 1 km
   a What is the distance, in metres, represented by 1 cm on the map?
   b What is the distance, in centimetres, represented by 1 cm on the map?
   c Express the scale as a ratio 1 : n.

10 On a road map the scale can be shown as 1 : 100 000 or 1 cm : 1 km
Match these scales.

1 : 50 000	1 cm : 1 km
1 : 100 000	2 cm : 25 km
1 : 1 250 000	10 cm : 1 km
1 : 10 000	1 cm : 4 km
1 : 4 000 000	2 cm : 1 km
1 : 400 000	1 cm : 40 km

11 What distance does 1 cm represent on a scale of:
   a 1 : 50 000   **b** 1 : 200 000
   c 1 : 5000   **d** 1 : 1 250 000
   e 1 : 5 000 000?

12 The scale on a road map is 1 : 25 000
   a What is the distance, in metres, represented by 3 cm?
   b What is the area of a field represented on the map by a rectangle 3 cm long and 4 cm wide?
   c What is the area of the field in hectares? ($10 000 \text{ m}^2 = 1 \text{ ha}$)

13 Sheldon makes a scale model of his bedroom.
The dimensions of the model are 8 cm long,
6 cm wide and 4 cm high. He uses a scale
of 1 : 50

a What is the area of the floor in the model?

b What is the actual area of Sheldon's
bedroom floor, in cm²?

c Copy and complete:
Area in the model : actual area
$= 48 : \square$
$= 1 : \square$

d What is the volume of the room in the
scale model?

e What is the actual volume of Sheldon's
bedroom?

f Copy and complete:
Volume in the model : actual volume
$= 192 : \square$
$= 1 : \square$

g How do the ratios of the areas in part **c**
and the volumes in part **f** compare to the
ratio of the lengths (1 : 50)?

In general, if the ratio of the lengths is 1 : a then the
ratio of the areas is 1 : a^2 and the ratio of the volumes
is 1 : a^3

3.3 Bearings

The position of an object relative to another object is
called its **bearing**.

The bearing is given as a three-figure angle, such as
045°. It is measured in a *clockwise* direction from North.

For example:

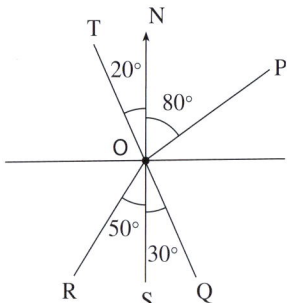

The bearing of P from O is 080°
The bearing of Q from O is 150°
The bearing of R from O is 230°
The bearing of T from O is 340°

When we write the bearing of A from B, this is short
for the bearing of A from the North line at B.

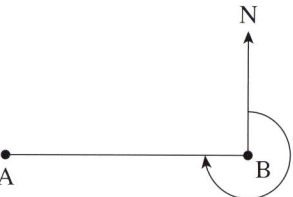

The bearing of B from A is different! Make sure you
understand the difference.

Exercise 3D

1 Which of these is not a bearing?

125°   048°   359°   12°

2

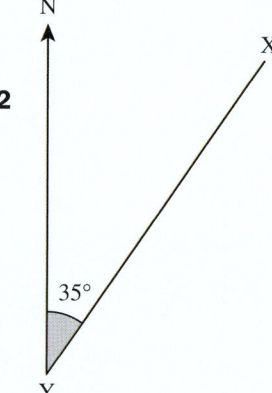

Jen says the bearing of X from Y is 35°. She
has not written this correctly. What mistake
has she made?

3

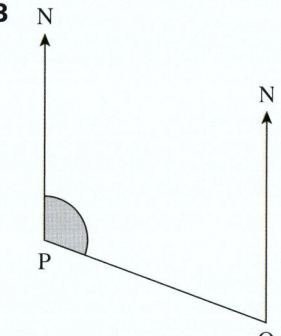

Which of these statements is correct?
The shaded angle shows:

● the bearing of P

● the bearing of P from Q

● the bearing of Q from P

● the bearing from Q to P.

4   Measure the bearing of B from A in each of
    these diagrams.

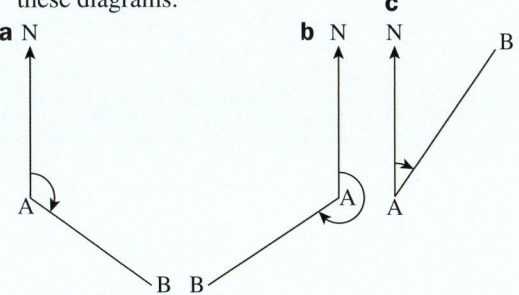

5   Draw each of these bearings.

   a   B from A is 035°

   b   C from D is 205°

   c   E from F is 345°

6   Here is a diagram showing compass points.

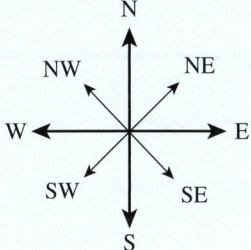

   Find the bearing of:

   a   SE

   b   S

   c   NE

   d   NW

   e   W

7   Measure the bearing of A from B in each of
    these diagrams. ('From B' means from the
    North line at B.)

   a

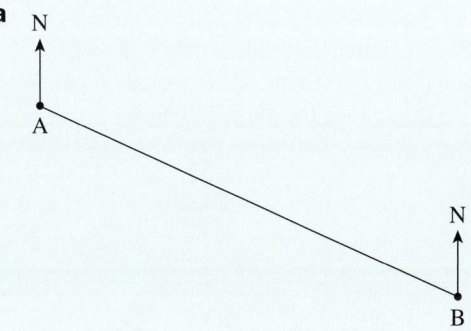

b

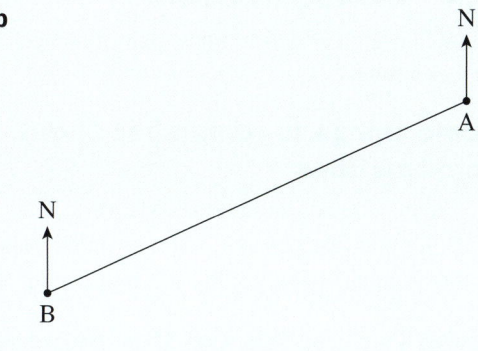

c

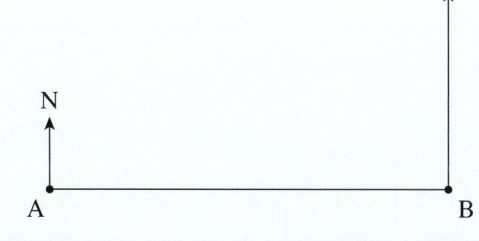

d

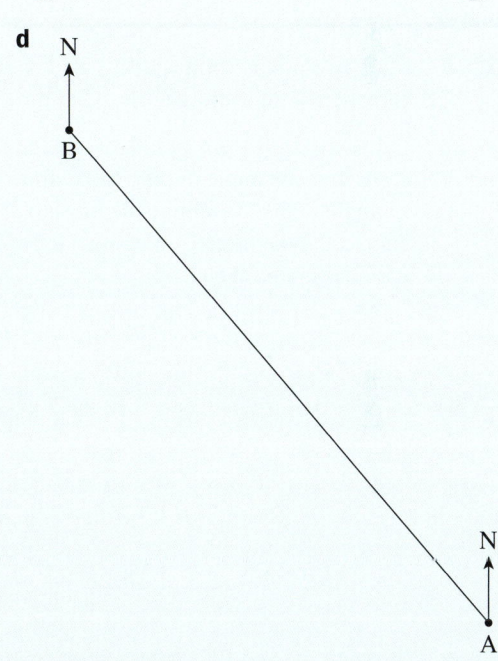

8   In each of the diagrams in question **7**, find
    the bearing of B from A.

9   Can you see a relationship between your
    answers to question **7** and your answers to
    question **8**?

3.4 Constructions

You will need a protractor and a pair of compasses as well as a ruler.

Some things to remember about constructions

- The pencil you use should be sharp.
- Don't press the pencil too heavily on the paper.
- Measure and draw lengths and angles carefully.

Given two angles and the included side (ASA)

You can draw triangles if you know two angles and the side length between them.

Exercise 3E

You need a protractor for this exercise to measure and draw the angles.

1 a Mark two points 7 cm apart.
Join them with a straight line. Call the line AB.

b At A, draw an angle of 50°. At B, draw an angle of 80°. Continue the lines to intersect at the point C, making a triangle ABC, as in this sketch.

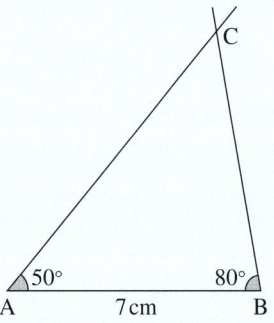

c Measure AC and BC. What sort of triangle have you drawn?

2 Make an accurate drawing of each of these triangles.

a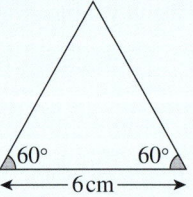

b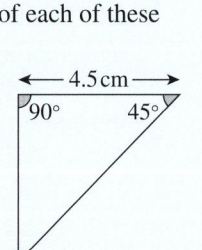

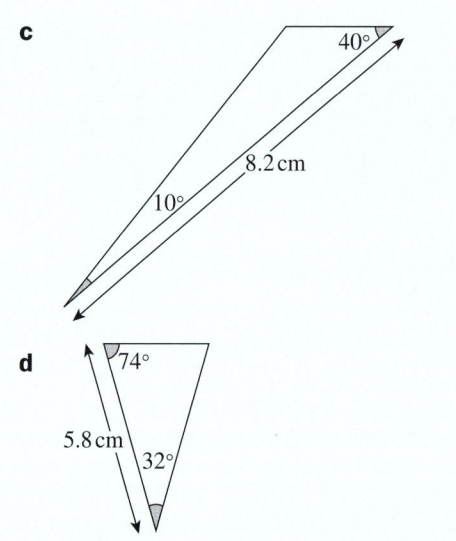

3 a Measure the unmarked angles in question **2**.
b Name the type of each triangle you drew.

4 Construct these triangles.
a Triangle ABC where AB = 6.8 cm, ∠BAC = 55° and ∠ABC = 40°
b Triangle DEF where DE = 7.4 cm, ∠EDF = 110° and ∠DEF = 42°

Using compasses

To draw other sorts of triangles you will need a pair of compasses.

First you need to know how to draw a line accurately using compasses.

Example 4

Draw the line AB exactly 6 cm long.

a First draw a line longer than 6 cm, say about 7.5 cm.
b Next, mark a point A near one end.
c Open the compasses to exactly 6 cm on your ruler, as in the diagram below.

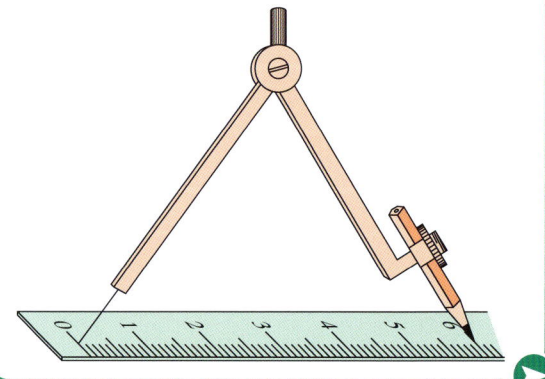

d An arc is part of a circle. Put the point of the compasses at A, and draw a small arc to cut the line.
Call the point where the arc intersects the line B.
You now have a line AB, 6 cm long.

Given two sides and the included angle (SAS)

You can construct a triangle if you know two side lengths and the angle between them.

Exercise 3F

1 Using compasses, draw the line:
  a MN, 6 cm   **b** PQ, 6.5 cm
  c ST, 5.8 cm   **d** CD, 7.2 cm
  e AB, 8.6 cm   **f** GH, 7.9 cm

2 a Using compasses, draw the line XY, 6.2 cm.
  b At X, draw an angle of 50° using your protractor.
  c Open the compasses to 5.1 cm. With the point of the compasses at X, draw an arc 5.1 cm along the second arm of the angle, as in the drawing below.
Call the point Z.

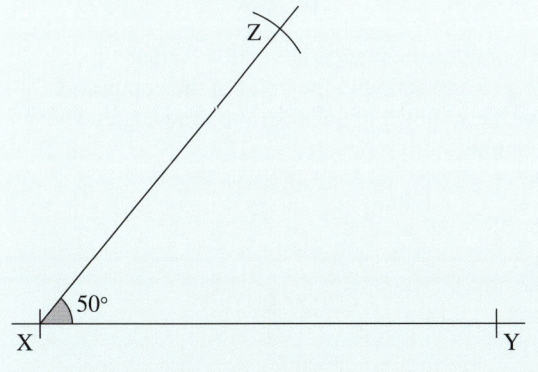

  d Join YZ. What sort of triangle have you constructed?

3 a Draw the line AB, 7.3 cm.
  b At B, draw an angle of 27°.

c With the point of the compasses at B, mark a point C, 5.9 cm from B, as in the drawing below.

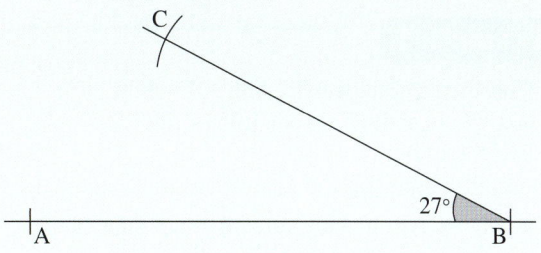

  d Join AC. What sort of triangle have you constructed?

4 In questions **2** and **3**, you could draw a triangle when you knew just two sides and the angle between them. Use the same method to make an accurate drawing of each of these triangles.

  a 4.7 cm 97° 5.3 cm
  b 4.9 cm 69° 6.8 cm
  c 3 cm 90° 4 cm
  d 136° 7.2 cm 7.2 cm

5 Construct these triangles.
  a Triangle ABC where AB = 7.9 cm, AC = 9.5 cm and ∠BAC = 45°
  b Triangle DEF where DE = 6.7 cm, DF = 7 cm and ∠EDF = 71°

Given three sides (SSS)

The construction of a triangle given the lengths of its sides is quite easy.

Example 5

Construct a triangle ABC with AB = 8 cm, AB = 6 cm and AC = 4 cm, using a ruler and compasses only.

....

a Draw AB, making sure it is 8 cm long.

A ——————————— B

b With your compasses, draw an arc of radius 6 cm centred at B.

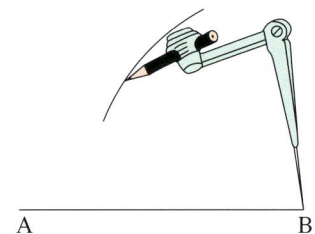

A   B

c Draw an arc of radius 4 cm centred at A.

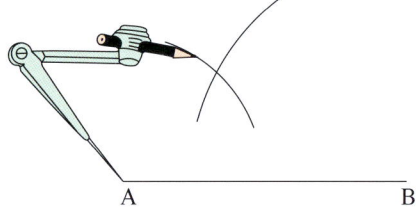

A   B

d Label the point where the two arcs intersect as C.

e Join AC and BC.

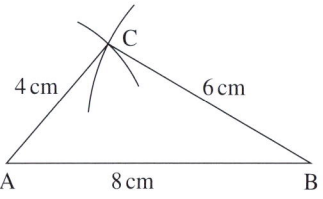

4 cm   C   6 cm

A   8 cm   B

Given a right angle, hypotenuse and one side (RHS)

In a right-angled triangle there are two perpendicular sides and a third side. The third side is opposite the right angle and is called the **hypotenuse**. The hypotenuse is the longest side of the right-angled triangle because it is opposite the largest angle in the triangle.

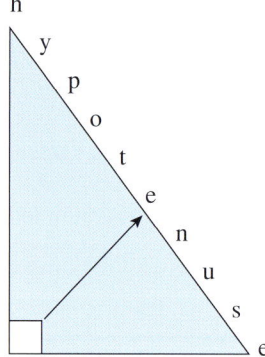

A right-angled triangle can be constructed using a pair of compasses.

Example 6

Construct the right-angled triangle PQR where PQ = 6 cm and the hypotenuse QR = 10 cm.

....

Draw a line longer than 6 cm. Mark P and Q on that line, 6 cm apart. You now need to construct the right angle at P.

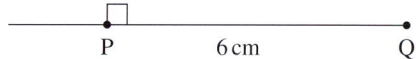

P   6 cm   Q

To construct the right angle, open the compasses to a short length (e.g. 2 cm), put the point of the compasses on P and draw arcs on the line either side of P. These arcs give you points A and B.

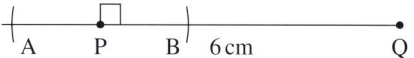

A   P   B   6 cm   Q

Then open the compasses a little further (e.g. 5 cm) and with the point of the compasses on A (as shown in the diagram) draw an arc above the line.

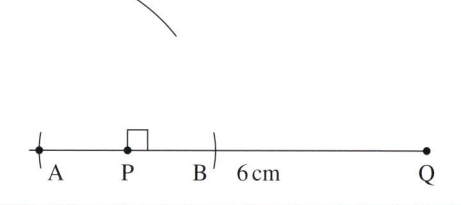

A   P   B   6 cm   Q

Keeping the same radius, and this time with the point of the compasses on B, draw another arc above the line.

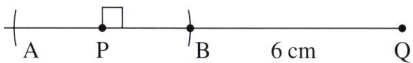

Draw a line up from P through the intersection of these two arcs, J. Remember, you do not know how long this line needs to be yet.

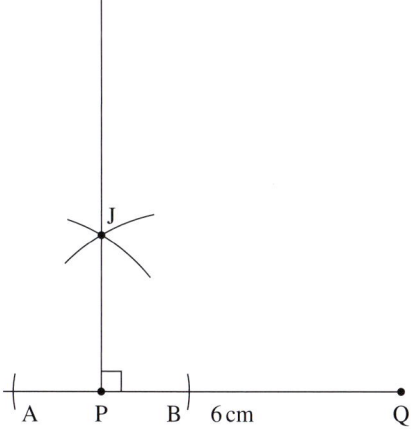

Put the point of the compasses on Q and with the radius set at 10 cm draw an arc on the line from P through J. Where the arc crosses the line is the point R.

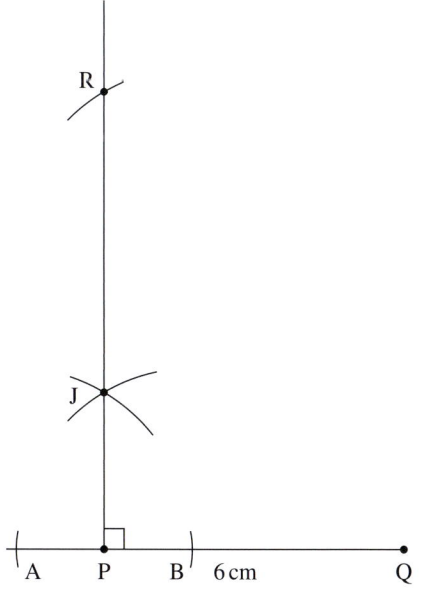

Finally, draw line QR to complete triangle PQR.

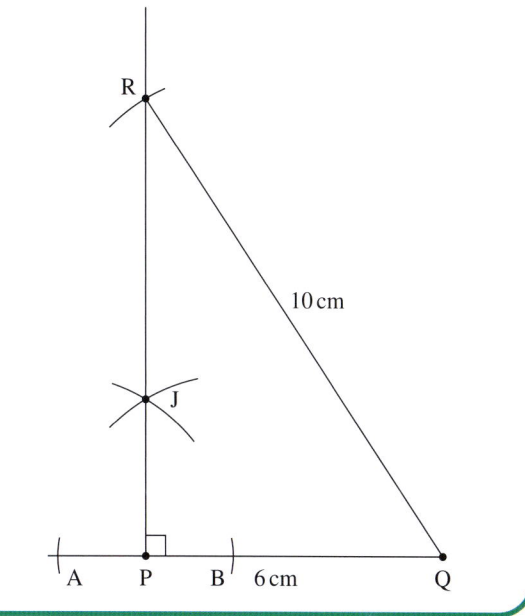

Exercise 3G

1   Construct triangles with sides:
   a   3 cm, 4 cm and 5 cm
   b   6 cm, 8 cm and 10 cm
   c   7 cm, 6 cm and 9 cm
   d   6.3 cm, 4.2 cm and 8.7 cm
   e   11.3 cm, 7.9 cm and 6.3 cm

2   Construct these right-angled triangles.
   a   ABC, where AB = 4 cm and the hypotenuse BC = 5 cm
   b   RST, where RS = 5 cm and the hypotenuse ST = 13 cm
   c   MNO, where MN = 7 cm and the hypotenuse NO = 8.5 cm

3   Draw the triangle ABC such that:
   a   AB = 8 cm , $\hat{CAB} = 60°$ and AC = 7 cm
   b   AB = 9 cm, $\hat{CAB} = 45°$ and AC = 6.8 cm
   c   AB = 4.8 cm, $\hat{CAB} = 30°$ and AC = 7.3 cm

You can use a protractor for this question.

4 Construct a right-angled isosceles triangle of your choice.

5

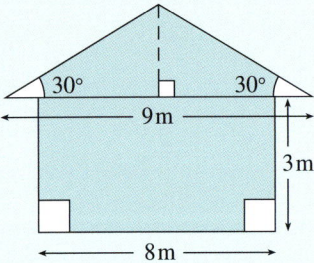

This diagram shows the side view of a house.

Using a scale of 1 cm to represent 1 m, make an accurate drawing of the side view.

6 In the triangle DEF, DF = 10 cm, FX = 6 cm and XE = 4 cm. Construct an accurate copy of triangle DEF.

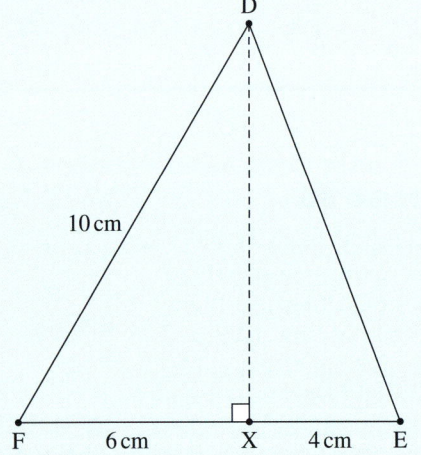

7 Explain why it is not possible to construct a triangle with sides 3 cm, 5 cm and 10 cm.

3.5 Bisecting angles and lines

The word **bisect** means to cut exactly in half. If you bisect a line segment you can find the halfway point, called the **midpoint**. If you bisect an angle you will be able to find half the angle. In this section you will learn how to bisect angles and lines.

In your diagrams you will draw lots of arcs to help you. These are called **construction lines** – do not rub them out, as they show the method you have used.

Perpendicular bisector of a line segment

You can bisect the line PQ with a pair of compasses and a ruler.

Draw an arc with centre P and radius more than $\frac{1}{2}$PQ. Draw another arc, centre Q, with the same radius.

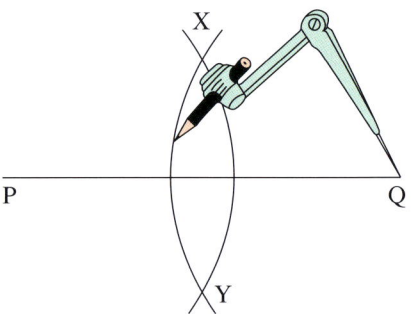

The arcs meet at X and Y. Join XY.

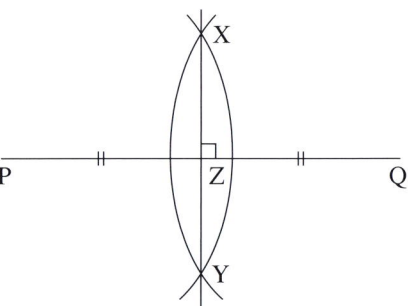

XY is **perpendicular** to PQ and bisects PQ at the midpoint Z. PZ = QZ. Perpendicular means at right angles.

Exercise 3H

1 Draw a line PQ = 8 cm.
  a Bisect the line.
  b Each part of your bisected line PQ should be 4 cm long. Check by measuring.

2 Copy the following lines and construct their perpendicular bisectors. (When you copy the lines keep them at approximately the angles shown.)

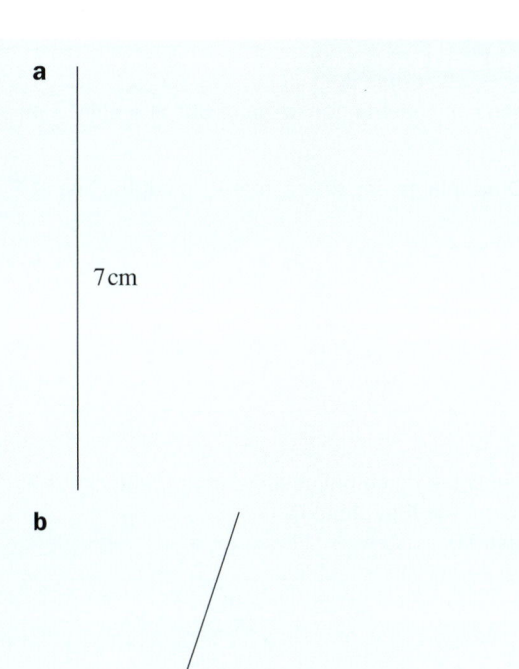

a

7 cm

b

9 cm

c

6 cm

3 a Draw four different triangles.
  b Construct the perpendicular bisectors of each side of the triangle.
  c What do you notice?

4 For your triangles in question **3**, are you able to draw a circle so that all three corners of the triangle lie on the circumference?
(**Hint:** use your answer to part **c**.)

5 Draw an approximate copy of this triangle, about twice as big.

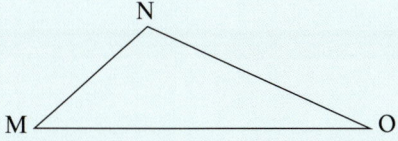

N

M   O

Construct a circle that passes through the three vertices M, N and O. This is called the **circumcircle** of the triangle.

Bisecting an angle

You can bisect the angle ABC with a pair of compasses and a ruler.

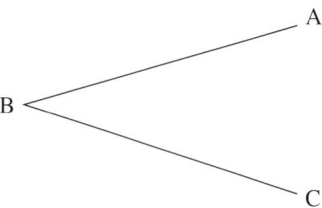

Draw an arc with centre B to cut AB at X and BC at Y.

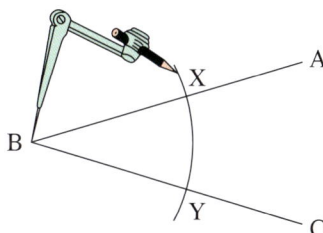

Draw two more arcs with the same radius, centred at X and Y. Label the point where they meet Z.

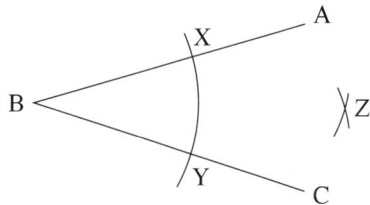

Join BZ. This line bisects the angle ABC.

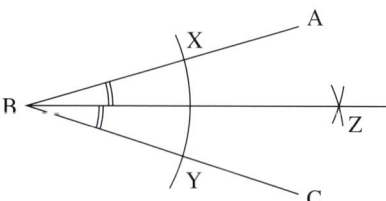

Exercise 3I

1 a Draw an angle of 90° using a protractor.
  b Bisect it to make an angle of 45°.

2 a Draw an angle of 60° using a protractor.
  b Bisect it to make an angle of 30°.

3 Draw an approximate copy of each angle, then bisect it.

a

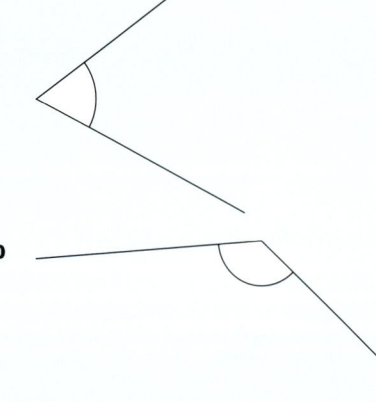

b

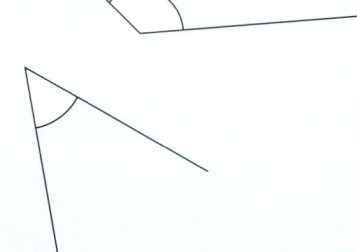

c

d

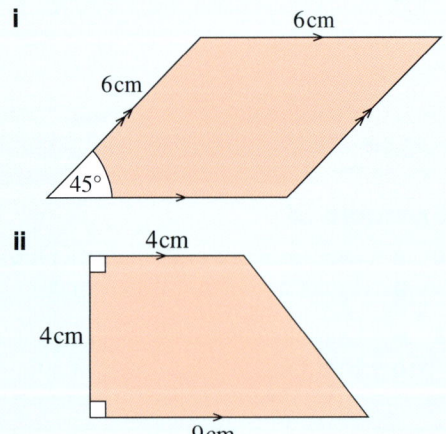

4 a Draw a straight line. This represents an angle of 180°.

b Bisect the angle of 180° to get an angle of 90°.

5 a Using only a ruler and pair of compasses, construct these shapes.

i

6cm
6cm
45°

ii

4cm
4cm
9cm

b What are the names of the shapes you constructed in part **a**?

Investigation

You can construct an angle of 60° at a point X on a line.

Draw a large arc, with centre X, to cut the line at P.

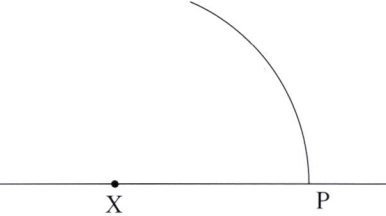

X   P

Using the same radius draw an arc, with centre P, to cut the first arc at Q.
Join XQ.

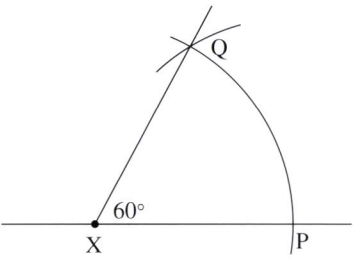

Q
60°
X   P

The angle QXP = 60°.
Investigate how to construct these angles using what you have learned: 45°, 30°, $22\frac{1}{2}°$, 15°, 75° and 135°.

Consolidation

Example 1

Write down four properties of:

a a parallelogram

b a square

....

a Parallelogram

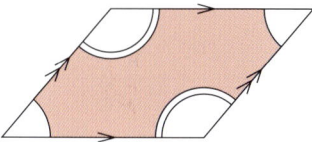

Two pairs of parallel sides
Two pairs of equal sides
Two pairs of equal angles
Diagonals bisect each other

b Square

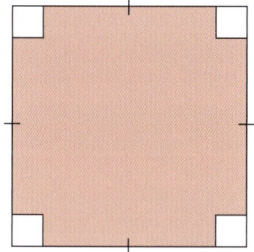

Two pairs of parallel sides
Four sides equal in length
Four right angles
Four lines of symmetry

Example 2

A polyhedron has 4 vertices and 6 edges. How many faces does this polyhedron have?

....

$V + F - E = 2$

$4 + F - 6 = 2$

$F = 2 + 6 - 4$

$F = 4$

It has four faces.

Example 3

This diagram shows the plan and elevations of a 3D shape. They are drawn to scale.

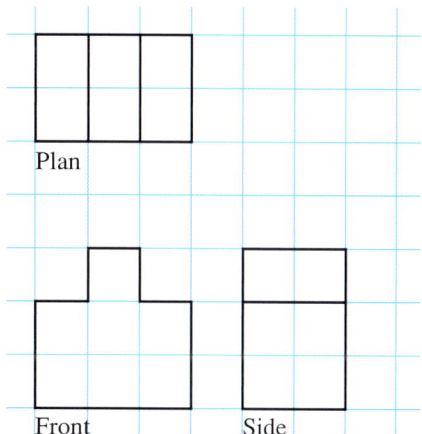

Here is a 3D diagram of this shape. Find the value of the missing side lengths (a and b).

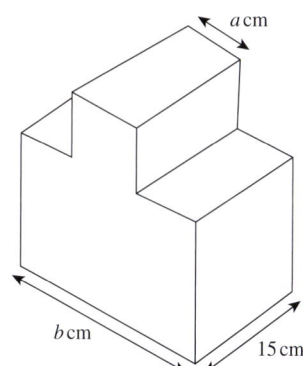

....

a 2 squares = 15 cm so 1 square = 7.5 cm

b 3 squares = $7.5 \times 3 = 22.5$ cm

Example 4

What is the bearing of B from A?

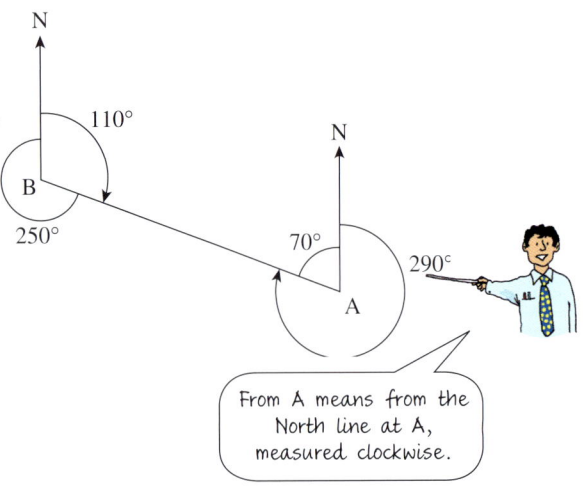

The bearing of B from A is 290°.

Example 5

Construct the triangle XYZ with

$XY = 8 \text{ cm}$, $Z\hat{X}Y = 35°$ and $Z\hat{Y}X = 65°$

1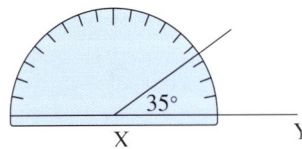

Draw base line XY 8 cm long.

2 Put the centre of your protractor at X then measure 35° at X.

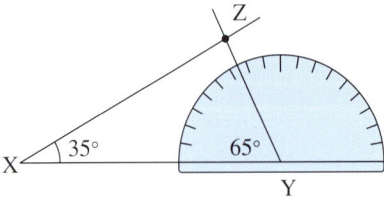

3 Repeat step 2, this time measuring 65° at Y.

4 Put point Z where these lines meet.

Example 6

Construct the triangle ABC with $B\hat{A}C = 90°$, $AB = 7 \text{ cm}$ and hypotenuse $BC = 10 \text{ cm}$.

First draw the line AB.

A ———————————————— B

Then construct a perpendicular at A.

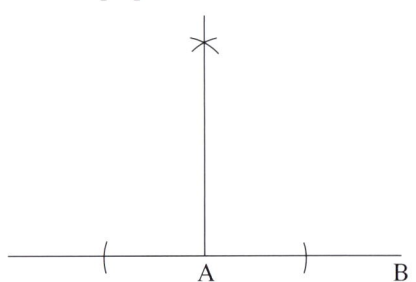

Open the compasses to 10 cm and, with the point of the compasses on B, draw an arc on the perpendicular line. Join B and C.

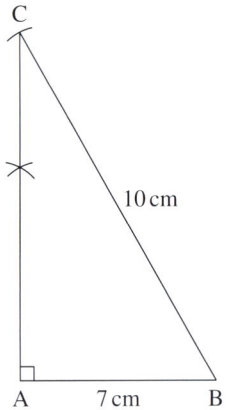

Exercise 3

1 Using ruler and compasses only, construct the triangle XYZ with:

    a   XY = 7 cm, YZ = 8 cm, XZ = 12 cm

    b   XY = 8 cm, XZ = 10 cm, XŶZ = 90°

2 Copy and complete this table.

Quadrilateral	Lines of symmetry	Number of right angles	Number of pairs of parallel sides	Do diagonals bisect each other?	Are diagonals equal in length?	Do diagonals bisect the angles?
Square						
Parallelogram						
Kite						
Trapezium						

3 What distance does 4.3 cm represent on a scale of:

    a   1 : 10000   **b**   1 : 25 000

    c   1 : 500 000   **d**   1 : 200?

4 A polyhedron has 5 faces and 8 edges.
How many vertices does this polyhedron have?

5 **a**   What is the bearing of A from B?   **b**   What is the bearing of C from D?

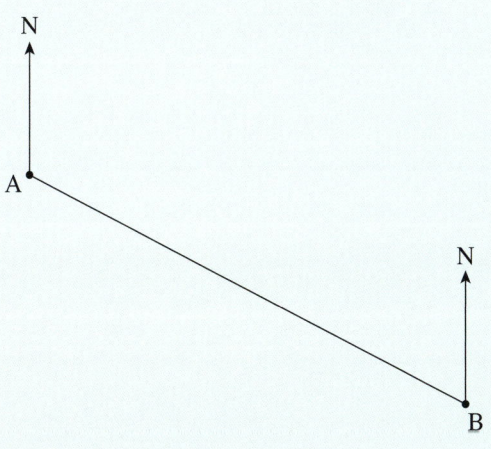
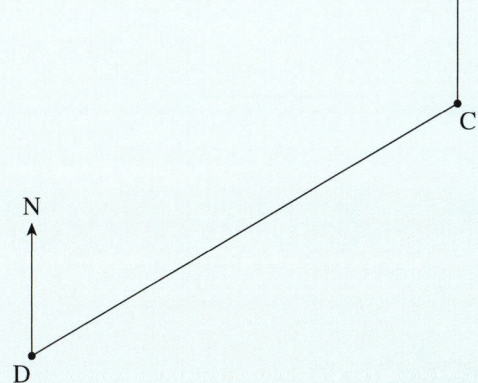

6 Which quadrilateral am I?

    a   I have four lines of symmetry.

    b   I have one line of symmetry. My diagonals are not equal in length.

    c   I have two lines of symmetry. My diagonals are equal in length.

    d   I have order of rotational symmetry 2 and four right angles.

    e   My diagonals bisect each other at right angles. I contain no right angles.

7 Construct these triangles.

    a   ABC, AB = 7 cm, BÂC = 40°, AB̂C = 40°

    b   JKL, JK = 7.5 cm, LĴK = 60°, JL = 6 cm

8 Draw a line 7 cm long.

    a   Construct the perpendicular bisector of the line.

    b   Label the midpoint of the line M.

9 This diagram shows the plan and elevations of a 3D shape. They are drawn to scale.

Plan

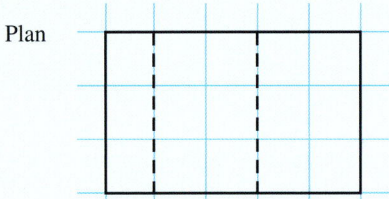

Front

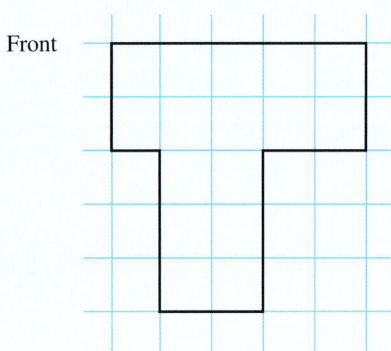

Side

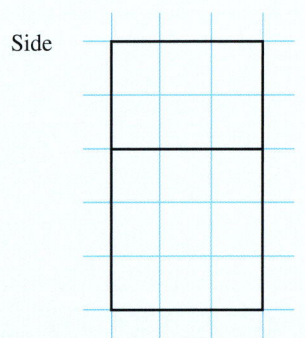

Here is a 3D diagram of this shape. Find the value of the missing side lengths.

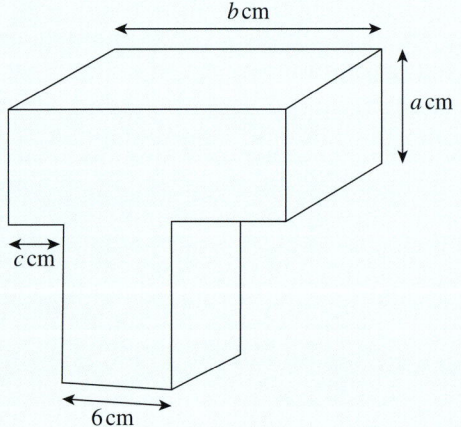

10 Draw an angle of 58°. Bisect the angle using only a pencil and compasses.

11 Describe as many properties as you can of a parallelogram. Include rotational symmetry, lines of symmetry and diagonals among the properties you consider.

12 A polyhedron has 8 vertices and 12 edges. How many faces does this polyhedron have?

13 a Construct the triangle ABC where AB = 4.5 cm, BC = 6 cm and AC = 7.5 cm.
 b What sort of triangle have you constructed?
 c What is the special name for side AC in this triangle?

14 The distance of 7 km would be represented by what length on a map with scale:
 a 1 : 50000 **b** 1 : 200 000
 c 1 : 1000 000 **d** 1 : 250 000?

15 Construct a triangle with:
 a one line of symmetry
 b three lines of symmetry

Summary

You should know ...

1 How to classify shapes in terms of:
- number of lines of symmetry
- number of right angles
- number of equal sides
- number of parallel sides
- properties of diagonals
- order of rotational symmetry.

2 Euler's formula for polyhedra.
For example:

A polyhedron has 12 faces and 30 edges.

How many vertices does this polyhedron have?

$V + F - E = 2$
$V + 12 - 30 = 2$
$V = 2 + 30 - 12$
$V = 20$

It has 20 vertices.

3 How to draw scale drawings.
For example:

This diagram shows a scale drawing of the plan and elevations of a cuboid.

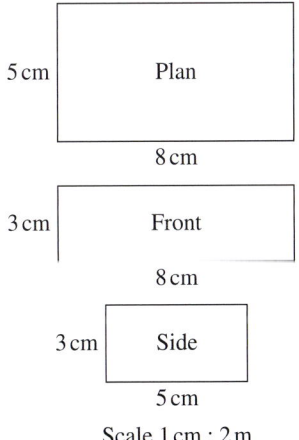

Scale 1 cm : 2 m

Check out

1 Write down three properties of:
  a a rhombus
  b a rectangle
  c a kite
  d a parallelogram

2 Copy and complete this table about polyhedra.

Number of vertices	Number of faces	Number of edges
	7	14
5	9	
7		15

3 The height of the real-life cuboid is $3 \times 2 = 6$ m.

Find the length and width of the real-life cuboid.

4 How to use bearings.
Bearings are three-digit angles measured clockwise from the
North direction.

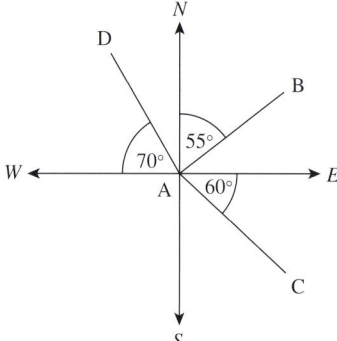

The bearing of B from A is 055°.

5 How to use a ruler and compasses to:
a construct a triangle
b construct the perpendicular bisector of a line
c bisect an angle.

4 Find these bearings.
a C from A
b D from A

5 a Construct triangle
ABC with AB = 7 cm,
AC = 12 cm and
AB̂C = 90°.
b Construct the triangle
with side lengths
5 cm, 7 cm and 8 cm.
c Draw a line 5 cm long.
Construct the
perpendicular bisector
of this line.
d Using a protractor,
draw an angle of 40°.
Construct the bisector
of this angle.

Sampling

Objectives

In this chapter you will learn about:

- justifying your choice of method of collecting data
- the advantage of one sampling method over another.

What's the point?

Scientists use statistical sampling all the time. For example, if they wanted to see how good a new vaccine was, they would choose a sample of the population to vaccinate to see how well the vaccine performed. They would also need to check that there were no serious side effects before going on to vaccinate lots more people. Choosing the correct sample size and the correct people to best represent the population is very important.

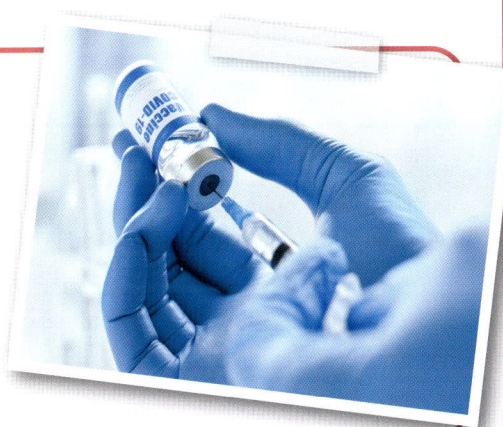

Before you start

You should know ...

How to make predictions before a statistical investigation and use these predictions to decide how to collect your data.

For example:
You may predict that school dinners are not very popular among students in your school. One way to collect data is to send out questionnaires to a sample of students in school.

Check in

What other methods of data collection do you know?

4.1 Methods of data collection

In the modern world huge amounts of data are collected every week by many different people. Some examples are shown in this table.

Person needing data	Purpose
Store manager	Stock control
Pollster	Find out public opinion on an issue
Teacher	Determine student progress
Company manager	Improve quality of product or service

To collect data you need a suitable method of data collection. Some methods are:

- questionnaires or surveys
- data collection sheets
- observation
- interviews
- focus groups
- tests
- case study
- opinion polls or online quizzes
- records and documents (that is, secondary data rather than primary).

The data you want to collect usually determines the method of data collection.

If you want to find out how popular certain clubs are in school, such as the football club or chess club, you would probably choose to give out a questionnaire.

There are advantages of choosing a questionnaire.

- Questionnaires are easy to use with a large sample size.
- They can be anonymous, so people may be more honest if sensitive issues are involved.
- Questionnaires are relatively cheap.
- More people can be involved.
- Questionnaires work well in combination with other methods, such as observation, case study.

There are also disadvantages of choosing a questionnaire.

- People may not complete it.
- Questions may be misunderstood. Only brief explanations can appear on the questionnaire and some questions could still be misunderstood by some individuals.

The next exercise is to get you to think about advantages and disadvantages of different kinds of data collection methods. You may want to do this exercise working in a group or a whole class so that you can discuss answers with other students.

Exercise 4A

1   Adam wants to find out what people think about his shop. He decides to interview people when they buy something.
    a   What advantages are there in conducting an interview?
    b   What disadvantages are there in conducting an interview?

2   Neeta says, 'A focus group is a good way to collect data because people in the group can help each other generate fresh ideas.' Write down some disadvantages of surveying a group of people all together in this way.

3   A school inspector decides to collect data on the quality of a school by observations. She observes a sample of teachers' lessons. Helga says, 'A disadvantage of observation is that the inspector may decide to join in with the lesson activity or help the teacher, then they won't be able to observe as well.' Write down some advantages of using observation as a method of data collection – it doesn't have to be in a classroom setting.

4   Samir is the manager of a company. His company own lots of shops. He wants to see which shops are doing well and to try to work out why some are doing better than others.

He decides to look at records of sales from a sample of different shops in his company. Give one advantage and one disadvantage of using secondary records and documents as a method of data collection.

5 | questionnaire | | observation | | tests |

| case study | | opinion polls |

| records and documents |

Use words from the list of different data collection methods to complete these sentences. Justify your choice in each case.

a A teacher wanting to see how much her students have understood this term should use . . . because . . . .

b A gym owner wants to know which exercise classes are most popular should use . . . because . . . .

c An . . . is a good method to use if you want to study people working in their own environment without them knowing that you are collecting data because . . . .

6 Lena wants to open a new takeaway restaurant. She wants to find out people's favourite kind of takeaway. She decides to collect data by interview. Give a reason why this could be a good choice of data collection method for Lena.

7 Javiera thinks that the most popular type of movie in the USA is action.
a Explain why a questionnaire would not be an appropriate method of data collection.
b What would be an appropriate method of data collection?

8 A new gym wants to find out some information about the health of people in the city. They interview people asking these two questions.
● How old are you?
● What is your weight?

a Discuss whether or not these are appropriate questions for an interview.
b Is there a better data collection method for this data? Justify your answer.

4.2 Sampling methods

There are many different ways you can choose a sample. These are some of the more popular methods.

● Random sampling – examples are: taking names out of a hat; using a random number generator.
● Systematic sampling – for example, you might use every 10th person in a list.
● Convenience sampling – for example, you might use the first 50 people you see.
● Stratified sampling – for example, if the population you are studying has twice as many females as males then the sample should have the same proportion: it should also have twice as many females as males.
● Cluster sampling – for example, you split the population into groups, where each group has similar characteristics to the whole population. Instead of sampling individuals from each group, you randomly select a few of the groups and survey everyone in the group.
● Quota sampling – for example, you split the population into groups and pick 10 people from each group.

There are many other types of sampling. While it is useful to know the names of sampling methods, you don't have to learn them at this stage.

Example 1

Adam wants to know whether girls or boys send the longest text messages in his year group at school. There are 120 girls and 60 boys in his year group. He gets a list of the names of all these students and chooses every 20th student from the list for his sample. He then plans to ask the students in his sample to write down the length of the last 10 text messages they wrote.

a What sort of sampling is this?
b Explain the advantages of using this method.
c If Adam decided to choose a sample size of 30 by stratified sampling, how many boys and how many girls would he choose?

a Systematic sampling
b It is simpler and quicker than other methods, such as random sampling. If the list is in an order, such as alphabetical order, this should

still make the sample representative. The first letter of your surname is unlikely to be related to the length of your text messages.

c Adam should choose a sample of 20 girls and 10 boys. As there are double the number of girls in the population for his survey, there should be double the number of girls in the sample.

Exercise 4B

1 Catalina wants to find out what students in her school think about the amount of homework set. She is considering the sampling methods in **a** to **d**. Name each method and the advantages and disadvantages of using it.

a Choose 10 students randomly from each age group in the school (which gives her a sample of 70).

b Choose 70 students names randomly from a list of all students in the school, for example using a random number generator to decide which students to pick.

c Choose every 10th student from an alphabetical list of all students in the school.

d Ask the first 70 students who arrive at school.

2 Farook wants to find out how people in a town feel about local recycling facilities. He has two possible methods for sampling.

Method 1 – sample the first 100 people he sees recycling at the local supermarket.
Method 2 – get a list of all the people in the town and use computer spreadsheet to randomize the names before he picks the first 100 names to sample.

a What is the name of sampling method 1?

b What is the name of sampling method 2?

c Give an advantage of each of these two sampling methods.

d Give a disadvantage of each of these two sampling methods.

3 Annie wants to buy a new apartment. She is looking at how the price of apartments relates to the number of bedrooms. There are 250 one-bedroom apartments for sale, 200 two-bedroom apartments for sale and 50 three-bedroom apartments for sale. She decides to sample 50 apartments to compare prices.

a Explain why a stratified sample would be a better choice for Annie than a random sample.

b There are 20 two-bedroom apartments in Annie's stratified sample. How many one-bedroom apartments will there be?

c Taking a cluster sample and sampling all 50 of the three-bedroom properties is not a sensible choice. Explain why.

4 Write down something you are interested in investigating. Decide how you might choose a sample and justify your choice of sampling method.

Consolidation

Example 1

To find out what people think about facilities at a local football ground you could send out a questionnaire.

a What advantages are there in sending out a questionnaire rather than conducting interviews?

b What advantages are there in conducting an interview rather than sending out questionnaires?

....

a A questionnaire can reach more people faster than an interview; it is likely to be cheaper; an interviewer can bias results with their own thoughts; people are likely to be less embarrassed if there are any personal questions; it is easier to reach a wider area (by post or online).

b In an interview, you can explain any questions that are not understood; there is likely to be a higher response rate; the interview can be more personalized.

Example 2

Explain the advantages and disadvantages of random sampling.

....

Random sampling means everyone has a chance of being chosen and the process should be relatively free from bias. It can be time consuming to sort out how to take the sample, for example by using random number generator, taking names out of a hat.

Exercise 4

1 Explain the advantages and disadvantages of collecting data by observation.

2 A company with 250 employees wants to make changes to its pay structure.
The managers want to find out what the employees think about the current pay structure.
They are thinking of using one of two data collection methods:
- interviewing each employee
- giving a questionnaire to each employee

Which method would you chose? Justify your answer.

3 Ravi owns a bookshop. He wants to know if people are more likely to buy books from his shop or from the internet.
  a Explain why interviewing his customers is not a good data collection method.
  b Explain why picking the first 50 people to walk past his shop might be a good sampling method.

4 Tomasz wants to open a café. He is interested to find out where people in the town would like his café to be located. He chooses between these two sampling methods.
Method 1 – group the town into 10 regions and select all the people in two of the regions.
Method 2 – group the town into 10 regions and select a sample of people from each region in proportion to the number of people living in the region.
  a What is the name of sampling method 1?
  b What is the name of sampling method 2?
  c Give an advantage for each of these sampling methods.
  d Give a disadvantage for each of these sampling methods.

Summary

You should know ...

1 The advantages and disadvantages of different methods of data collection.

For example:

These are some advantages of using a data collection sheet.
- It is cheap. (There is no need to photocopy lots of questionnaires.)
- Data collection sheets are useful if you do not need to find out lots of different things (for example if you can just complete tally marks).
- Data is already grouped for you so analysis afterwards is easier.
- There is a low chance of non-response.

This is a disadvantage of using a data collection sheet.
- It is not useful if lots of detailed answers are required.

2 The advantages and disadvantages of different sampling methods.
For example:
Here are some advantages of using convenience sampling,
(for example picking the first 50 people you see).
- It is easy to do.
- It is quicker and cheaper than other methods.

Here is a disadvantage of using convenience sampling.
- It is not random so can be affected by other factors, such as where you take the sample.

Check out

1 Write down the advantages and disadvantages of using a questionnaire.

2 Write down the advantages and disadvantages of random sampling.

5 Area, perimeter and volume

Objectives

In this chapter you will learn about:

- π and the circumference of a circle
- the area of a parallelogram and a trapezium
- the volume of a triangular prism
- the surface area of prisms and pyramids
- miles and kilometres.

What's the point?

Estate agents deal with the sale of properties. The key criteria for fixing the price of a piece of land are its location and its area. The larger the area, the more valuable the land. Precise measurements of land area are therefore very important!

Before you start

You should know ...

1 Area can be measured in mm^2, cm^2, m^2 and km^2.

$\square$ 1 mm
1 mm

1 cm
1 cm
1 cm

Check in

1 Estimate the area of:
  a a postage stamp
  b your thumbnail
  c a page of your exercise book

2 How to draw nets.
For example:
The net of this
triangular prism

is

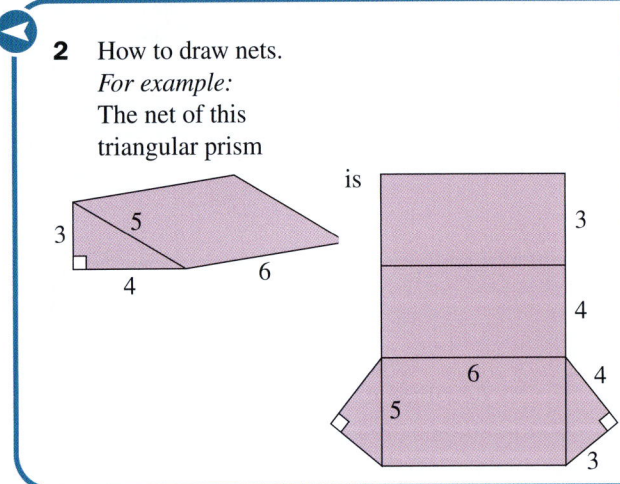

2 Draw the nets of these shapes.

a

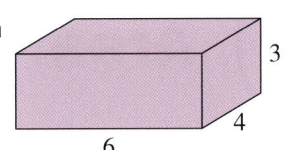

b

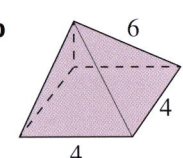

5.1 Circumference of a circle

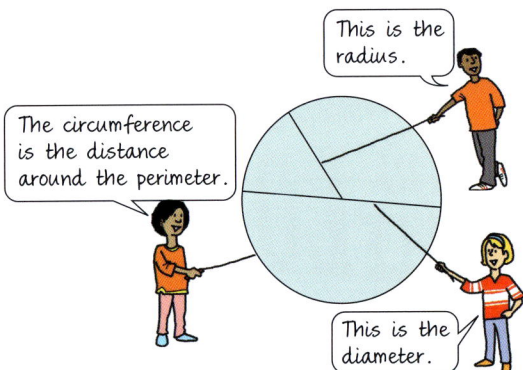

This is the radius.

The circumference is the distance around the perimeter.

This is the diameter.

- The **radius** of a circle is the length of a line drawn from the centre to a point on the circle.
- The **diameter** of a circle is the length of a line passing through the centre from one point on the circle to another (diameter = 2 × radius).
- The **circumference** of a circle is the distance around it.

Activity

Here is a good way to find the circumference of a cylindrical tin.

a Wrap a thin strip of paper around it, making sure the ends overlap. Stick a pin through the overlap.

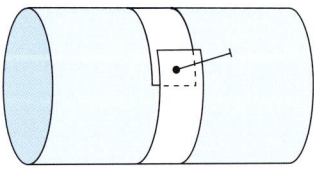

b Unwrap the paper and lay it flat on the table.

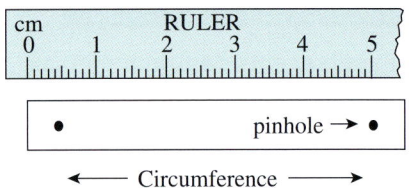

cm   RULER
0   1   2   3   4   5

•   pinhole → •

←——— Circumference ———→

c Measure the distance between the two pinholes. This gives the circumference of the tin.

d Use this method to find the circumference, C, of five cylindrical objects.

e With your ruler carefully measure the diameter, D, of each object.

f Copy and complete the table below.

Object	Circumfernce C cm	Diameter D cm	C ÷ D
tin	25.1 cm	8 cm	
plate			
cup			

What do you notice about the values of C ÷ D in your table?

If you were careful you should have found that

$C \div D$ is roughly 3.1

That is:

The circumference of a circle is just over three times the diameter of the circle.

The circumference of a circle is given more accurately by the relation

$$C = \pi \times D$$

The Greek letter π, or pi (pronounced 'pie'), cannot be found exactly. It is about 3.14 or $3\frac{1}{7}$.

Since

$$D = 2 \times \text{radius } (r)$$

You can also write

$$C = 2\pi r$$

- The circumference of a circle is approximately $3.14 \times$ diameter or $2 \times 3.14 \times$ radius.

Example 1

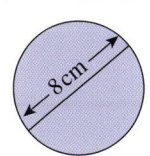

Find the circumference of a circle with diameter 8 cm. Take $\pi = 3.14$

Circumference $= \pi \times$ diameter
$$= 3.14 \times 8 \, \text{cm}$$
$$= 25.12 \, \text{cm}$$

Exercise 5A

Use 3.14 for π in this exercise.

1

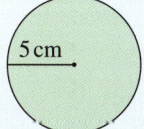

Anastasia works out the circumference of this circle. Here is her working.
$C = 3.14 \times 5 = 15.7 \, \text{cm}$
Is Anastasia correct? Explain your answer.

2 Calculate the circumference of a circle with a diameter of:
  a 2 cm   **b** 10 cm
  c 12 cm   **d** 21 cm

3 Calculate the circumference of a circle with a radius of:
  a 5 cm   **b** 8 cm
  c 13 cm   **d** 39 cm

4 The centre circle on a playing field has a radius of 7.5 m. Find its circumference.

5 A bicycle wheel has a diameter of 70 cm. What is the circumference of the wheel?

6 Each year the Earth goes around the Sun in a nearly circular path.

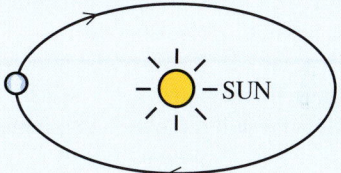

The Earth is about 150 000 000 km from the Sun. Assuming the path is a circle, how far does the Earth travel each year?

7 Here is Patrick's method to find the diameter of a circle with circumference 26 m.
$C = 3.14 \times d$
$26 = 3.14 \times d$
$26 \div 3.14 = d$
$d = 8.28025 \ldots$
$d = 8.3$ correct to 1 d.p.

Use Patrick's method to find the diameter of these circles. Give your answers correct to 1 d.p. The circumferences are:
  a 23 cm
  b 46 m
  c 12 mm

8 Find the radius of these circles. Give your answers correct to 1 d.p. The circumferences are:
  a 15 cm
  b 420 mm
  c 19 m

9 Work out the circumference of a circle with diameter 20 cm using each of these values of π. Give your answers correct to 1 d.p.

  a   3.14

  b   3.142

  c $\frac{22}{7}$

  d   the π key on your calculator

10 Repeat question **9** but round to 2 d.p. this time.

11 Look at your answers to questions **9** and **10**. The most accurate answer is part **d**, the one using the π key on your calculator.

Which of 3.14, 3.142 or $\frac{22}{7}$ gives the:

  a   least accurate approximation

  b   most accurate approximation?

12 The distance from the tip of the minute hand to the centre of a clock is 6 cm.

  a   How far will the tip of the minute hand move in one hour?

  b   How far will it move each minute?

13 A circular toy railway has a radius of 1.4 m. Calculate the time that a toy train will take to travel once round the track at a constant speed of 22 cm/s. Give your answer to the nearest second.

14 The diameter of the Earth is about 12 750 km. Find the distance around the equator.

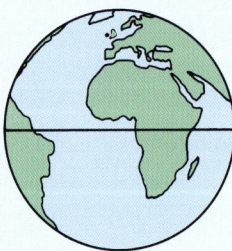

15 Find the perimeters of these shapes.

  a

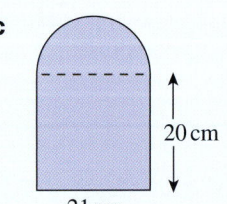

      14 cm

  b   1 cm   1 cm

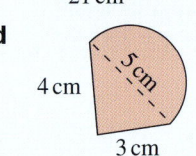

                 5 cm

        9 cm

  c

                    20 cm

        21 cm

  d

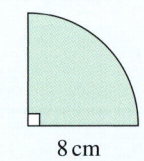

    4 cm   5 cm

         3 cm

  e

        8 cm

16 An athletics track consists of two equal semicircles joined by 100-metre straights.

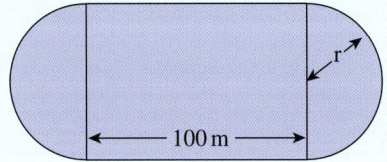

          100 m

The radius of each semicircle is 35 m. Find the perimeter of the track.

17 The distance around the inner circle of a rubber tyre is 220 cm.

  a   Find the inner radius.

  b   If the thickness of the tyre is 14 cm, find the distance around the outside of the tyre.

5.2 Areas of parallelograms and trapeziums

Area of a parallelogram

Activity

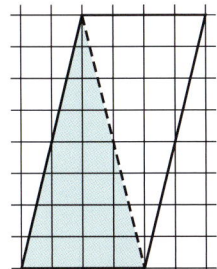 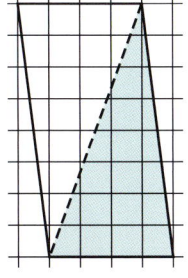

a Make a tracing of the shaded triangle in the first parallelogram above.
b Does your tracing fit exactly on to the white triangle in the first parallelogram?
c What can you say about the white triangle and the shaded triangle?
d Repeat parts **a** to **c** for drawing B.
e What is the connection between the area of each parallelogram above and the area of its shaded triangle?
f Can you suggest a quick way to find the area of a parallelogram?

A parallelogram is made up of two identical triangles. The area of each triangle is $\frac{1}{2}(b \times h)$. The area of the parallelogram is twice this.

- The area of a parallelogram is $A = b \times h$
  (b is base length, h is vertical height).

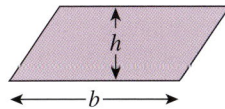

The area of a parallelogram can be shown another way.

Cut a triangle off the end of the parallelogram:

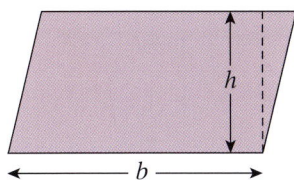

Put the triangle you cut off on the other end of the parallelogram:

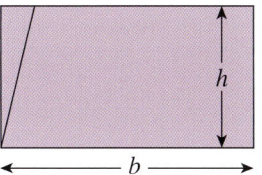

You can see it makes a rectangle.

So a parallelogram has the same area as a rectangle with the same base and height.

Example 2

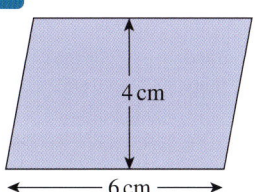

What is the area of the parallelogram?

Area of the parallelogram = base × height
$$= 6\,cm \times 4\,cm$$
$$= 24\,cm^2$$

Exercise 5B

1 Find the area of each parallelogram.

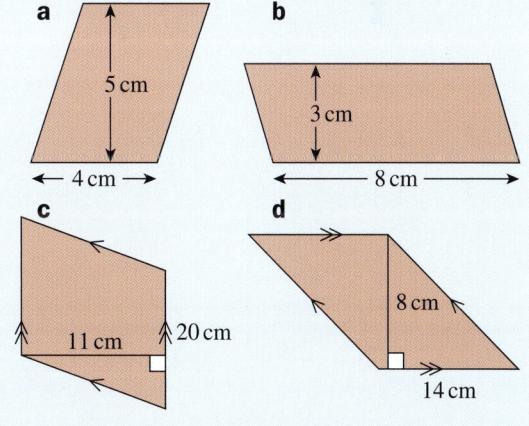

2 Find the area of the following parallelograms.

a

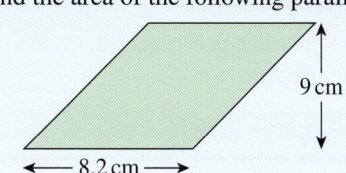

b

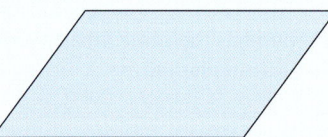

10.2 cm

← 9.8 cm →

3 By measuring carefully, find the area of this parallelogram.

4 Copy and complete this table for parallelograms.

	Base (cm)	Height (cm)	Area (cm²)
a	3.5	8	
b		16	144
c	6.5		52
d	2.3	3.2	
e		7.1	26.27

5 Find the missing side lengths.

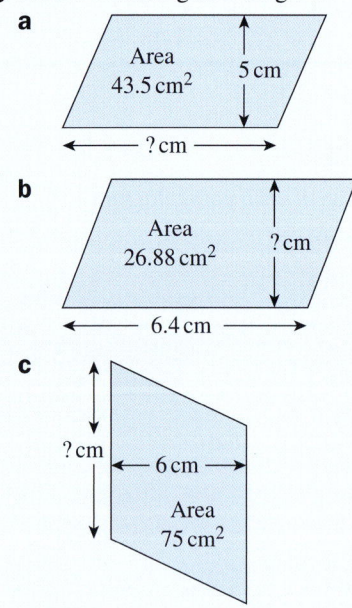

a

Area 43.5 cm²   5 cm

← ? cm →

b

Area 26.88 cm²   ? cm

← 6.4 cm →

c

? cm   ← 6 cm →

Area 75 cm²

Area of a trapezium

This shape has only one pair of parallel sides. It is called a **trapezium**.

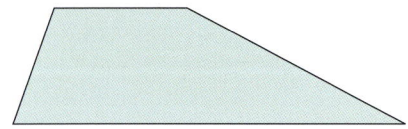

- A trapezium is a quadrilateral with one pair of sides parallel.

The area of a trapezium can be found by dividing it into two triangles.

Example 3

Find the area of this trapezium.

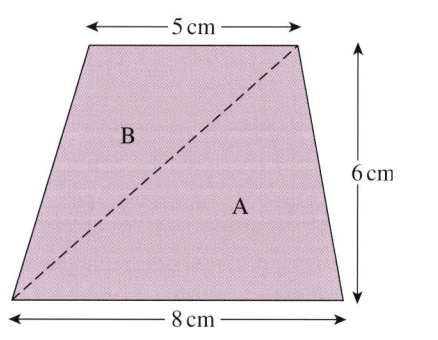

Divide the trapezium into two triangles, A and B.

Area of triangle A $= \frac{1}{2} \times 8\,\text{cm} \times 6\,\text{cm}$

$= 24\,\text{cm}^2$

Area of triangle B $= \frac{1}{2} \times 5\,\text{cm} \times 6\,\text{cm}$

$= 15\,\text{cm}^2$

Area of trapezium $= A + B$

$= 24\,\text{cm}^2 + 15\,\text{cm}^2$

$= 39\,\text{cm}^2$

Exercise 5C

1 For each diagram, find the area of:
   i the shaded triangle
   ii the unshaded triangle
   iii the trapezium

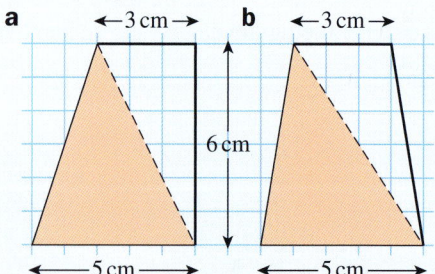

2 Draw a set of trapeziums each with a base of 5 cm, a height of 6 cm and a top edge of 3 cm. Do they all have the same area?

3 Find the area of this trapezium.

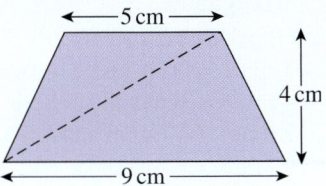

④ Work out the area of this parallelogram made up of two identical trapeziums.

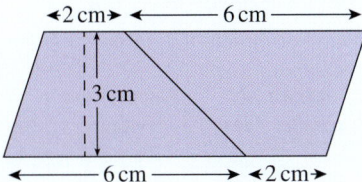

Look at this trapezium.

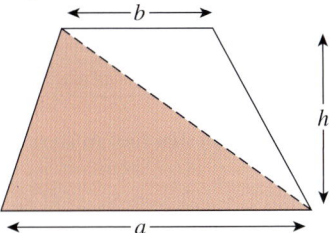

The area of the shaded triangle $= \frac{1}{2} a \times h$

The area of the white triangle $= \frac{1}{2} b \times h$

The area of the trapezium $= \frac{1}{2} a \times h + \frac{1}{2} b \times h$

$$= \frac{1}{2}(a+b) \times h$$

The formula for the area of a trapezium can be derived another way.
Look at this trapezium.

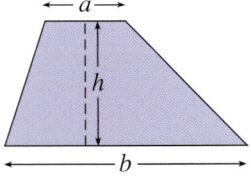

Rotate the trapezium round 180° and draw it touching the original trapezium:

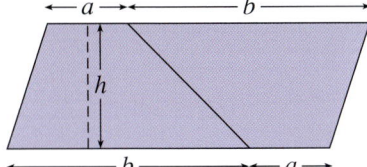

The new shape is a parallelogram with base $a + b$ and height h.

The area of the parallelogram is $(a + b) \times h$

The trapezium is half this shape and so the area is

$$\frac{1}{2}(a+b) \times h$$

- The area of a trapezium with two parallel sides of length a and b a perpendicular distance h apart is

$$A = \frac{1}{2}(a+b) \times h$$

Example 4

Find the area of the trapezium.

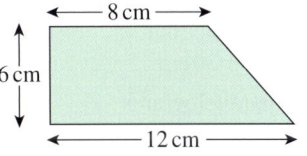

Area of trapezium $= \frac{1}{2}(a+b) \times h$

$$= \frac{1}{2}(12+8) \times 6$$

$$= \frac{1}{2} \times 20 \times 6$$

$$= 60 \text{ cm}^2$$

Compound shapes

A compound shape is what you get when two or more different shapes are combined to make a shape.

For example, this compound shape is made up of a triangle and a parallelogram.

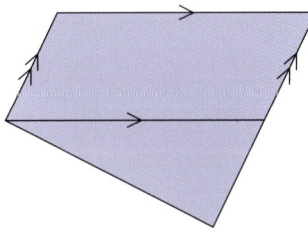

To work out the area of a compound shape, simply work out the area of the separate shapes then add them together to find the total area.

Some exercises you have completed so far have already used the idea of compound shapes. For example, see Exercise 5A, questions **15** and **16**.

Some areas may be worked out more easily using subtraction, rather than by adding up lots of smaller areas.

Example 5

Work out the shaded area in this compound shape.

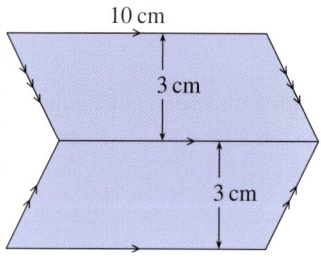

There are two parallelograms, each with base 10 cm and height 3 cm.

Area of one parallelogram
$= b \times h = 10 \times 3 = 30\,\text{cm}^2$

Area of both parallelograms
$= 30 \times 2 = 60\,\text{cm}^2$

Exercise 5D

1 Find the area of each trapezium.

a 3 cm 6.5 cm 7 cm

b 3 cm 5 cm 4.6 cm

c 8 cm 7 cm 4 cm

d 5 cm 7.3 cm 9 cm

e 12.5 cm 4.2 cm 8.5 cm

f 16 m 8 m 10 m

2 By measuring, find the area of this trapezium.

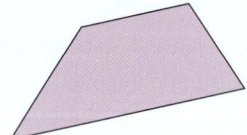

3 Using the formula $A = \frac{1}{2}(a+b) \times h$ find the area of a trapezium when:
   a $a = 7\,\text{cm}$, $b = 11\,\text{cm}$ and $h = 10\,\text{cm}$
   b $a = 4\,\text{cm}$, $b = 3\,\text{cm}$ and $h = 8\,\text{cm}$
   c $a = 2.6\,\text{cm}$, $b = 7.4\,\text{cm}$ and $h = 5.6\,\text{cm}$
   d $a = 9.3\,\text{m}$, $b = 6.3\,\text{m}$ and $h = 12.2\,\text{m}$

4 Find the area of a trapezium with parallel sides of length 24 cm and 16 cm and perpendicular height of 18 cm.

5 Copy and complete this table for trapeziums.

	Length of parallel sides		Perpendicular distance between PQ and RS	Area of trapezium
	PQ	RS		
a	9 m	15 m	7 m	
b	16.8 m	12.5 m	8.4 m	
c	23 cm	37 cm	40 cm	
d	12.4 m	6.8 m	4.5 m	
e	24 cm	16 cm	15.5 cm	

6 Draw a trapezium that has an area of $18\,\text{cm}^2$ and:
   a parallel sides of lengths 4 cm and 8 cm
   b a base of 5 cm and a height of 4 cm
   c a height of 6 cm

7 The area of a trapezium is $18\,\text{cm}^2$. Its parallel sides are 32 cm and 16 cm in length. Find the perpendicular height.

8 Find the area of this shape.

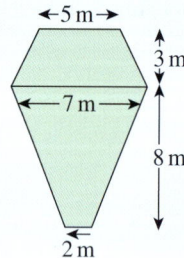

9 The diagram shows a piece of metal that is to be used to make the blade of a saw.

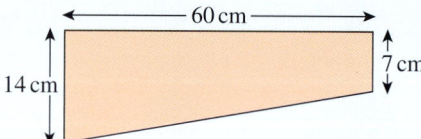

a Find the area of the blade.
b Metal costs $78 per square metre.
   What is the cost of metal contained in the blade?

10 The diagram shows a piece of wood that has been cut to make the deck of a toy boat. Find its area.

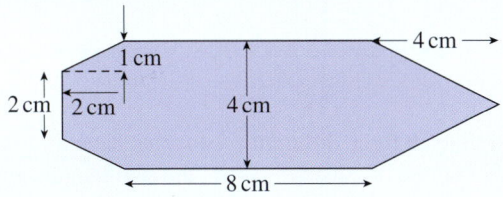

5.3 Volume of a triangular prism

A cuboid is one example of a **prism**.

Here are some more.

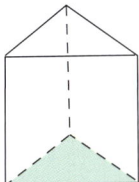 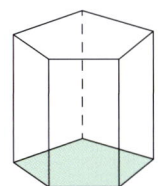

- A **prism** is a three-dimensional shape with a constant cross section.

Volume of a prism = area of cross section × height

- $V = A \times h$

Example 6

Find the volume of this prism.

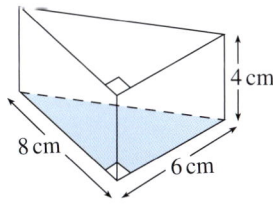

The base is a triangle.

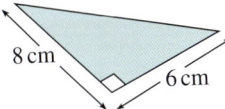

Area of base triangle, $A = \frac{1}{2}b \times h$

$= \frac{1}{2} \times 6\,\text{cm} \times 8\,\text{cm}$

$= 24\,\text{cm}^2$

Volume of prism $= A \times h$

$= 24\,\text{cm}^2 \times 4\,\text{cm}$

$= 96\,\text{cm}^3$

You can find missing side lengths for a prism when you know its volume and cross-sectional area.

Example 7

Calculate the missing side length, $x\,\text{cm}$, for this triangular prism with volume $336\,\text{cm}^3$.

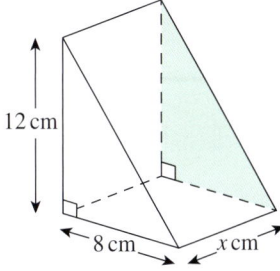

Cross-sectional area,

$A = \frac{1}{2}b \times h = \frac{1}{2} \times 8\,\text{cm} \times 12\,\text{cm} = 48\,\text{cm}^2$

Volume of prism, $V = A \times h$

Substituting $V = 336, A = 48, h = x$:

$336 = 48x$ [Solve to find x]

$x = 7$ [÷ 48]

so the missing side length is $7\,\text{cm}$.

Exercise 5E

1 Here is a triangular prism.

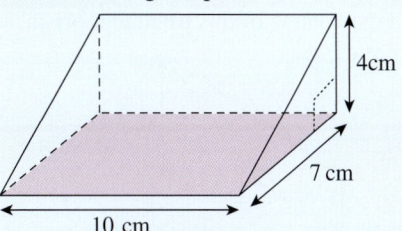

Maja and Manon each have a different way to work out the volume.

Manon's method

Area of cross section $= \dfrac{4 \times 7}{2} = 14\,\text{cm}^2$

Volume of prism $= 14 \times 10 = 140\,\text{cm}^3$

Maja's method

Volume of cuboid $= 4 \times 7 \times 10 = 280\,\text{cm}^3$

Volume of prism $= 280 \div 2 = 140\,\text{cm}^3$

Explain why both methods work.

2 Find the volumes of these triangular prisms.

a

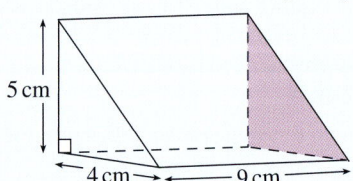

b

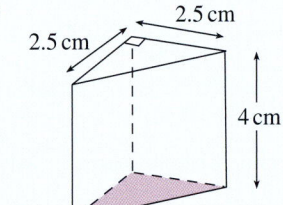

c

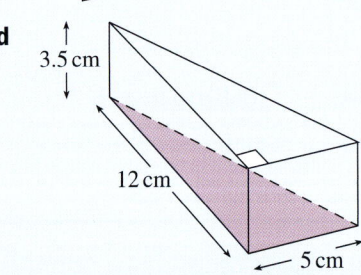

d

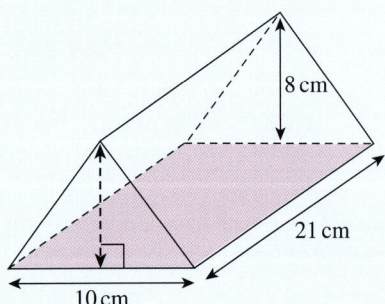

3 Find the volume of a triangular prism, with cross-sectional area 4 cm² and length 12 cm.

4 Find the volume of this triangular prism.

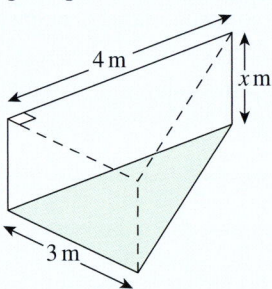

5 Calculate the missing side length for this triangular prism with volume 48 m³.

6 Sketch three different prisms that have a volume of 240 mm³. Write the side lengths on your sketches.

7 These two prisms have the same volume. Work out the missing side length in the triangular prism.

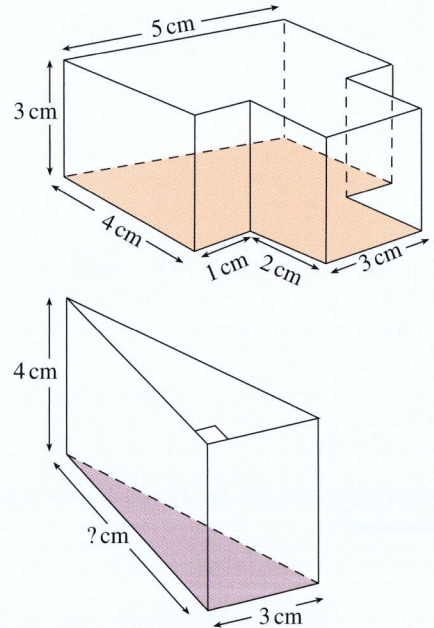

8 Find the volume of each prism.

a

First find the area of the parallelogram.

b

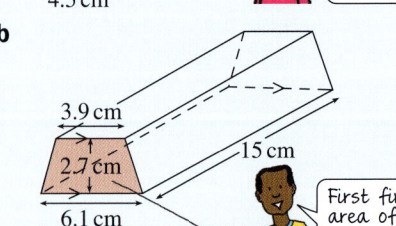

First find the area of the trapezium.

5.4 Surface area of prisms and pyramids

The total area of all faces of a solid is called the **surface area**. For a cuboid, this is the sum of the areas of its six rectangular faces. You can draw the net of the shape to help you work out the surface area.

Example 8

Find the surface area of this cuboid.

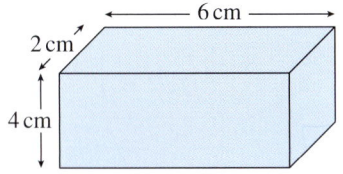

Draw the net and work out the areas of the six rectangular faces.

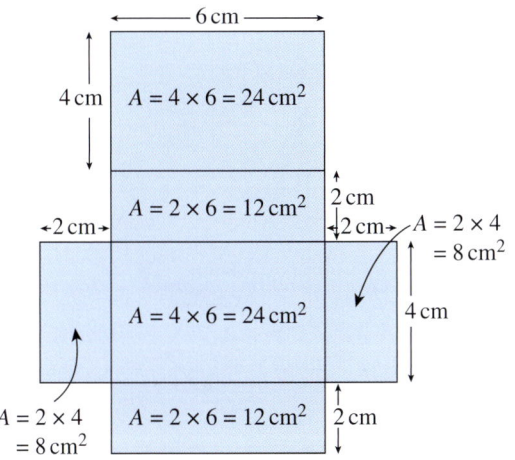

Add up the six areas:
$24 + 12 + 24 + 12 + 8 + 8 = 88$
So surface area = $88 \, cm^2$

The surface area of any prism is found by adding up the areas of the prism's faces.
For example, a triangular prism has five faces:

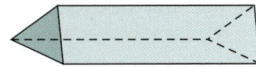

Two of its faces are triangles.
Three of its faces are rectangles.

To find the surface area you have to add the area of the two triangles to that of the three rectangles.

Example 9

Find the surface area of this triangular prism.

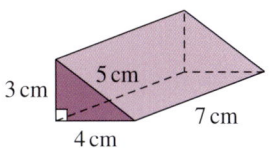

Area of triangular face $= \frac{1}{2} \times$ base $\times$ height
$$= \frac{1}{2} \times 4 \, cm \times 3 \, cm$$
$$= 6 \, cm^2$$

Area of base rectangle = length $\times$ width
$$= 7 \, cm \times 4 \, cm$$
$$= 28 \, cm^2$$

Area of upright rectangle $= 7 \, cm \times 3 \, cm$
$$= 21 \, cm^2$$
Area of sloping rectangle $= 7 \, cm \times 5 \, cm$
$$= 35 \, cm^2$$

Surface area
= area of triangular face $\times$ 2 + area of rectangles
$= 6 \, cm^2 \times 2 + 28 \, cm^2 + 21 \, cm^2 + 35 \, cm^2$
$= 12 \, cm^2 + 28 \, cm^2 + 21 \, cm^2 + 35 \, cm^2$
$= 96 \, cm^2$

You can draw a net if that helps.

Exercise 5F

1 Draw the net of a cube with side 2 cm. What is the surface area of a cube with side 2 cm?

2 Draw the net of this cuboid and work out its surface area.

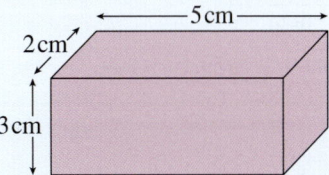

3 Angela said 'You don't need to draw the net to work out surface area. Just work out the area of the three faces you can see in the diagram, add them together and double this answer to include the three faces that you can't see.'

Use Angela's method to work out the surface area of this cuboid.

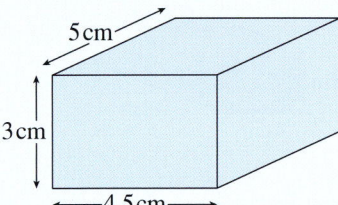

4 Each of the side lengths in cuboid Y are double those in cuboid X.

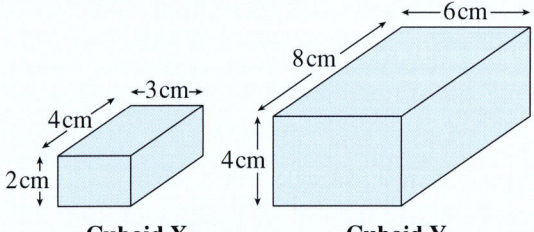

Cuboid X　　　　　**Cuboid Y**

a Find the surface area of:
　i cuboid X　　　**ii** cuboid Y
b How many times bigger is the surface area of cuboid Y than the surface area of cuboid X?
c Find the volume of:
　i cuboid X　　　**ii** cuboid Y
d How many times bigger is the volume of cuboid Y than the volume of cuboid X?

5 These diagrams show a prism and its net. Using the net, work out the surface area of the prism.

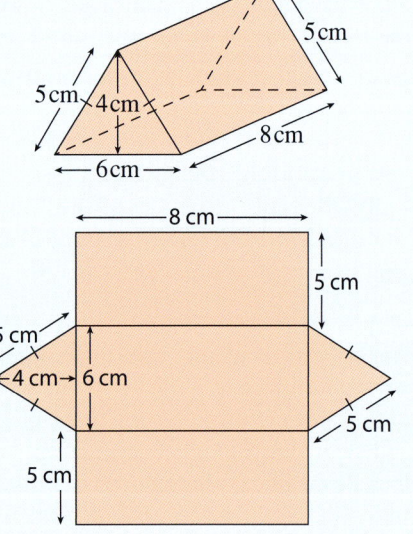

6 The total surface area of a cube is $150\,\text{cm}^2$. Find its volume.

7 Here is the net of a square-based pyramid. Find the surface area of the pyramid.

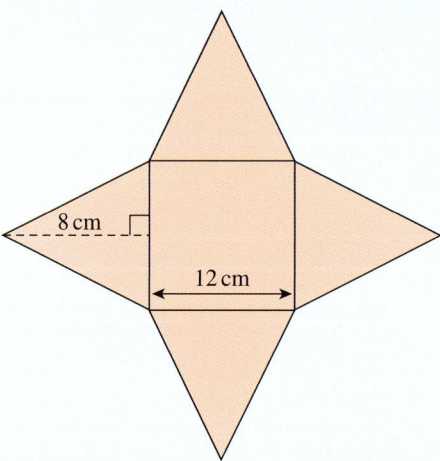

8 Find the surface area of these square-based pyramids.

a

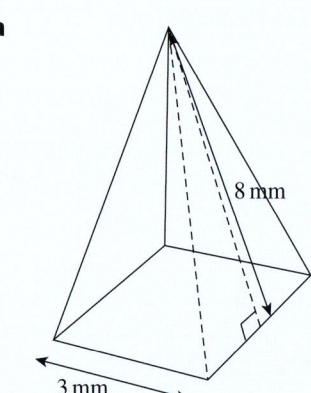

b

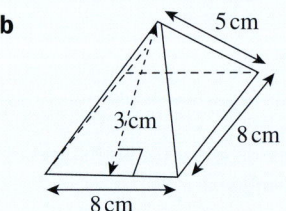

9 Find the surface area of each of these prisms.

a

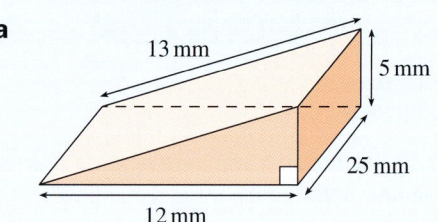

b
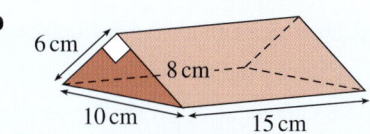

6 cm
8 cm
10 cm
15 cm

c
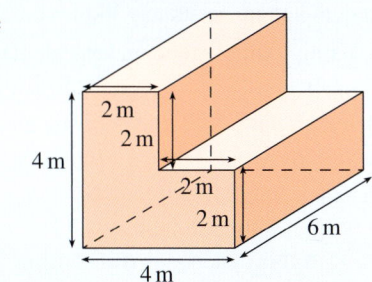

2 m
2 m
4 m
2 m
2 m
6 m
4 m

10 A cube has sides of length 7 cm.

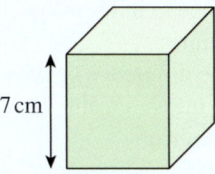

7 cm

Two of these cubes are stuck together to make a cuboid.

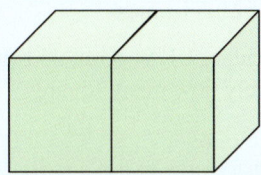

a Here is Amir's work to find the surface area of the cuboid made from two cubes.

Area of one face = 7 × 7 = 49

A cube has six faces so the surface area of one cube is 6 × 49 = 294

There are two cubes so 2 × 294 = 588

So the surface area of the cuboid made from two cubes is 588 cm²

Show why Amir is not correct.

b What would the surface area be if there were three cubes stuck together?

5.5 Imperial units

Some countries, including the USA and the UK, use a system for measuring called **imperial** measurements.

One imperial measurement of length is the mile. You can convert between miles and kilometres using the following approximation:

$$1 \, \text{km} \approx \frac{5}{8} \, \text{mile}$$

The word 'mile' comes from the Latin word *mille* (thousand), used over two thousand years ago by Romans to describe a distance of one thousand paces (*mille passuum*).

The symbol that looks like a wobbly equals sign ($\approx$) means 'approximately equal to'.

Example 10

Convert: **a** 48 km to miles **b** 45 miles to kilometres.

a $1 \, \text{km} \approx \frac{5}{8} \, \text{mile}$

$$48 \div 8 = 6$$

$$48 \, \text{km} \approx 48 \times \frac{5}{8} \, \text{miles} = 6 \times 5 \, \text{miles}$$
$$= 30 \, \text{miles}$$

$$48 \, \text{km} \approx 30 \, \text{miles}$$

b $1 \, \text{mile} \approx \frac{8}{5} \, \text{km}$

$$45 \div 5 = 9$$

$$45 \, \text{miles} \approx 45 \times \frac{8}{5} \, \text{km} = 9 \times 8 \, \text{km}$$
$$= 72 \, \text{km}$$

$$45 \, \text{miles} \approx 72 \, \text{km}$$

Some people forget whether they should be multiplying by $\frac{8}{5}$ or $\frac{5}{8}$. The most important thing to remember is that a mile is longer than a kilometre. So the number of kilometres will be larger than the number of miles if you have converted correctly.

Exercise 5G

1 Which of these conversions are clearly wrong?
  a 80 miles ≈ 50 km **b** 160 km ≈ 100 miles
  c 30 km ≈ 50 miles **d** 40 miles ≈ 25 km
  e 20 miles ≈ 32 km

2 Copy and complete:
  a 35 miles ≈ ☐ km **b** 32 km ≈ ☐ miles
  c 40 km ≈ ☐ miles **d** 65 miles ≈ ☐ km
  e 120 miles ≈ ☐ km

3 Sujatmi said that multiplying by $\frac{8}{5}$ was the same as multiplying by 1.6, so she wrote her method for changing 30 miles to kilometres as:

1.6 × 30 = 1.6 × 10 × 3 = 16 × 3

```
      16
   ×   3
   _____
      48
```

so 30 miles ≈ 48km

Use Sujatmi's method to convert:
  a 20 miles
  b 40 miles
  c 90 miles
    to kilometres.

4 The distance from Wellington in New Zealand to Jakarta in Indonesia is roughly 4800 miles. Approximately how many kilometres is this?

5 The distance from Abu Dhabi in UAE to Cairo in Egypt is roughly 2368 kilometres. Approximately how many miles is this?

6 Copy and complete:
  a 24 miles ≈ ☐ km   **b** 50km ≈ ☐ miles
  c 32 miles ≈ ☐ km   **d** 35km ≈ ☐ miles

⑦ Find the area of this piece of land that is in the shape of a trapezium.
Give your answer in km² correct to the nearest km².

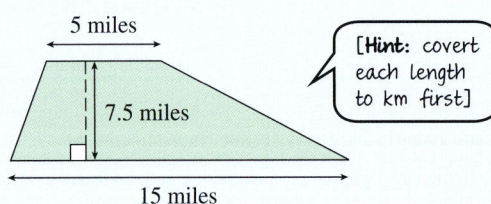

5 miles

7.5 miles

15 miles

[**Hint:** covert each length to km first]

⑧ A cube has side length 0.75 miles. Find the surface area of the cube. Give your answer in km² correct to the nearest km².

⑨ You may use a calculator for this question.
So far we have used the conversion 1km ≈ $\frac{5}{8}$ mile. As a decimal, $\frac{5}{8}$ = 0.625. The decimal conversion is more accurately 1 km ≈ 0.621371 miles. Using 1 km ≈ $\frac{5}{8}$ mile makes calculations quicker and easier when you haven't got a calculator. If you want more accurate answers then you need to use a more accurate decimal.

The distance from the Earth to the Sun is about 149.6 million kilometres.
  a Convert this distance to miles using
    1 km ≈ $\frac{5}{8}$ mile.
  b Convert this distance to miles using
    1 km ≈ 0.621371 miles.
  c Subtract to find the difference between the two converted measurements in **a** and **b**. This is the error.
  d Repeat parts **a** to **c** for a distance of 50 km
  e When does the greater error happen?
  f Can you find a more accurate conversion than 1 km ≈ 0.621371 miles?

10 The area of Cambridge, UK is about 45 square miles. Approximately how many square kilometres is this?

11 Jamil said that he was going to use a ratio method to convert between miles and kilometres. He said 8 km ≈ 5 miles, so the ratio of kilometres to miles is 8 : 5. So if you have the distance in miles and it is a multiple of 5 then the distance in kilometres will be the same multiple of 8. For example, 15 miles is the third multiple of 5 so the answer in kilometres is the third multiple of 8, which is 24 km.

Use Jamil's method to repeat question **2**.

Consolidation

Example 1

Find the circumference of this circle.

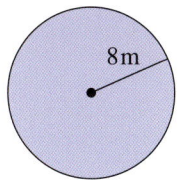

Circumference of circle

$= 2\pi \times$ radius

$= 2 \times 3.14 \times 8$

$= 50.24$ m

Example 2

Find the area of the trapezium.

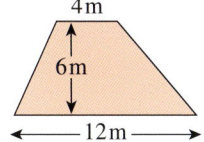

Area of trapezium $= \frac{1}{2}(a+b) \times h$

$\qquad = \frac{1}{2}(4\,\text{m} + 12\,\text{m}) \times 6\,\text{m}$

$\qquad = 48\,\text{m}^2$

Example 3

Find the area of this compound shape.

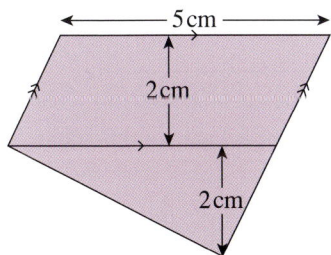

This shape is made up of a parallelogram and a triangle.

Area of the parallelogram
$= b \times h = 5\,\text{cm} \times 2\,\text{cm} = 10\,\text{cm}^2$

Area of the triangle
$= \frac{1}{2}b \times h = \frac{1}{2} \times 5\,\text{cm} \times 2\,\text{cm} = 5\,\text{cm}^2$

Total area $= 10\,\text{cm}^2 + 5\,\text{cm}^2 = 15\,\text{cm}^2$

Example 4

This diagram shows a triangular prism.

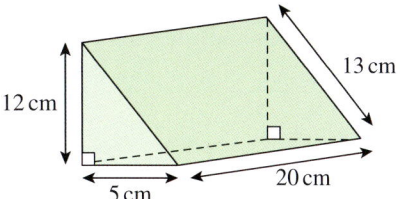

Calculate:

a the volume of the prism

b the surface area of the prism

a Volume = area × length

$\qquad = \frac{1}{2} \times 5 \times 12 \times 20$

$\qquad = 600\,\text{cm}^3$

b You can draw a net for the prism.

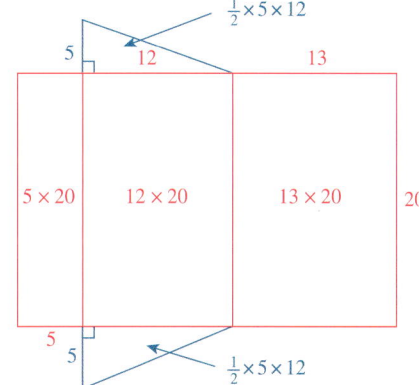

Surface area = the three rectangle faces +

$\qquad\qquad\qquad$ the two triangle faces

$\qquad = 5 \times 20 + 12 \times 20 + 13 \times 20 + \left(\frac{1}{2} \times 5 \times 12\right) \times 2$

$\qquad = 660\,\text{cm}^2$

Exercise 5

1 Work out the circumference of these circles.

a

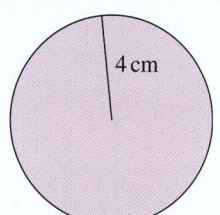

4 cm

b

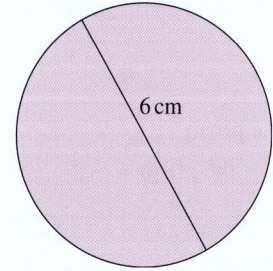

6 cm

2 The end face of a prism of length 10 cm is a right-angled triangle with sides 3 cm, 4 cm and 5 cm.
  a Draw a sketch of the prism.
  b Find the volume of the prism.

3 Find the area of these shapes.

a

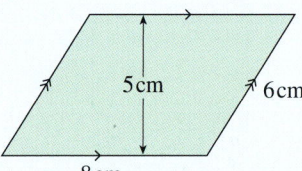

5 cm   6 cm

8 cm

b

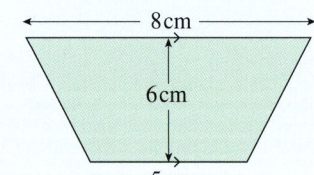

8 cm

6 cm

5 cm

c

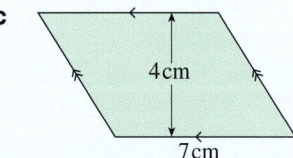

4 cm

7 cm

d

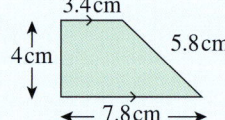

3.4 cm

4 cm   5.8 cm

7.8 cm

4 Copy and complete:
  a 32 km ≈ ☐ miles
  b 10 miles ≈ ☐ km

c 120 km ≈ ☐ miles
d 120 miles ≈ ☐ km

5 This diagram shows a triangular prism.

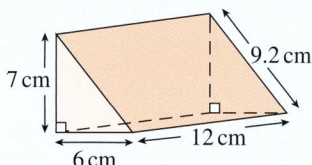

7 cm   9.2 cm

6 cm   12 cm

Calculate **a** the volume **b** the surface area of the prism.

6 What is the area of this field?

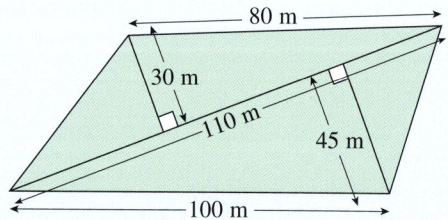

80 m

30 m

110 m

45 m

100 m

7 For this prism, find:
  a the volume
  b the total surface area

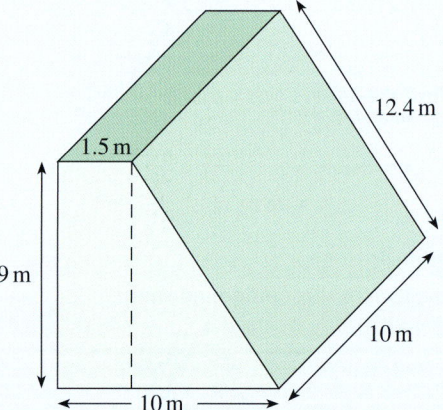

12.4 m

1.5 m

9 m

10 m

10 m

8 Find the surface area of this square-based pyramid.

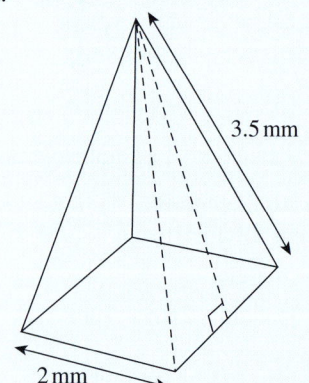

3.5 mm

2 mm

Summary

You should know ...

1 The circumference of a circle is πD or $2\pi r$, where $\pi = 3.14$

For example:

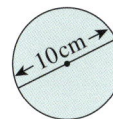

$$\begin{aligned} \text{Circumference} &= \pi D \\ &= 3.14 \times 10\,\text{cm} \\ &= 31.4\,\text{cm} \end{aligned}$$

2 The area of a parallelogram, A, is given by the formula

$A = b \times h$

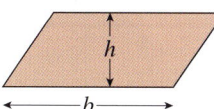

For example:

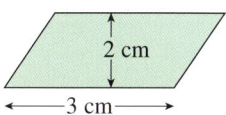

$A = 3\,\text{cm} \times 2\,\text{cm} = 6\,\text{cm}^2$

3 The area of a trapezium, A, is given by the formula

$A = \frac{1}{2}(a + b) \times h$

For example:

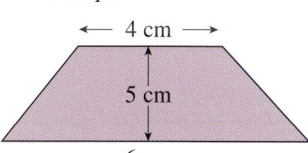

$A = \frac{1}{2}(4\,\text{cm} + 6\,\text{cm}) \times 5\,\text{cm}$

$= 25\,\text{cm}^2$

Check out

1 Find the circumference of these circles.

a

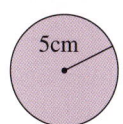

5cm

b

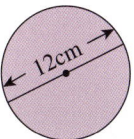

12cm

2 Find the area of these parallelograms.

a

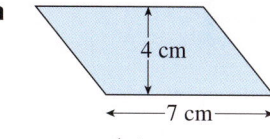

4 cm

7 cm

b

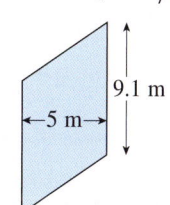

9.1 m

5 m

3 Find the area of these trapeziums.

a

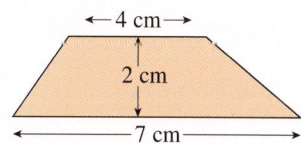

4 cm

2 cm

7 cm

b

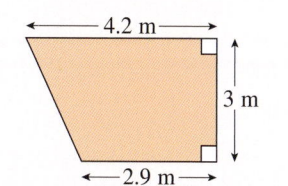

4.2 m

3 m

2.9 m

4 The volume of a triangular prism is the area of the end face × the length of the prism.

For example:

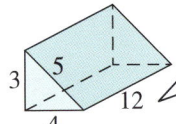

All measurements are in centimetres.

Volume $= \frac{1}{2} \times 4 \times 3 \times 12 = 72 \, \text{cm}^3$

5 The total surface area of a prism is the sum of the areas of all its faces.
For example:

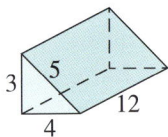

Surface area $= 2\left(\frac{1}{2} \times 4 \times 3\right) + (12 \times 4) + (12 \times 5)$

$\qquad\qquad + (12 \times 3) \, \text{cm}^2$

$\qquad = 156 \, \text{cm}^2$

6 That distances in the USA, the UK and some other countries are measured in miles, and that one kilometre is about $\frac{5}{8}$ of a mile.
For example: Convert 56 km into miles.

$1 \, \text{km} \approx \frac{5}{8} \, \text{mile}$

$56 \, \text{km} \approx 56 \times \frac{5}{8} \, \text{miles} = 7 \times 5 \, \text{miles}$

$\qquad\qquad\qquad = 35 \, \text{miles}$

$56 \, \text{km} \approx 35 \, \text{miles}$

4 Find the volume of this prism.

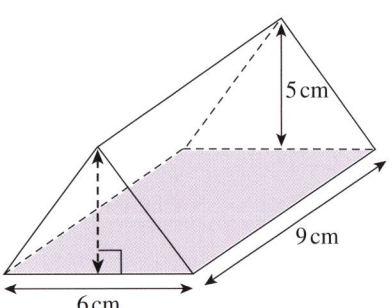

5 Find the total surface area of the prism. All measurements are in centimetres.

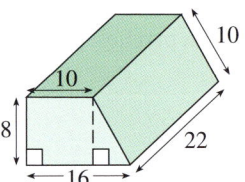

6 Copy and complete:
a 24 km ≈ ☐ miles
b 30 miles ≈ ☐ km
c 88 km ≈ ☐ miles
d 25 miles ≈ ☐ km

1 Find the circumference of these circles.

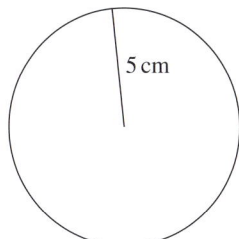
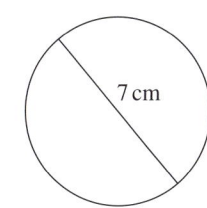

2 Work out:

a $\frac{15}{^-5} + \frac{^-12}{2}$ **b** $\frac{^-64}{^-8} + \frac{^-10}{5}$

c $\frac{^-20}{4} - \frac{^-50}{10}$

3 Decide whether each of the following is an equation, an expression or a formula.
a $C = \pi d$
b $2w + 2l$
c $5p - 1 = 9$
d $4y + 7$
e $I = PRT$
f $8t + 5 = 3t + 25$

4 A polyhedron has 11 faces and 20 edges.
How many vertices does this polyhedron have?

5 Jen says the bearing of X from Y is 25°. She has not written this correctly. What mistake has she made?

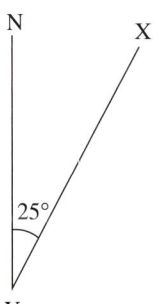

6 Draw triangle ABC such that
a AB = 8 cm, CÂB = 60° and AC = 6 cm
b AB = 7.5 cm, CÂB = 50° and AC = 6.3 cm

7 Work out:
a $4 \times {}^-3$ **b** ${}^-10 \times {}^-5$

c $2 \times {}^-8$ **d** ${}^-7 \times {}^-5$

e $12 \div {}^-3$ **f** ${}^-40 \div {}^-5$

g $16 \div {}^-8$ **h** ${}^-63 \div 7$

8 Sadie says she is going to collect data by observation because then she knows the data is relevant and first-hand. She is an expert in what she is observing and she can collect detailed data. Write down some disadvantages of collecting data in this way.

9 Find the area of a trapezium with parallel sides of length 7 cm and 13 cm and perpendicular height of 9 cm.

10 a Make an accurate drawing of this triangle.

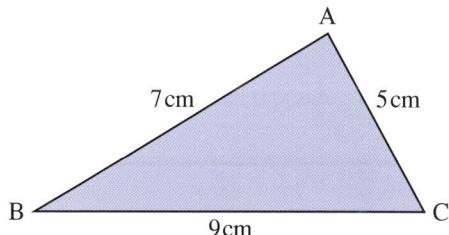

b Bisect CÂB.

11 a Use isometric paper to draw the 3D object represented by this plan and elevations. Each grid square is 1 cm².

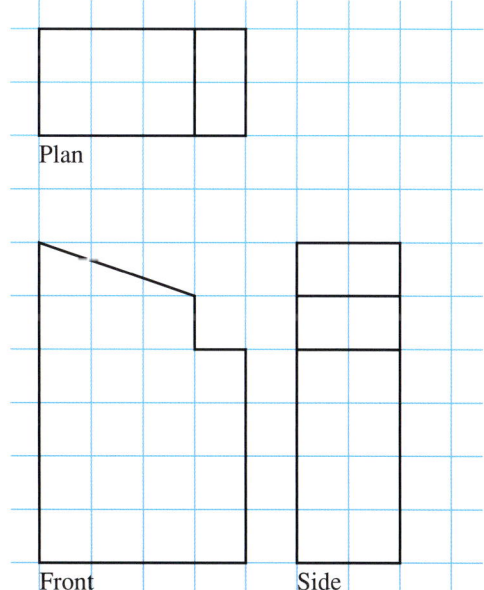

b If the plan was redrawn to measure 12 cm by 6 cm instead of 4 cm by 2 cm, what would be the height of the new side drawing?

12 Factorize:

a $5g + 10$ b $3c + 9v - 18$

c $20t + 16$ d $xy + xh$

e $4rf + 3r + rk$

13 Copy and complete:

a 75 miles $= \square$ km b 24 km $= \square$ miles

c 80 km $= \square$ miles d 85 miles $= \square$ km

14 Construct triangle ABC where $AB = 6.4$ cm, $AC = 5.2$ cm and $\angle BAC = 50°$

15 Find the value of:

a $x^3 + 4x$ when $x = {}^-2$

b $4x^2 - 4$ when $x = 3$

c $5y + y^3$ when $y = {}^-1$

d $3m^2 + 2m - 4$ when $m = {}^-3$

e $3p^3 - p^2 - 7$ when $p = 2$

16 Find the **a** volume **b** surface area of this prism.

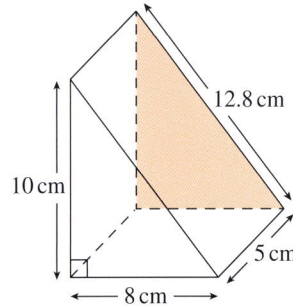

17 Rewrite using index notation:

a $3 \times 3 \times 3$

b $7 \times 7 \times 7 \times 7 \times 7$

c $5 \times 5 \times 5 \times 5 \times 5 \times 5$

18 Write these using algebra.

a I think of a number n, add 3 then multiply by 5.

b I think of a number p, double it then subtract 4.

c I think of a number x and a number y. I find the product of these two numbers, then multiply that answer by 3.

d I think of a number v, multiply it by itself, then double it.

19 Work out the following without using a calculator (write both possible answers where there are two).

a 8^2 b $\pm\sqrt{169}$

c $({}^-12)^2$ d $\pm\sqrt{10000}$

e $^-11^2$ f $\pm\sqrt{81}$

20 Find the value of each of the following when $t = 2$, $m = {}^-1$ and $c = 3$.

a $t(m + c)$ b $tc - m^2$

c $\dfrac{4m^2 - 2tc}{t + m + c}$

21 Find the area of each of these shapes.

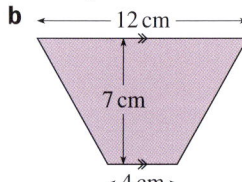

22 Which of these statements are true?

a A parallelogram has opposite sides equal and parallel.

b All rectangles are parallelograms.

c A trapezium has diagonals that bisect each other.

d All rectangles are squares.

e A kite has diagonals equal in length.

f The diagonals of a parallelogram are not equal in length.

23 Construct the right-angled triangle ABC, where $AB = 4$ cm and the hypotenuse $BC = 5$ cm.

24 Which of these is not a bearing?

$382°$ $027°$ $259°$ $92°$

25 Work out:

a $18 - 2 \times 6 + {}^-1$

b $^-2 \times {}^-9 + \sqrt[3]{8} \times 3^2$

c $8 \times 3 + ({}^-18 \div \sqrt{81})$

26 Draw a line 8 cm long. Construct the perpendicular bisector of that line.

27 The circumference of a circle is 19.5 m. Find the diameter of the circle. Use 3.14 for π. Give your answer correct to 1 d.p.

28 Omar wants to find out what people think about the new computer game he has written. He decides to do an online survey, which he emails to people after they buy his game. What are the advantages and disadvantages of this method of data collection?

29 Simplify:

a $8 - (x - 5) - 5(3 - x)$

b $4t + 6 \times 3t - 50t$

c $m - (4 - 2m) + 5 \times 3m$

30 Find the highest common factor of 90 and 225.

31 Draw a triangle with side lengths 4 cm, 5 cm and 7 cm.

32 Using the formula $A = 10d - 2w^2$, find A when:

    a $w = {}^-3$ and $d = 0.5$

    b $w = 4$ and $d = {}^-0.2$

33 Simplify:

    a $5(2x + y) - 4(x - 5y)$

    b $3x(y + z) - y(z + x) + 2z(x + y)$

34 Using your protractor, draw an angle of 80°. Construct the bisector of this angle.

35 Simplify:

    a $3^4 \times 3^6$ **b** $2^{15} \div 2^5$

    c $4^3 \times 4 \times 4^5$ **d** $\dfrac{7^{10}}{7^2}$

36 Measure the bearing of B from A in each of these diagrams.

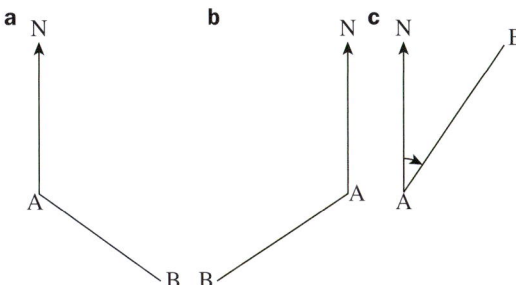

37 Write an expression for the total time taken to roast meat with a mass of k kilograms if it takes 15 minutes plus 40 minutes for each kilogram.

38 Expand and simplify:

    a $4x(2 - 3x) - x(6 - 2x)$

    b $2t^2(3t - 7x)$.

    c $2m(3m + 5x - 1)$

39 The circumference of a circle is 32 cm. Find the radius of the circle. Use 3.14 for π. Give your answer correct to 1 d.p.

40 If $\dfrac{1}{f} = \dfrac{1}{u} + \dfrac{1}{v}$ find f when:

    a $u = 2$ and $v = 4$ **b** $u = 8$ and $v = 12$

41 Use your ruler and protractor to draw this triangle accurately.

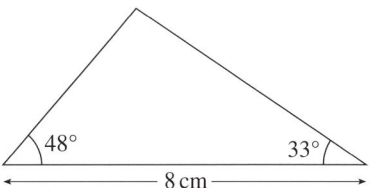

42 The distance from Wellington in New Zealand to Jakarta in Indonesia is roughly 7680 kilometres. Approximately how many miles is this?

43 Given $a = 2$, $b = {}^-1$, $c = 3$, evaluate:

    a $a + 2b$ **b** $ab - c$

    c $ac - 3b$ **d** $ab + bc$

    e $\dfrac{a}{b} - \dfrac{c}{b}$ **f** $\dfrac{abc}{-6}$

44 Catalina wants to find out what students in her school think about the school uniform. What are the advantages of each of these sampling methods?

    a   Quota sampling: for example, she chooses two students randomly from each class in the school.

    b   Random sampling: she chooses her sample using names taken randomly from a list of all students in the school, for example using names out of a hat.

    c   Convenience sampling: for example, she chooses the first 50 students who arrive in the school canteen.

45 Work out the shaded area in this shape.

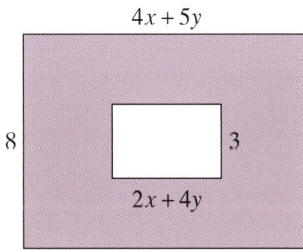

46 Find the area of each of these shapes.

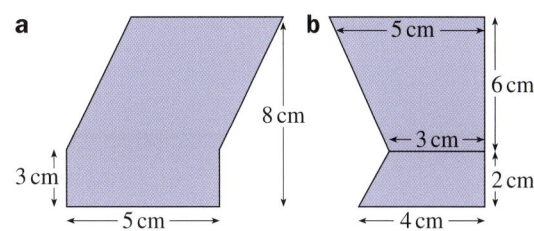

47 Using a protractor, draw each of these bearings.
   a B from A is 025°
   b C from D is 242°
   c E from F is 341°

48 Find the missing numbers.
   $2t(\square y + 8m) = 8ty + \square tm$

49 Find the total length of wire used to make a wire circle set inside a wire square of side 9 cm, as shown in this diagram.

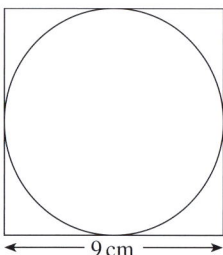

9 cm

50 Find the value of:
   a $3r^2 + r$ when $r = \frac{1}{2}$
   b $50 - 200b^2 - 10b^2$ when $b = {}^-0.4$

51 Draw an accurate net to scale for this square-based pyramid, constructing the triangle faces using compasses.

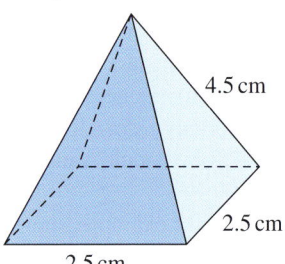

4.5 cm

2.5 cm

2.5 cm

52 The area of this rectangle is $24x^2 + 30x$. Find an expression for the perimeter of the rectangle in its simplest form.

4x + 5

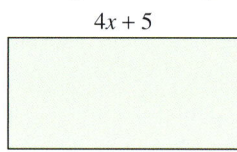

53 Using $s = \frac{1}{2}ut + at^2$ find u when $s = 80, t = 5$ and $a = 4$.

6 Fractions and decimals

Objectives

In this chapter you will learn about:

- multiplying and dividing by 0.1 and 0.01
- rounding to given significant figures
- recurring decimals
- subtracting mixed numbers
- dividing by a proper fraction
- multiplying by a mixed number
- simplifying calculations involving fractions.

What's the point?

Giving approximate answers and rounding to a certain number of significant figures is used a lot in the news. A newspaper would not report that there were 23 148 people at a concert. Instead they would write as the headline something like '23 000 attend concert'.

Before you start

You should know ...

1 Fractions can be simplified by dividing.

For example:

$$\frac{14}{35} \xrightarrow{\div 7} = \frac{2}{5} \xleftarrow{\div 7}$$

2 How to convert between mixed numbers and improper fractions.

For example:

$\frac{8}{5}$ means $8 \div 5 = 1$ remainder 3, so $\frac{8}{5} = 1\frac{3}{5}$

$3\frac{1}{4} = \frac{(3\times4)+1}{4}$, so $3\frac{1}{4} = \frac{13}{4}$

Check in

1 Simplify:

a $\frac{14}{20}$ **b** $\frac{16}{40}$ **c** $\frac{32}{48}$

d $\frac{15}{75}$ **e** $\frac{8}{96}$ **f** $\frac{18}{162}$

2 a Change these improper fractions to mixed numbers.

   i $\frac{23}{20}$ **ii** $\frac{18}{5}$ **iii** $\frac{73}{10}$

b Change these mixed numbers to improper fractions.

   i $1\frac{3}{10}$ **ii** $3\frac{7}{8}$ **iii** $5\frac{2}{3}$

3 How to round numbers.

For example:
Round:
a 23456 to the nearest 1000
b 4.7158 to 2 d.p.

....

a 23④56
The next number after the thousands column is a 4, so round down to 23000
b 4.71⑤8 to 2 d.p.
The next number after the second decimal place is a 5, so round up to 4.72

3 Round:
a 365 to the nearest 10
b 173.56 to the nearest whole number
c 490832 to the nearest 1000
d 0.6678 to 2 d.p.
e 0.412 to 1 d.p.
f 1689 to the nearest 100
g 0.21653 to 2 d.p.
h 197.483 to 1 d.p.
i 12.467 to the nearest whole number

6.1 Multiplying and dividing decimals

In Student Book 7 you learned how to multiply and divide whole numbers and decimals by powers of 10 where the power is a positive integer. For example, $100 = 10^2 = 10 \times 10$ and $1000 = 10^3 = 10 \times 10 \times 10$.

When the power gets higher the numbers are harder to say. In the USA and UK, commas are often used to help write large numbers (some calculators have commas on their display). We can also use spaces. To say large numbers, think what word goes in place of each comma or space.

$10^7 = 10000000$, or 10 million, is sometimes written 10,000,000.

Example 1

Work out:

a 12.4×10^2 **b** $21 \div 10^3$

....

a $12.4 \times 100 = 1240$

Move all digits two places to the left to make them worth 100 times their original value.

b $21 \div 1000 = 0.021$

Move all digits three places to the right to make them worth a 1000th of their original value.

Exercise 6A

1 Work out:

a 4.2×100 **b** 3.7×10

c 0.2×1000 **d** $53 \div 100$

e $2.8 \div 10$ **f** $350 \div 1000$

g 0.021×100 **h** 0.4×10

i 2.65×1000 **j** $42.4 \div 100$

k $0.34 \div 10$ **l** $92.1 \div 1000$

2 From your work on fractions and decimals in Student Book 7 you learned that

$$3 \times \frac{1}{10} = \frac{3}{10} = 0.3$$

Write the answers to these calculations as decimals.

a $7 \times \frac{1}{10}$ **b** $9 \times \frac{1}{10}$

c $4 \times \frac{1}{100}$ **d** $17 \times \frac{1}{100}$

e $18 \times \frac{1}{10}$ **f** $172 \times \frac{1}{100}$

g $63 \times \frac{1}{10}$ **h** $256 \times \frac{1}{100}$

3 Look at this pattern.
$25 \times 1000 = 25000$
$25 \times 100 = 2500$
$25 \times 10 = 250$
$25 \times 1 = 25$

$25 \times \frac{1}{10} = 2.5$

$25 \times \frac{1}{100} = 0.25$

Copy and complete this pattern.

$175 \times 1000 =$

$175 \times 100 =$

$175 \times 10 =$

$175 \times 1 =$

$175 \times \frac{1}{10} =$

$175 \times \frac{1}{100} =$

4 Work out:

a $7 \times \frac{1}{10}$ **b** $7 \div 10$

c $21 \times \frac{1}{100}$ **d** $21 \div 100$

e $24 \div 10$ **f** $24 \times \frac{1}{10}$

g $315 \div 100$ **h** $315 \times \frac{1}{100}$

Comment on your answers.

5 Copy and complete these sentences.

Multiplying by $\frac{1}{10}$ is the same as dividing by ... .

Multiplying by $\frac{1}{100}$ is the same as dividing by ... .

Multiplying integers and decimals by 0.1

You know that $0.1 = \frac{1}{10}$ and $0.01 = \frac{1}{100}$.

- Multiplying by 0.1 is the same as multiplying by $\frac{1}{10}$ or dividing by 10.
- Multiplying by 0.01 is the same as multiplying by $\frac{1}{100}$ or dividing by 100.

Example 2

Work out:

a 24×0.1

b 173.4×0.01

c 0.08×0.1

....

a $24 \times 0.1 = 24 \div 10 = 2.4$

b $173.4 \times 0.01 = 173.4 \div 100 = 1.734$

c $0.08 \times 0.1 = 0.08 \div 10 = 0.008$

Exercise 6B

1 Work out:

a 32×0.1 **b** 256.1×0.01

c 0.1×0.04 **d** 356×0.1

e 28×0.01 **f** 5×0.1

g 0.01×4.1 **h** 7×0.01

i 0.2×0.1 **j** 2300×0.01

k 0.1×4560 **l** $30900 \div 0.01$

2 A glass holds 0.1 litres of water. How many litres would there be in 9 glasses of water?

3 A sheet of paper is 0.1 mm thick. How thick would 360 sheets of paper be:

     a in mm   **b** in cm?

4 A box has a mass of 0.01 kg. What is the mass of 7 of these boxes:

     a in kg   **b** in g?

5 Look at this pattern.

$3200 \div 1000 = 3.2$

$3200 \div 100 = 32$

$3200 \div 10 = 320$

$3200 \div 1 = 3200$

$3200 \div \frac{1}{10} = 32000$

$3200 \div \frac{1}{100} = 320000$

Copy and complete this pattern.

$410 \div 1000 =$

$410 \div 100 =$

$410 \div 10 =$

$410 \div 1 =$

$410 \div \frac{1}{10} =$

$410 \div \frac{1}{100} =$

6 Copy and complete these sentences.

Dividing by $\frac{1}{10}$ is the same as multiplying by ... .

Dividing by $\frac{1}{100}$ is the same as multiplying by ... .

7 Put the results of these calculations in order, starting with the smallest.

54×0.1	$504 \div 10$	$50.4 \div 100$
5.4×10	$54 \div 100$	5.4×0.01

8 Complete these calculations.

a $456 \times \square = 4.56$

b $23 \times \square = 2.3$

c $\square \times 0.01 = 0.65$

d $\square \times 0.1 = 480$

9 Find the area of this rectangle.

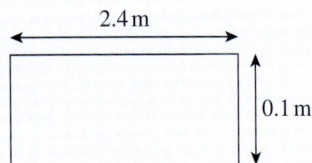

2.4 m

0.1 m

10 Solve these equations.

a $\dfrac{x}{0.1} = 550$ **b** $\dfrac{y}{0.01} = 2000$

Dividing integers and decimals by 0.1 and 0.01

Since $0.1 = \frac{1}{10}$ and $0.01 = \frac{1}{100}$:

- dividing by 0.1 is the same as dividing by $\frac{1}{10}$ or multiplying by 10
- dividing by 0.01 is the same as dividing by $\frac{1}{100}$ or multiplying by 100.

Example 3

Work out:

a $12 \div 0.1$

b $0.17 \div 0.1$

c $0.43 \div 0.01$

....

a $12 \div 0.1 = 12 \times 10 = 120$

b $0.17 \div 0.1 = 0.17 \times 10 = 1.7$

c $0.43 \div 0.01 = 0.43 \times 100 = 43$

Exercise 6C

1 Work out:

a $59 \div 0.1$ **b** $35.21 \div 0.01$

c $0.07 \div 0.1$ **d** $249 \div 0.1$

e $76 \div 0.01$ **f** $8 \div 0.1$

g $3.7 \div 0.01$ **h** $15 \div 0.01$

i $0.5 \div 0.1$ **j** $4100 \div 0.01$

k $7290 \div 0.1$ **l** $19300 \div 0.01$

2 A 0.1 kg piece of cheese costs \$0.48. What is the cost of cheese per kg?

3 A rectangle has an area of $0.45\,\text{m}^2$. The width of the rectangle is 0.1 m. Find the length of the rectangle.

4 Anna uses this method to work out $45.6 \div 0.01$

$$\frac{45.6 ^{\times 100}}{0.01 _{\times 100}} = \frac{45.600}{1} = 45.6$$

Explain the mistake Anna has made.

5 Copy and complete, filling in the blanks with 0.01, 0.1, 10 or 100.

a $74 \div \square = 740$ **b** $6.43 \times \square = 0.643$

c $0.8 \div \square = 0.08$ **d** $329 \div \square = 3.29$

e $85 \times \square = 8.5$ **f** $2.1 \div \square = 210$

g $1.23 \times \square = 12.3$ **h** $41 \div \square = 0.41$

6 Complete these sentences.

a $2 \div 0.1$ is the same as $20 \div \square$

b $3.2 \div 0.1$ is the same as $\square \div 1$

c $16 \div 0.01$ is the same as $160 \div \square$ or $1600 \div \square$

d $43.21 \div 0.01$ is the same as $\square \div 1$

7 Work out:

a 0.1^2

b 0.01^2

8 Farook says, 'Dividing by 0.1 is the same as multiplying by 10, so when you divide by a decimal, the answer will always be bigger than the number you started with.' Is Farook correct? Explain your answer.

6.2 Significant figures

Rounding to one or two decimal places doesn't always give a simple approximation of a number.

For example:
6 381 278.68
rounded to 1 d.p. is
6 381 278.7
which is not much simpler!

In such cases it is more useful to round to a given number of **significant figures**.

Significant figures show the relative importance of the digits in a number, with the first non-zero digit from the left being the most important.

For example, to one significant figure (1 s.f.), 2914 is 3000.

Notice the place value:

Th	H	T	U
2	9	1	4

The first non-zero digit from the left is the 2, which is the number of thousands. In this case rounding to 1 s.f. is the same as rounding to the nearest thousand.

Example 4

Write 83 562 to:

a 2 s.f.

b 1 s.f.

....

The column headings are

TTh	Th	H	T	U
8	3	5	6	2

a The second digit heading is thousands so round to the nearest thousand:
83 562 = 84 000 (2 s.f.)
b The first digit heading is tens of thousands so round to the nearest ten thousand:
83 562 = 80 000 (1 s.f.)

> Don't forget to include the zeros in 84 000 to maintain the place value.

Here are some other examples.

Number	67 341	23.478	0.03286
3 s.f.	67 300	23.5	0.0329
2 s.f.	67 000	23	0.033
1 s.f.	70 000	20	0.03

Notice that in the number 0.03286 the first non-zero digit from the left is 3 which is in the hundredths place.

So 0.03286 = 0.03 (1 s.f.).

When rounding 0.0040791 correct to 3 significant figures note that the blue zero is significant as it follows a non-zero digit. So 0.0040791 correct to 3 s.f. is 0.00408.

Exercise 6D

1 Write the following correct to 1 s.f.
  a 420 **b** 8314 **c** 396
  d 48 **e** 4.81 **f** 91 265
  g 10 034 **h** 0.0032 **i** 7.0036

2 Decide whether these statements are true or false.
  a Leading zeros at the start of a decimal are not significant. For example, in 0.000432, the 4 is the first significant figure.
  b Zeros in the middle of a decimal are not significant. For example, in 0.04032, the 3 is the second significant figure.
  c Zeros at the end of a decimal are significant. For example, if a rounded figure is given as 0.0230, then you know it is rounded to 4 d.p. or 3 s.f.

3 Write the following correct to 2 s.f.
  a 962 **b** 489 **c** 6183
  d 17 638 **e** 17.34 **f** 26.92
  g 0.00389 **h** 133.68 **i** 8.035

4 Copy and complete:

Number	613 752	1.6831	0.004753
3 s.f.			
2 s.f.			
1 s.f.			

5 Round:
  a 0.0046572 to 3 s.f.
  b 0.0189 to 1 s.f.
  c 0.00022278 to 2 s.f.
  d 0.399 to 2 s.f.

6 Ramsingh wrote the following.
- **i** 6384 = 6000 (1 s.f.)
- **ii** 6384 = 6400 (2 s.f.)
- **iii** 816952 = 8 (1 s.f.)
- **iv** 0.0356 = 0.03 (1 s.f.)
- **v** 0.08942 = 0.089 (2 s.f.)
- **vi** 899 = 800 (2 s.f.)

a Which of these are correct?

b Write down the correct answers for any Ramsingh got wrong.

7 Round 82 736 to:
a 1 s.f.　**b** 2 s.f.　**c** 3 s.f.　**d** 4 s.f.

8 Round 0.003565 to:
a 1 s.f.　**b** 2 s.f.　**c** 3 s.f.　**d** 4 s.f.

9 A number rounded to 1 s.f. is 60. Write down five examples of what the number could be.

10 Imagine your job is to write the headlines for a newspaper. For each fact below, decide whether you would write an *approximate* number or the *exact* number in your headline. If *approximate*, write down the number you would use instead.

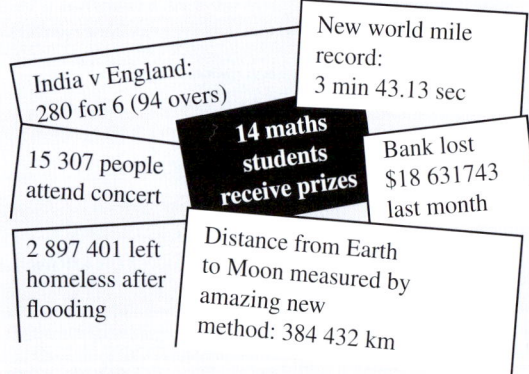

India v England: 280 for 6 (94 overs)

New world mile record: 3 min 43.13 sec

14 maths students receive prizes

Bank lost $18 631743 last month

15 307 people attend concert

2 897 401 left homeless after flooding

Distance from Earth to Moon measured by amazing new method: 384 432 km

11 Round 8 134 767 to:
a 4 s.f.　**b** 3 s.f.
c 2 s.f.　**d** 1 s.f.

12 You need to be sensible when rounding. Say what the problem is in these situations.
a A theatre has 956 seats. The head of the theatre rounded this correct to 2 s.f. and sold that number of tickets.
b A bridge can safely carry a weight of 14.65 tonnes. A sign shows this amount correct to 2 s.f.

13 A whole number rounded to 1 s.f. is 5000. What is:
a the largest number it could have been
b the smallest number it could have been?

14 Calculate 0.053 × 1.4, giving your answer:
a exactly
b correct to 2 decimal places
c correct to 1 significant figure

15 Calculate 0.6234 ÷ 17.2 and write your answer:
a correct to 2 decimal places
b correct to 3 significant figures

Rounding appropriately

Often a question doesn't require an exact answer, particularly when that answer is a decimal with a large or infinite number of decimal places. A question will not always tell you how to round your answer, however, and you need to think carefully about how to do this. The guidelines below should help. Remember, these are not rules but sensible suggestions, for use only if the question doesn't tell you how to round.

1 It is a good idea to write lots of digits, 5 or 6 significant figures or more, in your working before rounding your final answer. If you over-round or make an error in rounding your answer, your working will still show the correct numbers and you can still gain marks.

2 Do not round part way through a calculation. Where there are two parts to a question, use your unrounded answer to the first part in your calculation for the second part. Only round your final answers.

3 If an answer is a short, terminating decimal do not round it.

4 If an answer is a recurring decimal you don't need to round it. You can write it with recurring dots to show an exact number.

> You will learn more about recurring decimals in section 6.3.

5 Answers involving money should be given to 2 d.p. For example, answers in dollars will be to the nearest cent.

6 If you are doing an exam, check the front of the exam paper. It may contain instructions about the number of decimal places or significant figures to which you should write your answers. If you are preparing for an exam with rules for rounding, get into the habit of using these rules in all your maths lessons.

7 Use the same accuracy as the question, or slightly better. For example, if all numbers in the question are given to 1 d.p. then giving answers to 1 d.p. or 2 d.p. is sensible.

In Exercise 6E use these guidelines to decide whether to round your answers, and, if so, how to round them. In each case, write which guideline you have used.

Exercise 6E

You may use a calculator for this exercise.

1 Change these fractions to decimals.

a $\frac{3}{20}$ **b** $\frac{4}{11}$

c $\frac{3}{7}$ **d** $\frac{3}{16}$

2 Work out:

a $2.74 + 3.681$ **b** 1.748×3.26

c $7.4 \div 1.7$

3 Work out the missing angle, x.

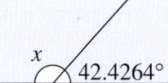

4 a What is the area of a rectangle with length 2.73 cm and width 4.68 cm?

    b What is the width of a rectangle with area 14 cm² and length 7.6 cm?

5 Cloth costs $13.17 per metre. What is the cost of

    a 7 m   **b** 2.3 m

    c 0.41 m?

6 a What area of metal is required to make the road sign shown in the diagram?

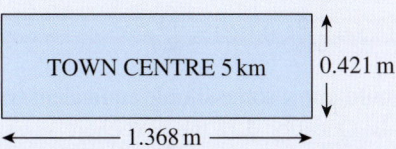

    b If the metal for the road sign costs $15.19 per square metre, what is the cost of the metal required?

6.3 Recurring decimals

Any fraction can be written as a decimal. For example:

$$\frac{1}{2} = 1 \div 2 = 1.0 \div 2 = 0.5$$

$$\frac{3}{4} = 3 \div 4 = 3.00 \div 4 = 0.75$$

To convert fractions to decimals you will need to divide.

Example 5

Change these fractions to decimals.

a $\frac{1}{4}$ **b** $\frac{3}{8}$

....

a $\frac{1}{4} = 1 \div 4 = 1.00 \div 4$

$$4\overline{)1.0\,{}^2 0}\quad 0.2\,5$$

So $\frac{1}{4} = 0.25$

b $\frac{3}{8} = 3 \div 8 = 3.000 \div 8$

$$8\overline{)3.0\,{}^6 0\,{}^4 0}\quad 0.3\,7\,5$$

So $\frac{3}{8} = 0.375$

Exercise 6F

Do not use a calculator for question **1**.

1 Change these fractions to decimals.

a $\frac{2}{5}$ **b** $\frac{3}{5}$ **c** $\frac{4}{5}$ **d** $\frac{1}{4}$

e $\frac{3}{4}$ **f** $\frac{3}{8}$ **g** $\frac{5}{8}$ **h** $\frac{7}{8}$

i $\frac{1}{16}$ **j** $\frac{5}{16}$

What do you notice?

These decimals are called terminating decimals because they contain a finite number of digits after the decimal point (they don't go on forever).

2 Change these fractions to decimals.

a $\frac{1}{3}$ **b** $\frac{2}{3}$ **c** $\frac{1}{6}$ **d** $\frac{1}{7}$

What do you notice?

These decimals are called **non-terminating decimals** because they go on for ever.

3 The fraction $\frac{1}{3} = 0.333... = 0.\dot{3}$

The digits in the decimal repeat themselves. Decimals like this are called **recurring decimals**. The dot above the 3 shows it repeats. Find four other recurring decimals, using your calculator to help you.

4 Look at this diagram. It shows ten identical strips of paper. A fraction of each is shaded.

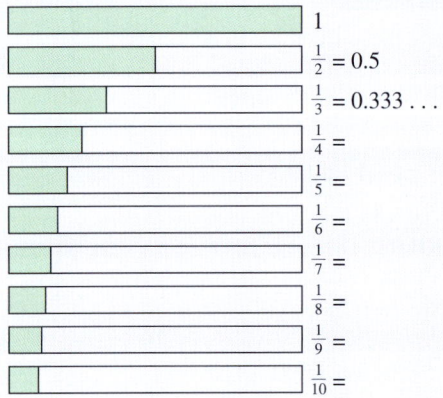

1

$\frac{1}{2} = 0.5$

$\frac{1}{3} = 0.333...$

$\frac{1}{4} =$

$\frac{1}{5} =$

$\frac{1}{6} =$

$\frac{1}{7} =$

$\frac{1}{8} =$

$\frac{1}{9} =$

$\frac{1}{10} =$

The fractions are written on the right-hand side. The first two have been converted to decimals.

a Why are there dots after the fraction for $\frac{1}{3}$?

b Copy the list of fractions and convert the rest of them to decimals.

c Which of these decimals are
   i terminating   **ii** recurring?

5 Convert the fraction to a decimal. If the decimal is non-terminating, find out whether it recurs.

a $\frac{5}{6}$ **b** $\frac{2}{7}$ **c** $\frac{4}{7}$ **d** $\frac{2}{9}$

e $\frac{5}{9}$ **f** $\frac{2}{11}$ **g** $\frac{1}{14}$ **h** $\frac{1}{12}$

6 In question **5**, did you find that all the decimals were recurring? When you change a fraction to a decimal, does the decimal always either terminate or recur?

7 Change these fractions to decimals. Check each answer on your calculator.

a $\frac{1}{16}$ **b** $\frac{11}{16}$ **c** $\frac{3}{20}$ **d** $\frac{7}{20}$

e $\frac{11}{20}$ **f** $\frac{3}{200}$ **g** $\frac{7}{200}$ **h** $\frac{1}{250}$

8 a Change these fractions to decimals.

   i $3\frac{2}{5}$ **ii** $7\frac{3}{8}$ **iii** $2\frac{1}{3}$ **iv** $12\frac{3}{16}$

b Change these decimals to fractions, giving your answers as mixed numbers.

   i 2.8   **ii** 7.625
   iii 2.4   **iv** 15.4375

9 The decimal $0.333... = \frac{1}{3}$

We use the recurring dot symbol to show a **recurring decimal**.

For example, $\frac{2}{3}$ will be written as $0.\dot{6}$ rather than 0.666…

Sometimes two or more digits recur.

The next two examples will show why 0.53… is not a good way to represent a recurring decimal.

$\frac{53}{99} = 0.\dot{5}\dot{3}$; the 5 and the 3 both recur so this would be 0.53535353…

$\frac{8}{15} = 0.5\dot{3}$; just the 3 recurs so this would be 0.53333…

$\frac{533}{999} = 0.\dot{5}3\dot{3}$; the 5 and two 3s recur so this would be 0.533533533…

Notice in the last case that we usually put the recurring dot above the first and the last digit that recur, not over the top of all three digits.

All recurring decimals can be written as fractions. This means they are rational numbers (you learned about this in Chapter 1).

Look at the examples below. Try to spot the pattern.

$$0.\dot{4} = \frac{4}{9} \qquad 0.\dot{6} = \frac{6}{9} = \frac{2}{3}$$

$$0.\dot{5}\dot{6} = \frac{56}{99} \qquad 0.\dot{1}\dot{2} = \frac{12}{99} = \frac{4}{33} \qquad 0.\dot{0}\dot{8} = \frac{8}{99}$$

$$0.\dot{1}7\dot{5} = \frac{175}{999} \qquad 0.\dot{0}6\dot{1} = \frac{61}{999}$$

You should know some simple equivalent fractions and decimals. Question **1** in the next exercise expects you to know some fractions and decimal equivalents without having to do any calculations.

However, at this stage you don't need to know *how* to convert recurring decimals to fractions. Questions **2–6** and the Challenge on the following page are included as extension material for you to try if you have time.

Exercise 6G

1 Copy and complete this table of equivalent fractions and decimals. Cancel each fraction to its simplest form.

Decimal	Fraction
0.6	
	$\frac{3}{4}$
0.$\dot{6}$ (or 0.666 . . .)	
	$\frac{7}{10}$
0.25	
0.03	
	$\frac{27}{100}$
	$\frac{2}{5}$
0.8	

Questions **2–6** are extension material.

2 Change these recurring decimals into fractions. Cancel them to their simplest from if necessary.

a 0.$\dot{2}$ **b** 0.$\dot{3}$

c 0.$\dot{2}\dot{8}$ **d** 0.$\dot{0}\dot{9}$

e 0.$\dot{0}\dot{7}$ **f** 0.$\dot{3}\dot{6}$

g 0.1$\dot{4}\dot{3}$ **h** 0.00$\dot{7}$

i 0.0$\dot{1}\dot{2}$

3 Make up five recurring decimals of your own and change them to fractions.

4 Dhanesh wrote this working.

$$\frac{2}{9} = 0.222$$

$$\frac{7}{9} = 0.777$$

$$0.222 + 0.777 = 0.999$$

So $\frac{2}{9} + \frac{7}{9} = 0.999$

What mistake has he made?

5 Look at this pattern.

$$0.0\dot{7} = \frac{7}{90}$$

$$0.0\dot{6} = \frac{6}{90} = \frac{1}{15}$$

$$0.00\dot{6} = \frac{6}{900} = \frac{1}{150}$$

$$0.0\dot{1}\dot{3} = \frac{13}{990}$$

Notice that not all numbers after the decimal point repeat.

For example, 0.0$\dot{1}\dot{3}$ means 0.013131313 … and not 0.013013013013 … . Look at how this affects the denominator of the fraction.

Change these recurring decimals into fractions. Cancel them to their simplest form if necessary.

a 0.0$\dot{2}$ **b** 0.0$\dot{5}$

c 0.0$\dot{8}$ **d** 0.0$\dot{2}\dot{9}$

e 0.00$\dot{7}$

6 Change these recurring decimals into fractions. Cancel them to their simplest form if necessary.

a 0.00$\dot{1}\dot{8}$

b 0.0$\dot{2}0\dot{3}$

c 0.000$\dot{7}$

Challenge

To convert a decimal that begins with a non-repeating part, such as 0.48888 . . . (or 0.4$\dot{8}$) to a fraction, follow these steps.

Step 1
Write it as the sum of the non-repeating part and the repeating part.
0.4 + 0.088888 . . .

Step 2
Convert each of these decimals to fractions.
(See the pattern in Exercise 6G, question **5** to help you.)

$$\frac{4}{10} + \frac{8}{90}$$

Step 3
Write these as fractions with a common denominator.

$$\frac{36}{90} + \frac{8}{90}$$

Step 4
Add the fractions.

$$\frac{36}{90} + \frac{8}{90} = \frac{44}{90}$$

> **Step 5**
> Cancel the result to its simplest form (if necessary).
>
> $$\frac{44}{90} = \frac{22}{45}$$
>
> Your challenge is to convert these recurring decimals to fractions.
>
> **1** $0.7\dot{1}$ **2** $0.3\dot{7}$
>
> **3** $0.24\dot{5}$ **4** $0.01\dot{6}$
>
> **5** $0.48\dot{2}\dot{3}$

Ordering fractions

One of the reasons for converting fractions to decimals is so that you can write fractions more easily in order of size. Another way of ordering fractions is to write them with common denominators. You will practise both methods in the next exercise.

Example 6

Which fraction is larger:

a $\frac{4}{5}$ or $\frac{7}{9}$ **b** $\frac{1}{3}$ or $\frac{2}{7}$?

----

a $\frac{4}{5} = \frac{8}{10} = 0.8$

$\frac{7}{9} = 7.000 \div 9 = 9)\overline{7.000...}^{\,0.777...}$

Make both decimals the same number of decimal places, for example 2 d.p., to compare them more easily. 0.80 is larger than 0.78, so the larger fraction is $\frac{4}{5}$.

b $\frac{1}{3} = \frac{7}{21}$

$\frac{2}{7} = \frac{6}{21}$

So $\frac{1}{3}$ is larger.

Write both fractions with a common denominator.

Exercise 6H

1 By converting to fractions with a common denominator, which is larger:

     a $\frac{1}{4}$ or $\frac{3}{10}$ **b** $\frac{2}{5}$ or $\frac{3}{8}$ **c** $\frac{7}{20}$ or $\frac{2}{5}$?

2 By converting to decimals, which is smaller:

     a $\frac{4}{9}$ or $\frac{3}{7}$ **b** $\frac{2}{9}$ or $\frac{3}{20}$ **c** $\frac{2}{15}$ or $\frac{1}{8}$?

3 Choose two pairs of fractions of your own, with different denominators. By writing equivalent fractions with a common denominator for each pair, write which fraction is larger.

4 Write down which of these is bigger than $\frac{3}{5}$.

     0.61 $\frac{59}{100}$ $\frac{5}{8}$ $\frac{6}{10}$ 0.597 $\frac{13}{20}$ $\frac{5}{9}$

5 Are these statements true or false?

     a $\frac{4}{15} < 0.2$ **b** $\frac{14}{30} = \frac{21}{45}$ **c** $\frac{5}{9} > 0.6$

6 Choose two pairs of fractions of your own, with different denominators. Convert each fraction to a decimal by division, then write which fraction is larger for each pair.

7 Choose from <, = or > to complete these.

     a $\frac{4}{7} \square \frac{5}{9}$ **b** $\frac{4}{3} \square 1\frac{2}{5}$ **c** $1.8 \square \frac{9}{5}$

8 Complete these sentences using 'larger' or 'smaller'. The numerators and denominators are all positive.

     a Two fractions have the same numerator. For these fractions, the larger the denominator the . . . the fraction.

     b Two fractions have the same denominator. For these fractions, the larger the numerator the . . . the fraction.

9 Use > or < to compare these fractions.

     a $\frac{5}{13} \square \frac{6}{13}$ **b** $2\frac{19}{47} \square 2\frac{19}{43}$

     c $\frac{17}{15} \square 1\frac{1}{15}$ **d** $\frac{^-1}{6} \square \frac{^-1}{7}$

     e $\frac{^-11}{19} \square \frac{^-12}{19}$

10 Decide whether these statements are true or false.

     a $60\% < \frac{2}{3}$ **b** $\frac{42}{55} > 80\%$

     c $3\frac{8}{15} \neq 3.5\dot{3}$ **d** $^-4.2 > ^-4\frac{2}{9}$

11 Find a possible value for x in each of these. x must be a fraction.

     a $4\frac{7}{9} < x < 4.8$ **b** $1\frac{2}{11} < x < 1.2$

     c $0.2 < x < \frac{2}{7}$

6.4 Subtracting mixed numbers

You can always add or subtract amounts of the same object.

For example, 7 tables − 2 tables = 5 tables

In the same way:

7 eighths − 2 eighths = 5 eighths

or $\quad \dfrac{7}{8} \; - \; \dfrac{2}{8} \; = \; \dfrac{5}{8}$

If you are adding or subtracting mixed numbers you can add or subtract the whole numbers first.

Example 7

Work out:

$3\dfrac{3}{4} - 2\dfrac{1}{4}$

$3\dfrac{3}{4} - 2\dfrac{1}{4} = 1\dfrac{2}{4}$ — Subtract the whole numbers (3 − 2)

$\qquad = 1\dfrac{1}{2}$ — Subtract the fractions $\left(\dfrac{3}{4} - \dfrac{1}{4}\right)$

Always simplify fractions.

You don't have to add or subtract the whole numbers first; you can change to improper fractions instead. You may wish to do this if, by subtracting the proper fraction parts, the result is a negative answer.

Example 8

Work out:

$3\dfrac{2}{5} - 1\dfrac{3}{4}$

$3\dfrac{2}{5} - 1\dfrac{3}{4} = \dfrac{17}{5} - \dfrac{7}{4}$

$\qquad = \dfrac{68}{20} - \dfrac{35}{20}$ — Change to improper fractions.

$\qquad = \dfrac{33}{20}$ — Common denominator is 20

$\qquad = 1\dfrac{13}{20}$ — Change to a mixed number.

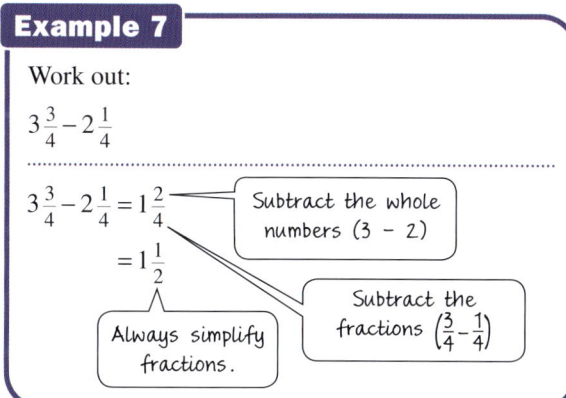

Exercise 6I

1. Work out:

   a $1\dfrac{3}{4} - \dfrac{1}{4}$ **b** $3\dfrac{2}{3} - 1\dfrac{1}{3}$

   c $2\dfrac{2}{5} - 1\dfrac{1}{5}$ **d** $1\dfrac{3}{5} - \dfrac{1}{5}$

   e $4\dfrac{3}{8} - 2\dfrac{2}{8}$ **f** $3\dfrac{7}{12} - 1\dfrac{5}{12}$

2. Somnath and Waqar are both working out the answer to $5\dfrac{1}{5} - 3\dfrac{4}{5}$

   Here is Waqar's method.

   $5 - 3 = 2$

   $\dfrac{1}{5} - \dfrac{4}{5} = -\dfrac{3}{5}$

   $5\dfrac{1}{5} - 3\dfrac{4}{5} = {}^-2\dfrac{3}{5}$

   Here is Somnath's method.

   $5\dfrac{1}{5} - 3\dfrac{4}{5} = 4\dfrac{6}{5} - 3\dfrac{4}{5} = 1\dfrac{2}{5}$

   a Who is correct? Explain why.

   b How could the person who is incorrect improve their method so that it works.

3. Work out:

   a $3\dfrac{1}{2} - 1\dfrac{1}{4}$ **b** $4\dfrac{3}{5} - 2\dfrac{2}{7}$

   c $8\dfrac{1}{2} - 3\dfrac{2}{3}$ **d** $5\dfrac{2}{3} - 2\dfrac{3}{4}$

   e $5\dfrac{1}{4} - 1\dfrac{4}{5}$ **f** $4\dfrac{1}{5} + 2\dfrac{3}{4} - 2\dfrac{1}{2}$

   g $10\dfrac{3}{5} - 4\dfrac{3}{8} + 2\dfrac{7}{20}$

4.

   Aakesh bought $3\dfrac{1}{4}$ kg of oranges. He gave his sister $1\dfrac{2}{3}$ kg.

   What was the mass of the oranges he had left?

5. A plank of wood is 3 m in length. How long will it be if I cut $\dfrac{5}{8}$ m of wood from it?

6. A container holds $4\dfrac{1}{2}$ litres of water.

   How much water is left in the container if Jerry drinks $2\dfrac{7}{8}$ litres?

7 Work out the missing numbers in this linear sequence.

$?, 4\frac{1}{3}, 6\frac{5}{6}, ?$

8 Fill in the missing numbers.

a $\square + 3\frac{1}{3} - 2\frac{1}{5} = 5\frac{4}{15}$

b $4\frac{1}{2} - \square + 3\frac{1}{10} = 2\frac{9}{10}$

9 Find the value of the letters in the following.

a $1\frac{a}{5} + 2\frac{3}{10} = 4\frac{1}{10}$ **b** $\frac{1}{3} + b\frac{5}{6} = 4\frac{1}{6}$

c $6\frac{c}{10} + 1\frac{3}{5} = 8\frac{1}{2}$ **d** $3\frac{7}{9} + \frac{7}{d} = 4\frac{1}{6}$

Investigation

A **unit fraction** is a fraction with a numerator of 1, e.g. $\frac{1}{2}, \frac{1}{6}, \frac{1}{13}$.

The ancient Egyptians didn't write fractions with a numerator greater than 1. Instead they would write them as a sum of two or more *different* unit fractions. For example, they wouldn't write $\frac{2}{3}$; instead they would write $\frac{2}{3}$ as $\frac{1}{2} + \frac{1}{6}$. (They wouldn't use $\frac{1}{3} + \frac{1}{3}$ as this involves repeating a unit fraction.) Investigate different non-unit fractions. Can all non-unit fractions be written as a sum of different unit fractions?

6.5 Multiplying and dividing fractions

You will need 24 identical quarter-circles, made from paper or thin card.

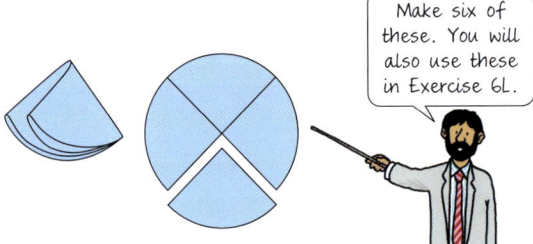

Make six of these. You will also use these in Exercise 6L.

Multiplying fractions by integers

Example 9

What is $3 \times \frac{3}{4}$?

Take three $\frac{3}{4}$-circles:

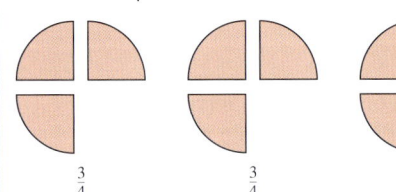

$\frac{3}{4}$ $\frac{3}{4}$ $\frac{3}{4}$

There are nine $\frac{1}{4}$-circles $= \frac{9}{4}$, so $3 \times \frac{3}{4} = \frac{9}{4} = 2\frac{1}{4}$

Exercise 6J

1 **a** Take two $\frac{3}{4}$-circles. How many quarter-circles have you used?

    b Copy and complete: $2 \times \frac{3}{4} = \frac{\square}{4}$

2 **a** Make up four $\frac{3}{4}$-circles. How many quarters have you used?

    b Copy and complete: $4 \times \frac{3}{4} = \frac{\square}{4}$

3 Use your $\frac{1}{4}$-circles to make up $\frac{3}{4}$-circles.

    a Copy and complete this table.

Number of $\frac{3}{4}$-circles made	Number of $\frac{1}{4}$-circles used	The multiplication
1	3	$1 \times \frac{3}{4} = \frac{3}{4}$
2	6	$2 \times \frac{3}{4} = \frac{6}{4}$
3	9	$3 \times \frac{3}{4} = \frac{9}{4}$
4		
5		
6		
7		
8		

    b Look at the last column in your table. Can you see a pattern?

    c What numbers multiplied together give the numerator in the second fraction?

    d What do you notice about the denominator in both fractions?

    e Can you see a way of multiplying any fraction by an integer?

- To multiply a fraction by an integer you multiply the numerator by the integer.

Example 10

Find: **a** $\frac{3}{5} \times 4$ **b** $1\frac{3}{4} \times 4$ **c** $^-2 \times 5\frac{1}{3}$

....

a $\frac{3}{5} \times 4 = \frac{3 \times 4}{5} = \frac{12}{5} = 2\frac{2}{5}$

b $1\frac{3}{4} \times 4 = \frac{7}{4} \times 4$

$\qquad = \frac{28}{4} = 7$

c $^-2 \times 5\frac{1}{3} = ^-2 \times \frac{16}{3}$

$\qquad = \frac{^-32}{3}$

$\qquad = ^-10\frac{2}{3}$

> First change to an improper fraction

Exercise 6K

1 Work out:

a $\frac{1}{6} \times 13$ **b** $4 \times \frac{5}{8}$ **c** $\frac{4}{9} \times 3$

d $\frac{3}{11} \times 8$ **e** $6 \times \frac{7}{10}$ **f** $10 \times \frac{5}{12}$

g $\frac{2}{5} \times 14$ **h** $20 \times \frac{7}{12}$ **i** $\frac{4}{5} \times 72$

j $96 \times \frac{4}{17}$

2 Work out:

a $1\frac{2}{3} \times 6$ **b** $7 \times 2\frac{1}{6}$

c $5 \times 3\frac{1}{4}$ **d** $2\frac{1}{2} \times 8$

e $6\frac{1}{9} \times 12$ **f** $14 \times 5\frac{7}{8}$

g $1\frac{2}{7} \times 5$ **h** $2 \times 4\frac{3}{8}$

i $5\frac{1}{3} \times 4$ **j** $2 \times 7\frac{1}{8}$

k $2\frac{3}{5} \times ^-3$ **l** $^-3 \times 1\frac{4}{15}$

3 Akila studies for $1\frac{3}{4}$ hours each day.

How long does she study in one week?

4 A box holds $1\frac{1}{2}$ kg of nails. What weight of nails do seven such boxes hold?

5

A kilogram is $2\frac{1}{5}$ pounds. A baby has a mass of 4 kg.
What is 4 kg in pounds?

6 What is the area of a rectangular garden that is $6\frac{1}{4}$ m long and 5 m wide?

7 Here is Aiko's method for multiplying an integer by a mixed number using the distributive law.

$10 \times 12\frac{5}{18}$

$10 \times 12 = 120$

$10 \times \frac{5}{18} = \frac{10 \times 5}{18} = \frac{50}{18} = \frac{25}{9} = 2\frac{7}{9}$

$10 \times 12\frac{5}{18} = 120 + 2\frac{7}{9} = 122\frac{7}{9}$

> Multiply 10 by the integer first then multiply 10 by the fraction part then add the two answers.

Aiko's method means you don't have to do a long multiplication to work out 12×18.

Use Aiko's method to repeat question **2**. Is it easier than using the improper fractions method from Example 10?

8 Put these in ascending order.

$$3 \times 1\frac{5}{12} \quad \ ^-3 \times 2\frac{1}{5} \quad 4 \times 1\frac{1}{3} \quad \left(4 \times 1\frac{1}{4}\right)^2 \quad 2 \times 3\frac{1}{8}$$

9 Work out:

a $\ ^-4\frac{2}{7} \times {}^-5$ **b** $\ ^-2 \times {}^-5\frac{3}{8}$

10 Two-thirds of the children in a class are girls. Of these, $\frac{1}{4}$ wear glasses. If there are 24 children in the class, how many girls wear glasses?

11 In the village of Lowcroft there are 630 people. Two-thirds of the villagers are under 16 and $\frac{4}{7}$ of these are girls.

a How many girls are under 16 in Lowcroft?
b How many boys are under 16?

Dividing integers by fractions

You can divide a number by a fraction.

For example:

what is $2 \div \frac{1}{4}$?

One way of working this out is to ask how many quarter-circles make 2 circles.

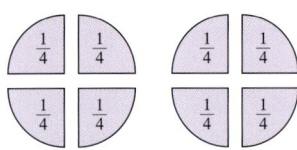

8 quarters = 2, so $2 \div \frac{1}{4} = 8$

Exercise 6L

1 a Using quarter-circles, make up 6 circles. How many quarter-circles are in 6 circles? Copy and complete: $6 \div \frac{1}{4} = \square$

b From your 6 circles make $\frac{3}{4}$ circles. How many $\frac{3}{4}$ circles can you make from 6 circles? Copy and complete: $6 \div \frac{3}{4} = \square$

2 a Using quarter-circles, make up 9 circles. How many quarter-circles are in 9 circles? Copy and complete: $9 \div \frac{1}{4} = \square$

b From your 9 circles make up $\frac{3}{4}$ circles. How many $\frac{3}{4}$ circles can you make from 9 circles? Copy and complete: $9 \div \frac{3}{4} = \square$

3 Look at your answers to question **1**.

$$6 \div \frac{1}{4} = 24$$

$$6 \div \frac{3}{4} = 8$$

a There are 24 quarter-circles in 6 circles. What number could you have *multiplied* 6 by, to get 24? Copy and complete:

$$6 \div \frac{1}{4} = 6 \times \square = 24$$

b What number should you *divide* 24 by, to find how many $\frac{3}{4}$ there are in 24 quarter-circles? Copy and complete:

$$6 \div \frac{3}{4} = \frac{6 \times 4}{\square} = 8$$

4 Look at your answers to question **2**.

$$9 \div \frac{1}{4} = 36 \qquad 9 \div \frac{3}{4} = 12$$

a There are 36 quarter-circles in 9 circles. What number could you have *multiplied* 9 by, to give 36? Copy and complete:

$$9 \div \frac{1}{4} = 9 \times \square = 36$$

b To find how many $\frac{3}{4}$ are in 36 quarter-circles, what number should you *divide* 36 by? Copy and complete:

$$9 \div \frac{3}{4} = \frac{9 \times 4}{\square} = 12$$

5 Look at some of the answers to questions **3** and **4**:

$$6 \div \frac{1}{4} = 6 \times 4 = 24 \quad 6 \div \frac{3}{4} = \frac{6 \times 4}{3} = 8$$
$$9 \div \frac{1}{4} = 9 \times 4 = 36 \quad 9 \div \frac{3}{4} = \frac{9 \times 4}{3} = 12$$

Can you see the pattern?
Can you see a way to divide a whole number by a fraction? Explain.

To divide an integer by a fraction you turn the fraction upside down and multiply. This is multiplying by the **reciprocal** of the fraction.

Example 11

Work out:
$2 \div \frac{3}{4}$

$2 \div \frac{3}{4} = 2 \times \frac{4}{3}$

$= \frac{8}{3}$

$= 2\frac{2}{3}$

Division by $\frac{3}{4}$ = multiplication by $\frac{4}{3}$

Exercise 6M

1   Copy and complete:

   a $4 \div \frac{2}{3} = 4 \times \frac{3}{2} = \frac{12}{2} = \square$

   b $10 \div \frac{2}{5} = 10 \times \frac{5}{2} = \frac{\square}{\square} = \square$

   c $10 \div \frac{4}{5} = 10 \times \frac{5}{4} = \frac{\square}{\square} = \square$

2   Do the division and write the answer in its simplest form.

   a $5 \div \frac{1}{9}$ **b** $4 \div \frac{3}{4}$ **c** $6 \div \frac{2}{3}$

   d $8 \div \frac{5}{6}$ **e** $9 \div \frac{3}{4}$ **f** $7 \div \frac{7}{9}$

   g $^-3 \div \frac{2}{5}$ **h** $^-2 \div \frac{5}{18}$ **i** $^-10 \div \frac{5}{7}$

3   How many half-litre bottles of juice can you get from a 10-litre container?

4   I have 12 oranges.
   How many people can I give $\frac{2}{3}$ of an orange to?

5   Andrew uses $\frac{2}{5}$ of a bag of fertilizer each week. How long will one bag of fertilizer last?

6   The area of this rectangle is $2\,\text{m}^2$. Find the missing side length.

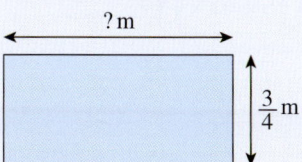

?m

$\frac{3}{4}$ m

7   Find the missing fraction.

   $\square\frac{\square}{\square} \times \frac{4}{5} = 3$

8   Sahil draws this picture to represent $4 \div \frac{1}{2} = 8$

Draw a picture to represent $4 \div \frac{1}{3} = 12$

9   Change the mixed number to an improper fraction, then do the division.

   a $6 \div 1\frac{1}{4}$ **b** $4 \div 1\frac{1}{2}$ **c** $9 \div 1\frac{5}{6}$

   d $7 \div 2\frac{3}{8}$ **e** $10 \div 3\frac{4}{5}$ **f** $8 \div 2\frac{5}{11}$

   g $^-6 \div 2\frac{3}{5}$ **h** $^-4 \div 1\frac{1}{7}$ **i** $^-10 \div 4\frac{2}{3}$

10   Which is better value:
   $\frac{3}{4}$ kg of soap powder for \$9,
   or $\frac{2}{3}$ kg of soap powder for \$8?

Investigation

Scientific calculators usually have a fraction key that looks like:

 or

This key can be used to simplify fractions and carry out calculations. It will allow you to convert fractions to decimals and some calculators will convert mixed numbers to improper fractions and improper fractions to mixed numbers.

Here are some examples.

To simplify $\frac{25}{35}$ key in

   [2] [5] [a b/c] [3] [5] [=]

You should see an answer of $\frac{5}{7}$.

To do $7 \times \frac{3}{14}$ key in

You should see an answer of $\frac{3}{2}$ or $1\frac{1}{2}$ or 1.5.

(Some calculators can be set up so that the answer will be an improper fraction, a mixed number or a decimal – see what yours can do.)

Your calculator will add and subtract fractions too. There are many different makes of calculator. Find out how to do calculations with fractions on your calculator. You may need to look at the instructions book if you have it, or you can ask your teacher. Repeat some of the questions from the previous exercises using your calculator.

Investigation

In the calculation $4 \div 2$, if you replace the division sign with a subtraction sign you get the same answer. So $4 \div 2 = 4 - 2$.

Here are two more calculations involving fractions where this is also true.

$$4\frac{1}{2} \div 3 = 4\frac{1}{2} - 3$$

$$6\frac{1}{4} \div 5 = 6\frac{1}{4} - 5$$

Investigate this further.

Consolidation

Example 1

Work out:

a 17×0.01 **b** $53 \div 0.1$

a $17 \times 0.01 = 17 \div 100 = 0.17$
b $53 \div 0.1 = 53 \times 10 = 530$

Example 2

a   Write the following to 1 significant figure (s.f.).

   i   20 871   **ii**   4175   **iii**   0.01652

b   Write the following to 2 significant figures (s.f.).

   i   20 871   **ii**   18 381
   iii   0.01652   **iv**   0.0003841

a   **i**   20 871 $= 20\,000$ (1 s.f.)
   ii   4175 $= 4000$ (1 s.f.)
   iii   0.01652 $= 0.02$ (1 s.f.)

b   **i**   20 871 $= 21\,000$ (2 s.f.)
   ii   18 381 $= 18\,000$ (2 s.f.)
   iii   0.01652 $= 0.017$ (2 s.f.)
   iv   0.0003841 $= 0.00038$ (2 s.f.)

Example 3

Change these fractions to recurring decimals.

a $\dfrac{7}{9}$ **b** $\dfrac{3}{7}$

a $9\overline{)7.^70^70...}^{\;0.\,7\,7...} = 0.\dot{7}$

b $7\overline{)3.^30^20^60^40^50^10^30...}^{\;0.\,4\,2\,8\,5\,7\,1\,4...} = 0.\dot{4}2857\dot{1}$

Example 4

Work out: $2\frac{3}{4} - 1\frac{2}{3}$

$$2\frac{3}{4} - 1\frac{2}{3} = \frac{11}{4} - \frac{5}{3}$$
$$= \frac{33}{12} - \frac{20}{12}$$
$$= \frac{13}{12} = 1\frac{1}{12}$$

Example 5

Calculate: $6 \times 3\frac{2}{5}$

$$6 \times 3\frac{2}{5} = \frac{6}{1} \times \frac{17}{5}$$
$$= \frac{102}{5}$$
$$= 20\frac{2}{5}$$

Example 6

Calculate: $^{-}5 \div \frac{3}{4}$

$$^{-}5 \div \frac{3}{4}$$
$$= {}^{-}5 \times \frac{4}{3}$$
$$= \frac{^{-}20}{3}$$
$$= {}^{-}6\frac{2}{3}$$

$\div \frac{3}{4}$ is the same as $\times \frac{4}{3}$

Example 7

Using the distributive law, calculate $6 \times 15\frac{5}{24}$

$6 \times 15\frac{5}{24} = 6 \times 15 + 6 \times \frac{5}{24}$

$= 90 + \frac{30}{24}$

$= 90 + 1\frac{6}{24}$

$= 91\frac{1}{4}$

Exercise 6

1 Work out:
   a 29×0.01 **b** 8×0.1
   c $97 \div 0.1$ **d** $3 \div 0.01$
   e 3200×0.01 **f** 900×0.1
   g $0.4 \div 0.1$ **h** $0.035 \div 0.01$

2 Copy and complete using decimals or integers:
   a $61 \div \square = 6100$
   b $\square \times 0.1 = 0.23$
   c $\square \div 10 = 7.1$
   d $\square \times 100 = 408$
   e $234 \times \square = 2.34$
   f $\square \div 0.01 = 780$

3 Write correct to 1 significant figure:
   a 850 **b** 9317 **c** $41\,290$
   d $43\,651$ **e** 0.763 **f** 0.072
   g 0.0065 **h** 8.93 **i** 0.000398

4 Write correct to 2 significant figures:
   a $537\,000$ **b** $41\,562$ **c** $0.037\,206$
   d $0.000\,215$ **e** 11.931 **f** 1.076
   g 0.00476 **h** 1432 **i** 0.3061

5 Write correct to 3 significant figures:
   a 18.403 **b** $157\,683$ **c** $10\,467$
   d 0.030084 **e** 15.092 **f** 113.387
   g $21\,547$ **h** 13.406 **i** 21.004

6 Here are some numbers written to 3 significant figures. Some of them are wrong. Pick out the wrong ones and write them correctly.
   a $539\,010$ $\rightarrow$ $539\,000$
   b 0.005706 $\rightarrow$ 0.00570
   c 0.050621 $\rightarrow$ 0.0506
   d $508\,716$ $\rightarrow$ $508\,700$
   e $15\,480\,000$ $\rightarrow$ $15\,500\,000$
   f 0.01006 $\rightarrow$ 0.0100

7 Complete these calculations.
   a $216 \div \square = 2160$
   b $240 \div \square = 24\,000$
   c $\square \div 0.01 = 20$
   d $\square \div 0.1 = 0.7$

8 Write as decimals:
   a $\frac{3}{8}$ **b** $\frac{2}{7}$ **c** $\frac{2}{9}$ **d** $\frac{5}{6}$
   e $\frac{6}{13}$ **f** $1\frac{2}{3}$

9 Write these numbers in order of size, starting with the smallest, by:
   a converting to fractions
   b converting to decimals.

   i $0.\dot{2}, \frac{1}{8}, \frac{3}{10}, 0.25, \frac{1}{5}$

   ii $\frac{2}{3}, \frac{7}{9}, 0.8, \frac{5}{8}, 0.75$

10 Work out:
   a $1\frac{1}{3} - \frac{2}{3}$ **b** $2\frac{3}{4} - 1\frac{1}{4}$ **c** $4\frac{5}{8} - 1\frac{7}{8}$
   d $6\frac{1}{12} - 4\frac{7}{12}$ **e** $2\frac{1}{3} - 1\frac{5}{9}$ **f** $4\frac{11}{12} - 1\frac{5}{6}$

11 Calculate:
   a $3\frac{3}{4} - 1\frac{3}{5}$ **b** $2\frac{1}{2} - 1\frac{3}{4}$ **c** $3\frac{1}{3} - \frac{3}{4}$
   d $2\frac{2}{5} - 1\frac{3}{4}$ **e** $4\frac{2}{7} - 1\frac{7}{8}$

12 Kimani's home is $6\frac{1}{2}$ km from school. She walks $\frac{5}{8}$ km to the bus stop and takes a bus for the rest of the journey. How far is her school from the bus stop?

13 Olive has 3 parcels to post. The first parcel has a mass of $2\frac{1}{3}$ kg. The second parcel has a mass of $\frac{3}{4}$ kg. The parcels have a mass of 5 kg in total. What is the mass of the third parcel?

14 Calculate:
   a $12 \times \frac{2}{5}$ **b** $^{-}5 \times \frac{2}{3}$ **c** $4\frac{3}{7} \times {}^{-}3$
   d $6 \times 2\frac{1}{5}$ **e** $^{-}2 \times 3\frac{7}{10}$ **f** $2\frac{6}{7} \times 34$

15 Work out :
   a $10 \div \frac{1}{2}$ **b** $-6 \div \frac{3}{4}$ **c** $8 \div \frac{4}{9}$
   d $7 \div \frac{2}{3}$ **e** $15 \div \frac{10}{11}$ **f** $^{-}10 \div \frac{5}{7}$

Summary

You should know ...

1 How to multiply and divide decimals.
For example:
a $92 \times 0.1 = 92 \div 10 = 9.2$
b $0.31 \div 0.01 = 0.31 \times 100 = 31$

2 How to round numbers to a given number of significant figures.
For example:

$2348 = 2000 \,(1 \text{ s.f})$
$0.005796 = 0.0058 \,(2 \text{ s.f. })$
$46\,891.47 = 46\,900 \,(3 \text{ s.f. })$
$46\,891.47 = 47\,000 \,(2 \text{ s.f.})$
$46\,891.47 = 50\,000 \,(1 \text{ s.f. })$

3 How to order fractions by writing them with common denominators or by dividing and converting them to decimals.
For example:

Write $\frac{3}{5}, \frac{2}{3}$ and $\frac{11}{20}$ in order of size, starting with the smallest.

$\frac{3}{5} = 3 \div 5 = 0.6 = 0.60$ to 2 d.p.

$\frac{2}{3} = 2 \div 3 = 0.\dot{6} = 0.67$ to 2 d.p.

$\frac{11}{20} = \frac{55}{100} = 55 \div 100 = 0.55$ to 2 d.p.

In size order they are: $\frac{11}{20}, \frac{3}{5}, \frac{2}{3}$

4 To add or subtract fractions their denominators must be the same.
For example:

$2\frac{3}{4} - 1\frac{2}{5}$

$= 2\frac{15}{20} - 1\frac{8}{20}$

$= 1\frac{7}{20}$

You can't subtract fifths from quarters

but you can subtract 8 twentieths from 15 twentieths.

Check out

1 Work out:

   a 18×0.01
   b $0.27 \div 0.1$
   c $3 \div 0.01$
   d 51.6×0.1

2 **a** Write correct to 3 s.f.
     i 8712   **ii** 0.004864

   b Write correct to 2 s.f.
     i 48 487 **ii** 0.0195

   c Write correct to 1 s.f.
     i 58   **ii** 0.000783

3 Write $\frac{7}{9}, \frac{17}{20}$ and $\frac{3}{4}$ in order of size, starting with the smallest.

4 **a** Work out:

     i $1\frac{4}{7} - \frac{2}{9}$

     ii $3\frac{3}{4} - 1\frac{7}{8}$

   b Beverly drinks $\frac{2}{3}$ litre of milk from a $1\frac{1}{2}$ litre bottle. How much milk is left in the bottle?

5 How to multiply by fractions.
For example:

$$^-3 \times 3\frac{3}{8} = {}^-3 \times \frac{27}{8} = \frac{^-81}{8} = {}^-10\frac{1}{8}$$

6 To divide an integer by a fraction, you turn the fraction over and then multiply.
For example:

$$5 \div \frac{3}{8} = 5 \times \frac{8}{3} = \frac{40}{3} = 13\frac{1}{3}$$

5 Work out:

a $4 \times 2\frac{3}{5}$

b $3\frac{1}{6} \times {}^-5$

c $4 \times 5\frac{2}{3}$

6 a Work out:

i $6 \div \frac{3}{8}$ **ii** $^-7 \div \frac{2}{3}$

b How many $\frac{5}{8}$ m pieces of material can be cut from a piece of material 25 m long?

Equations, formulae and inequalities

Objectives

In this chapter you will learn about:

- deriving and using formulae
- changing the subject of a formula
- deriving an equation
- solving equations with unknowns on both sides

- representing an inequality on a number line and understanding equivalent inequalities.

What's the point?

Different parts of the world use different currencies and different measuring scales. It is sometimes necessary to convert between currencies, for example when travelling or in business. Some parts of the world measure temperature in degrees Celsius, while some measure it in degrees Fahrenheit (both scales are named after the person who invented them). You have also learned about metric and imperial measurements. It is important to know which measuring scale is being used and how to convert between them – algebra and formulae help with this.

Before you start

You should know ...

1 About the order of operations, BIDMAS.

$(10 \times 6) - 5 \times 3^2$ Bracket first

$= 60 - 5 \times 3^2$ then Indices

$= 60 - 5 \times 9$ then Multiplication

$= 60 - 45$ then Subtraction.

$= 15$

2 How to expand brackets and use the order of operations with algebra.

For example:

$6 - 2(3x - 4) + 10x$

$= 6 - 6x + 8 + 10x$

$= 14 + 4x$

3 How to multiply an integer by a fraction.

For example:

$25 \times \frac{8}{5} = (25 \div 5) \times 8$

$\qquad\quad = 5 \times 8 = 40$

Find $\frac{1}{5}$ of 25 first, then $\times$ 8

Check in

1 Work out:

a $7 \times 3^2 - (5 \times 3)$

b $2 \times 10 - 12 \div (1 + 2)$

c $(3^3 + 4) - 2^2 \times 5$

d $\dfrac{4^2 + 5}{7}$

2 a Expand:

i $4(2p - 5)$

ii $^-5(3 - 2x)$

b Expand and simplify:

i $4(3p - 5m) + 5(2p + 7m)$

ii $60 - 10(3x + 2) + 80x$

3 Work out:

a $64 \times \dfrac{5}{8}$

b $55 \times \dfrac{4}{5}$

c $60 \times \dfrac{5}{12}$

7.1 Deriving formulae

You have already learned that an expression does not contain an equals sign, but a formula does. A **formula** describes a relationship between two variables. **Formulae** is the plural of formula.

To derive a formula means to use known information to write a formula connecting variables.

Example 1

An electrician charges a call out fee of $30 and then $25 for each hour that he works. Write a formula for the total cost, T dollars, of a job that takes h hours.

....

h hours at $25 dollars for each hour is:

$25 \times h$ (or $25h$)

Add to that the call out fee so:

$25h + 30$

Then change the expression to a formula with an equals sign:

$T = 25h + 30$

In Example 1 you will have seen there are different parts to the formula. The 25 and the 30 are **fixed** parts of the formula. They do not change. The **variable** parts of the formula are the T and the h, which can take different values. T is the **subject** of the formula and the formula is said to be **in terms of** h as h is the variable used on the other side of the equals sign.

Some countries measure distances in miles. One kilometre is about $\frac{5}{8}$ of a mile. You can derive a formula to convert between miles and kilometres.

To convert 8 km into miles do $8 \times \frac{5}{8} = 5$

To convert 24 km into miles do $24 \times \frac{5}{8} = 15$

To convert 56 km into miles do $56 \times \frac{5}{8} = 35$

To convert k km into miles do $k \times \frac{5}{8} = \frac{5k}{8}$

$\frac{5k}{8}$ is an expression for the number of miles in k km. To turn it into a formula you need to have an equals sign. If m = the number of miles then the formula is $m = \frac{5k}{8}$

Some countries measure temperature using degrees Fahrenheit and some use degrees Celsius. To compare the two scales, consider the boiling point and freezing point of water:

	Celsius	Fahrenheit
Freezing temperature of water	0°C	32°F
Boiling temperature of water	100°C	212°F

You can see that when the temperature in Celsius increases by 100°C the temperature in Fahrenheit increases by 180°F.

So an increase of 1°C will be an increase of 1.8°F.

Since the freezing point of water in Fahrenheit is 32°F we need to add 32°F as well.

If F = the temperature in Fahrenheit and C = the temperature in Celsius, then

$$F = 1.8C + 32$$

You could write this using fractions to make calculations easier to do without a calculator:

$$F = \frac{9}{5}C + 32$$

Exercise 7A

1 Use the formula $m = \frac{5k}{8}$ to convert the following into miles.

    a 48 km   **b** 80 km   **c** 32 km

2 **a** Derive a formula to convert m miles into k kilometres.

    b Use your formula to convert the following into kilometres.

       i 50 miles   **ii** 75 miles

3 **a** Use the formula $F = \frac{9}{5}C + 32$, without a calculator, to convert the following into Fahrenheit.

       i 30°C   **ii** 80°C   **iii** 12°C

    b Some people find this formula hard to remember and use. Did you find part **a iii** difficult without a calculator? Could you do it in your head or did you have to write down working?

    Instead of using the formula, people can find an approximation by doubling the degrees in Celsius then adding 32 to find the degrees in Fahrenheit. Derive the formula for this.

c Without using a calculator, use your formula from part **b** to convert the following into Fahrenheit.

    i 30°C   **ii** 80°C   **iii** 12°C

    Was part **iii** easier this time?

4 Here is a rectangle.

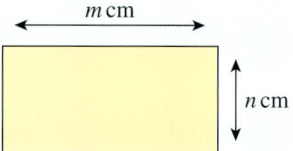

    a Write down a formula for the area, A, of the rectangle.

    b Write down a formula for the perimeter, P, of the rectangle.

    c When $m = 3$ and $n = 15$ use your formulae to find the area and perimeter of the rectangle.

5 Here is a cuboid.

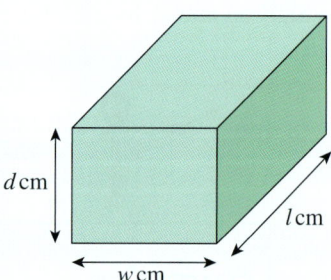

Write down a formula for the volume, V, of the cuboid.

6 Pier says, 'Anything with an equals sign is an equation.'

    a Explain why Pier is wrong.

    b Which of these are equations?

       $2a - 4 = 3$ $v = u + at$ $a^2 + b^2 = c^2$

7 A plumber charges a call out fee of $20 and then $30 for each hour that he works. Write a formula for the total cost, T dollars, of a job that takes h hours.

8 Fatimah says the following is a formula to find the total cost T, of x jumpers at $20 per jumper and $3 delivery charge.

$$T = 20(x + 3)$$

Is she correct? Explain your answer.

9 *n* cubes are stacked in a tower like the one below, then placed on a table.

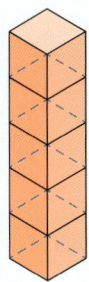

Faces that touch each other or the table when the cubes are stacked are hidden from view and cannot be seen.

a Derive a formula for the number of faces, *f*, that can be seen.

b Derive a formula for the number of faces, *c*, that cannot be seen.

c Amira thinks that this formula shows the total number of faces: $f + c = 6n$
Do you think she is correct? Use your answers to parts **a** and **b** to check.

10 A rectangle has a width of *x* cm. The length is 4 cm more than the width.
a Find the area of the rectangle when the width is 5 cm.
b Find the area of the rectangle when the width is 10 cm.
c Find a formula for the area, *A*, of the rectangle in terms of *x*.

11 A company charges $2 for a 2GB memory stick and $5 for an 8GB memory stick. The company also charges $3 for post and packing (which is the same no matter how many memory sticks you buy). Find a formula for the total order cost, *T* dollars, of *x* 2GB memory sticks and *y* 8GB memory sticks.

12 Amil receives $12 per hour for her job and receives a travel allowance of $3 each day.
a How much does Amil receive in 4 days when she works 8 hours per day?
b Write a formula for the total, *T* dollars, Amil receives in *d* days working *h* hours per day.

13 Sanjit bought 4 more oranges than apples. He bought *n* apples. Write a formula for *T*, the total number of pieces of fruit he bought, in terms of *n*, in its simplest form.

14 Which of the following formulae correctly show the relationship between the area *A*, base *b* and perpendicular height *h* of a triangle?

$$A = \frac{1}{2} \times b \times h \qquad \frac{2A}{b} = h \qquad A = \frac{1}{2}bh$$

$$2A = b \times h \qquad \frac{2A}{h} = b$$

15 Cubes are stacked in a tower like the one below, with 4 cubes in each layer. The tower is placed on a table. It is *n* cubes high.

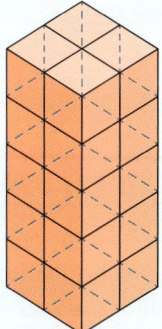

a Derive a formula for *t*, the total number of cubes used in the tower.
b Derive a formula for *f*, the number of faces that can be seen.
c Use your formula from part **b** to find the total number of faces that can be seen when the tower is 12 cubes high.
d Use your formula from part **b** to find the height of the tower when the number of faces that can be seen is 76.

16 Write a formula to find the side length, *l*, of the cuboid in question **5** in terms of *V*, *w* and *d*.

17 Write a formula to find the side length, *m*, of the rectangle in question **4** in terms of *P* and *n*.

18 Derive a formula to convert degrees Fahrenheit into degrees Celsius.

Investigation

To convert from Celsius to Fahrenheit we use the formula $F = \frac{9}{5}C + 32$

This is not so easy to remember or use so many people use the approximation $F \sim 2C + 32$

Try a few conversions using both formulae. When is the approximation most accurate?

7.2 Deriving and solving equations

In Year 7, you learned how to derive and solve equations such as $2x + 3 = 13$. The first exercise is a recap of that work.

Example 2

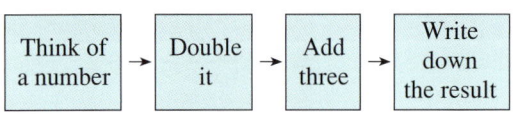

Using the given number machine:

a find the result if you start with 5

b find the result if the starting number is x

c find the starting number if the result is 15

---

a

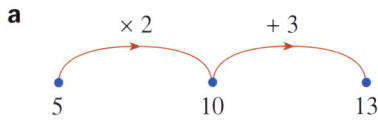

The result is 13.

b

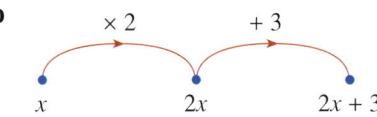

The result is $2x + 3$.

c To do this, use the machine backwards:

-3 is the inverse of $+3$

$\div 2$ is the inverse of $\times 2$

So subtract 3 and then divide by 2:

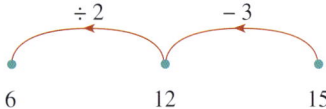

The starting number was 6.

Exercise 7B

1 For each machine, write down the result when the starting number is:

i 5

ii 12

a

b

c

2 Draw a diagram to show the reverse machine for each part of question **1**.

3 For each machine in question **1**, find what starting number will give 26 as the result. Use your machines from question **2** to help you.

4 For each machine in question **1**, write down the result when the starting number is x.

Simple equations can be solved using number machines.

Balancing equations

Can you use a number machine to solve

$$2x = x + 5?$$

You cannot because there are xs on both sides of the equation.

The idea of a balance can be used to solve equations like these.

Look at the balance. On one side it has 2 tins of paint. On the other side it has 1 tin of paint and a 5 kg weight.

Each tin of paint weighs x kg.

The masses on each side of the balance are equal. If you remove one tin of paint from each side of the balance …

… you will see that $x = 5$.

Algebraically you can write:

$$2x = x + 5$$

so, subtracting x from each side:

$$x = 5$$

Example 3

Use the balance idea to solve:

a $w - 5 = 9$ **b** $4w = w + 3$

c $\dfrac{x}{3} + 4 = 8$

....

a $w - 5 = 9$

Add 5 to both sides:

$w - 5 + 5 = 9 + 5$

$\qquad w = 14$

Add 5 to **both** sides!

b $4w = w + 3$

Subtract w from both sides:

$4w - w = w + 3 - w$

$\qquad 3w = 3$

Divide both sides by 3:

$\dfrac{3w}{3} = \dfrac{3}{3}$

$\qquad w = 1$

c $\dfrac{x}{3} + 4 = 8$

Subtract 4 from both sides:

$\dfrac{x}{3} + 4 - 4 = 8 - 4$

$\qquad \dfrac{x}{3} = 4$

Multiply each side by 3:

$3\left(\dfrac{x}{3}\right) = 3(4)$

$\qquad x = 12$

Exercise 7C

1 a Look at this diagram.

Do you agree that it shows the equation $5x + 1 = 2x + 10$?

b Can you explain how the diagram below was obtained from the diagram in part **a**? Do the two sides still balance?

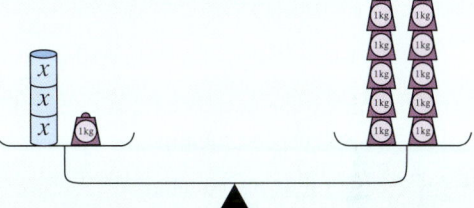

c Write an equation to show the information in the diagram in part **b**.

d Draw a diagram like the one in part **b**, but this time remove 1 kg from each side. Do the two sides still balance?

e Work out the mass of one tin, x.

2 Complete each of these statements.

a Given that $5x + 8 = 2x + 14$, then $3x + 8 = \ldots$

b Given that $3x - 7 = 3 - 2x$, then $5x - 7 = \ldots$

c Given that $6x + 19 = 10x - 5$, then $19 = \ldots$

d Given that $5x + 3 = 9x - 13$, then $5x + 16 = \ldots$

3 Kevin tried to solve the equation

$5x - 4 = 3x + 6$

Here is his working.

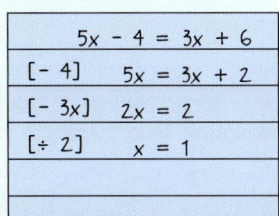

$5x - 4 = 3x + 6$	
[- 4]	$5x = 3x + 2$
[- 3x]	$2x = 2$
[÷ 2]	$x = 1$

a What was Kevin's mistake?

b Solve the equation correctly for Kevin.

4 Each of **a** to **g** are correct equivalent statements for solving $10x - 9 = 6x + 3$ Which ones are helpful in finding a solution? Which would you use to find the value of x?

   a $8x - 9 = 4x + 3$
   b $^-9 = ^-4x + 3$
   c $0 = 6x + 3 - 10x + 9$
   d $10x = 6x + 12$
   e $10x - 12 = 6x$
   f $4x - 12 = 0$
   g $4x - 9 = 3$

5 Given that $7x + 15 = 9x + 5$:

   a which of these are equivalent statements?

      $^-2x + 15 = 5$ $15 = 2x + 5$

      $8x + 15 = 10x + 5$ $7x + 10 = 9x$

   b which is the most helpful starting point for solving the equation?

6 Solve the following equations.
   a $7x + 3 = 3x + 11$
      (**Hint:** subtract $3x$ from both sides)

   b $2x + 19 = 5x + 4$
      (**Hint:** subtract $2x$ from both sides)

   c $5x + 1 = 8 - 2x$
      (**Hint:** add $2x$ to both sides)

   d $6x + 2 = 16 - x$
      (**Hint:** add x to both sides)

7 Use the balance idea to solve:

   a $9x + 3 = 4x + 8$

   b $3x + 7 = 5x + 1$

   c $2x + 3 = 28 - 3x$

   d $4x - 7 = 13 - x$

   e $6x - 7 = 2x + 3$

   f $5x - 12 = 3 + 2x$

   g $9x - 11 = 3x + 13$

   h $8x - 12 = 5x - 3$

8 Use the balance idea to solve the following equations. Write down what you are doing to both sides of the equation, as in Example 3.

   a $x + 3 = 5$ **b** $x - 3 = 7$

   c $6 = 2x$ **d** $\dfrac{x}{3} = 4$

   e $2x + 3 = 9$ **f** $2x - 3 = 9$

   g $5p - 7 = 8$ **h** $\dfrac{x}{3} + 4 = 8$

   i $7 + 6x = 7x$ **j** $4 - 2x = 2x$

   k $\dfrac{x}{5} - 7 = 2$ **l** $7x = 20 - 3x$

   m $3 - x = 2$ **n** $3x = x + 6$

9 Hannah answers question **6a** and gets the answer $x = 2$
She uses this method of substitution to check her answer is correct.
$7x + 3 = 3x + 11$
$7 \times 2 + 3 = 3 \times 2 + 11$
$14 + 3 = 6 + 11$
$17 = 17$ ✓

Use this method of substitution to check your answers to questions **6**, **7** and **8**. Make sure the left-hand side of the equation equals the right-hand side of the equation.

10 Solve these equations.

   a $\dfrac{x}{3} + 4 = 8$ **b** $\dfrac{x}{5} - 6 = 1$

   c $\dfrac{x}{4} + 8 = 17$

11 Solve these equations.

   a $\dfrac{(x + 2)}{3} = 10$ **b** $\dfrac{(x + 3)}{2} + 3 = 7$

   c $\dfrac{(x - 3)}{2} = 42$

12 Explain why there is no solution to $3x + 5 = 3x$

13 Show that ABCD is a parallelogram.

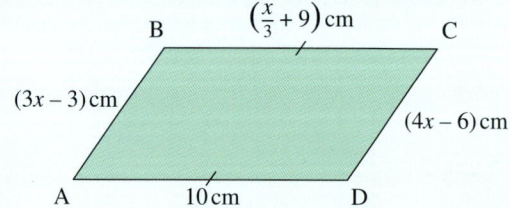

Sometimes you have to simplify equations by first removing brackets.

Example 4

Solve $5(x+3) = 2x + 27$

Expand brackets:
$$5x + 15 = 2x + 27$$

Subtract $2x$ from both sides:
$$5x + 15 - 2x = 2x + 27 - 2x$$
$$3x + 15 = 27$$

Subtract 15 from both sides:
$$3x + 15 - 15 = 27 - 15$$
$$3x = 12$$

Divide both sides by 3:
$$x = 4$$

Sometimes you need to expand brackets and collect terms.

Example 5

Solve $7(2p+3) + 6(p-2) = 10p + 29$

Expand brackets:
$$14p + 21 + 6p - 12 = 10p + 29$$

Simplify by collecting like terms:
$$20p + 9 = 10p + 29$$
$[-10p]$ $10p + 9 = 29$
$[-9]$ $10p = 20$
$[\div 10]$ $p = 2$

Exercise 7D

1 Solve:

   a $3(x+3) = 21$

   b $10(t-3) = 40$

   c $60 = 4(2x-5)$

   d $42 = 3(3m-4)$

   e $3(4+x) = 24$

   f $2(x-3) = 1$

   g $3(x+4) = 2(x+7)$

   h $5(x+2) = 3(x+10)$

   i $4(x-5) = 2(x+3)$

   j $7(p+3) = 4(p+6)$

   k $4(2x+1) = 6(x+3)$

   l $7d = 5(3+d)$

2 Manon and Sukhraj are each solving $7(x+4) = 49$. Here are their methods.

Manon:
$$7(x+4) = 49$$
$$7x + 28 = 49$$
$$7x = 21$$
$$x = 3$$

Sukhraj:
$$7(x+4) = 49$$
$$x + 4 = 7$$
$$x = 3$$

Both methods are correct.

   a   Sort the equations below into two categories **A** or **B**.

      A – where it is easier to multiply the bracket out as the first step

      B – where it is easier to **not** multiply the bracket out as the first step.

      $7(x-3) = 35$ $5(x-2) = 6$

      $6(x+3) = 36$ $4(x+3) = 17$

   b   Give a reason for the way you have sorted the equations.

3 Given that $6(x+3) = 4(5x+8)$, which of these are true and which are false?

   a $6x + 3 = 20x + 8$

   b $6x + 9 = 9x + 12$

   c $6x + 18 = 4(5x+8)$

   d $6x + 3 = 20x + 8$

   e $6(x+3) = 20x + 32$

   f $6x + 18 = 20x + 32$

   g $2(3x+9) = 2(10x+16)$

4 Solve these equations.

   a $4x + 2(x+1) = 16$

   b $3(x+1) + 2(x+2) = 17$

   c $2(2x-7) + x = 2x + 6$

   d $4 + 3(x-5) = 10$

   e $x + (x+1) + (x+2) = 63$

   f $2(x+7) - 6 = x + 15$

5 Samina and Ben each try to solve the equation $\frac{5}{16}x = 15$

Samina's method, dividing by the numerator first, is:

$\frac{5}{16}x = 15$ [÷ 5]

$\frac{1}{16}x = 3$ [× 16]

$x = 48$

Ben's method, multiplying by the denominator first, is:

$\frac{5}{16}x = 15$ [× 16]

$5x = 240$ [÷ 5]

$x = 48$

a Both methods are correct. Which method is easier and why?

b Is it easier to use Samina's method of dividing by the numerator of the fraction first, or Ben's method of multiplying by the denominator first to solve this equation?

$\frac{4}{15}x = 2$

Give a reason for your answer.

6 Solve these equations.

a $\frac{2}{3}x = 16$ **b** $\frac{3}{16}x = 30$

c $\frac{4}{15}x = 24$ **d** $\frac{3}{5}x - 2 = 10$

e $2\left(\frac{3}{8}x + 2\right) = x$ **f** $6\left(\frac{3}{4}x + 5\right) = 7x$

Solving problems

Creating an equation and solving it is a powerful way of working on many types of problem.

There are four steps to solving a problem.

1 Understand the problem.

2 Devise a plan.

3 Carry out your plan.

4 Look back.

Example 6

10.5 cm

A rectangle with length 10.5 cm has perimeter 36 cm. What is its width?

Understand the problem
What is to be found?
The width, call it w.

Devise a plan
Form an equation:
$w + 10.5 + w + 10.5 = 36$

Carry out the plan
Solve the equation.
First, collect like terms:
$w + w + 10.5 + 10.5 = 36$

Simplify:
$2w + 21 = 36$

Subtract 21 from both sides:
$2w = 15$

Divide both sides by 2:
$w = 7.5$

Look back
The width of the rectangle is 7.5 cm.
This is correct, since:
$10.5 + 7.5 + 10.5 + 7.5 = 36$

Exercise 7E

For all of these questions, first write down an equation, then solve it.

1 The result when adding a number to 6 is 41. What is the number?

2 A certain number when multiplied by 2 and then added to 5 gives a result of 97. What is the number?

3 The sum of two consecutive whole numbers is 91. What are the numbers?

4 The perimeter of this isosceles triangle is 29 cm. What is the missing side length, t?

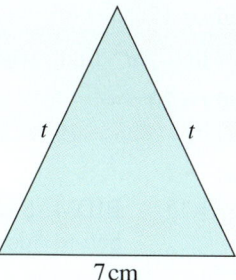

7 cm

5 Find the width of rectangles with:

a length 5 cm, perimeter 24 cm

b length 8.1 cm, perimeter 28.5 cm

c length 17.3 m, perimeter 41.7 m

6 This diagram shows an equilateral triangle with perimeter 54 cm.

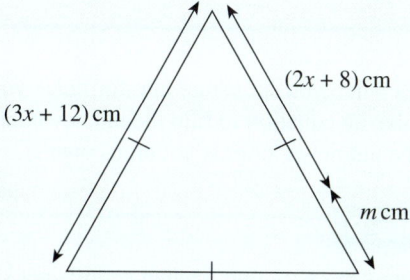

Find the value of m.

7 Work out the size of the obtuse angle in this isosceles triangle.

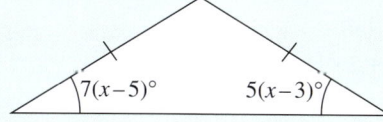

8 In this diagram these shapes made using rectangles have the same area. Find the value of x. The drawings are not to scale.

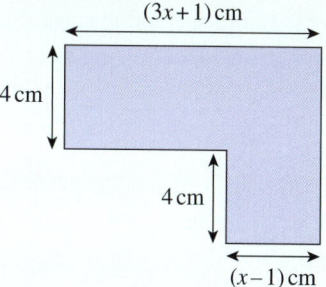

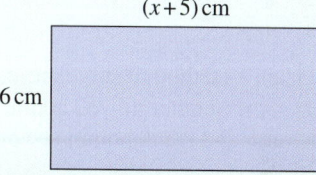

9 The perimeter of a triangle is 34 cm. What are the lengths of the sides if the first side is twice the length of the second side and the third side is 2 cm longer than the second side?

10 Janet is 6 years younger than Safiya. How old is each girl if the sum of their ages is 26?

11 Anton has three times as many marbles as Kamil. Kamil has 4 more marbles than Abdul. How many marbles does each boy have if there are 96 marbles altogether?

12 In 16 years' time Jim will be three times his current age. How old will Jim be in 4 years' time?

13 The sum of two consecutive even numbers is 214. What are the numbers?

14 The sum of three consecutive odd numbers is 243. What are the numbers?

15 What same number do you add to the numerator and the denominator of $\frac{4}{7}$ to make $\frac{2}{3}$?

Here is Hani's method to solve this question.

$\frac{4+n}{7+n} = \frac{2}{3}$ $\qquad$ [× 3]

$\frac{3(4+n)}{7+n} = 2$ $\qquad$ [× (7 + n)]

$3(4+n) = 2(7+n)$ $\qquad$ [expand the brackets]

$12 + 3n = 14 + 2n$ $\qquad$ [− 2n]

$12 + n = 14$ $\qquad$ [− 12]

$n = 2$

Check:

$\frac{4+2}{7+2} = \frac{6}{9} = \frac{2}{3}$ ✔

Use Hani's method to solve these questions. What same number do you add to the numerator and the denominator of:

a $\frac{3}{11}$ to make $\frac{5}{9}$

b $\frac{2}{7}$ to make $\frac{2}{3}$

c $\frac{1}{14}$ to make $\frac{1}{2}$

d $\frac{3}{10}$ to make $\frac{2}{3}$?

Substituting into formulae

Example 7

In electricity, Ohm's law states that $V = IR$ where V = voltage, I = current and R = resistance.
Find V when $I = 0.5$ and $R = 6$

. . . .

$V = 0.5 \times 6$
$V = 3$

When something moves in a straight line with constant acceleration, the following formulae apply:

$v = u + at$ $\qquad$ and $\qquad$ $s = ut + \frac{1}{2}at^2$

where
a = acceleration $\qquad\qquad$ u = initial velocity

t = time taken $\qquad\qquad$ s = displacement (distance
v = final velocity $\qquad\qquad\qquad$ travelled from a point)

These are called equations of motion. They are used in mechanics and physics.

Example 8

a Using $v = u + at$, find v when $u = 10$, $a = 2$ and $t = 15$

b Using $s = ut + \frac{1}{2}at^2$, find s when $u = 15$, $a = 3$ and $t = 20$

. . . .

a $v = u + at$
$v = 10 + 2 \times 15$ $\qquad$ BIDMAS says Multiply
$\qquad\qquad\qquad\qquad\qquad$ before Adding
$v = 10 + 30$
$v = 40$

b $s = ut + \frac{1}{2}at^2$

$s = 15 \times 20 + \frac{1}{2} \times 3 \times 20^2$ $\qquad$ BIDMAS says
$\qquad\qquad\qquad\qquad\qquad\qquad\quad$ Indices first

$s = 15 \times 20 + \frac{1}{2} \times 3 \times 400$ $\qquad$ then Multiply

$s = 300 + 600$ $\qquad\qquad\qquad\qquad$ then Add.

$s = 900$

Sometimes when you substitute into formulae you need to solve an equation to find the unknown letter because the unknown letter is not on its own.

Example 9

a Using $v = u + at$, find u when $v = 50$, $a = 3$ and $t = 15$

b Using $v = u + at$, find t when $v = 40$, $u = 10$ and $a = 2$

. . . .

a $v = u + at$
$50 = u + 3 \times 15$
$50 = u + 45$
Subtract 45 from both sides:
$50 - 45 = u + 45 - 45$
$\qquad 5 = u$

b $v = u + at$
$40 = 10 + 2t$
Subtract 10 from both sides:
$40 - 10 = 10 - 10 + 2t$
$\qquad 30 = 2t$

Divide both sides by 2:
$\frac{30}{2} = \frac{2t}{2}$

$15 = t$

Note that all of the formulae used in the following exercises are used in real life.

Exercise 7F

1 Using $v = u + at$, find v when:
 a $u = 20$, $a = 2$ and $t = 25$
 b $u = 50$, $a = 23$ and $t = 10$

2 Using $V = IR$, find V when:
 a $I = 2$ and $R = 5$
 b $I = 0.5$ and $R = 4$

3 Using $v = u + at$, find u when:
 a $v = 100$, $a = 5$ and $t = 10$
 b $v = 20$, $a = {}^-5$ and $t = 9$

4 Using $V = IR$, find I when:
 a $V = 8$ and $R = 4$
 b $V = 5$ and $R = 10$

5 Using $v = u + at$, find t when:
 a $v = 60$, $a = 5$ and $u = 20$
 b $v = 40$, $a = {}^-2$ and $u = 68$

6 Using $F = ma$, find m when:
 a $F = 24$ and $a = 8$
 b $F = 15$ and $a = 6$

7 Here is Angelique's working using $v = u + at$ to find a when $v = 70$, $t = 10$ and $u = 20$

 $70 = 20 + 10a$ $\quad [\div 10]$
 $7 = 20 + a$ $\quad [-20]$
 $^-13 = a$

 It is an acceptable method to divide by 10 first, if this is done correctly. Explain the mistake that Angelique has made.

8 Using $Ft = mv - mu$, find v when $F = 42$, $u = 10$, $t = 5$ and $m = 7$

9 Which of the following formulae are equivalent to $s = ut + \frac{1}{2}at^2$?

 $s - ut = \frac{1}{2}at^2$ $\qquad s + ut = \frac{1}{2}at^2$

 $s - \frac{1}{2}at^2 = ut$ $\qquad s + \frac{1}{2}at^2 = ut$

 $2s = ut + at^2$ $\qquad 2s = 2ut + at^2$

10 Using $s = ut + \frac{1}{2}at^2$, find u when $s = 78$, $t = 3$ and $a = 4$

11 Using $v^2 - u^2 = 2as$, find:
 a s when $v = 20$, $u = 8$ and $a = 4$
 b a when $v = 15$, $u = 5$ and $s = 200$

12 Hooke's Law states that $F = kx$ where F = force, x = displacement and k is the spring's constant (which tells you how powerful the spring is). Find k when $F = 20$ and $x = 0.1$

13 Using $v^2 - u^2 = 2as$, find v when $s = 21$, $u = 4$ and $a = 2$

14 Using $s = ut + \frac{1}{2}at^2$, find t when $s = 96$, $u = 0$ and $a = 3$

15 Using $v^2 - u^2 = 2as$, find u when $v = 35$, $a = 3$ and $s = 200$

7.3 Changing the subject of a formula

The formula for the circumference of a circle is $C = 2\pi r$

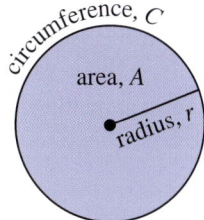

How would you find the radius, r, given the circumference, C?

You can find the circumference of a circle using a function machine:

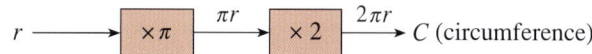

You can use the reverse machine to find the radius:

$$r \xleftarrow{\frac{C}{2\pi}} \boxed{\div\,\pi} \xleftarrow{\frac{C}{2}} \boxed{\div\,2} \xleftarrow{} C$$

That is, $r = \dfrac{C}{2\pi}$

r is now the **subject** of the formula.

Example 10

Make P the subject of the formula
$I = P \times R \times T$

....

The formula can be shown as a function machine:

$$P \longrightarrow \boxed{\times R} \longrightarrow \boxed{\times T} \longrightarrow I$$

P can be found by reversing the machine:

$$P \longleftarrow \boxed{\div R} \longleftarrow \boxed{\div T} \longleftarrow I$$

which gives the formula
$$P = \frac{I}{T \times R}$$

Notice that in Example 10 we don't write $I \div T \div R$ or $\frac{I \div T}{R}$. It is common for students to write 'two-tiered' fractions such as $\frac{\frac{I}{T}}{R}$, which are messy and should be avoided. It is worth remembering that dividing by x is the same as multiplying by $\frac{1}{x}$. This is why, in Example 10, dividing $\frac{I}{T}$ by R is shown as multiplying the T by R : $\frac{I}{T \times R}$.

Exercise 7G

1 Using function machines to help you, rearrange each formula to make the letter in brackets the subject.
   a $t = u + v$ (v) **b** $d = h + y$ (y)
   c $x = f - 3$ (f) **d** $m = 2c + g$ (c)
   e $\frac{b}{n} = s$ (b) **f** $\frac{de}{f} = a$ (d)
   g $q = 3(x - f)$ (x)

2 The simple interest formula is $I = \frac{P \times R \times T}{100}$
   a Draw a function machine for finding I when starting with T.
   b Draw the reverse machine, and complete the equation $T = \frac{\square \times \square}{\square \times \square}$

3 The formula for converting temperature from Fahrenheit to Celsius is $C = \frac{5}{9}(F - 32)$
   a Draw a function machine to show how you can find C, starting with a value for F.
   b Use your machine to find C when F is:
      i 41°   **ii** 59°
      iii 86°   **iv** 212°

4 **a** Draw the reverse machine for question **3a**.
   b Use this to write down the rearranged formula with F as its subject.
   c Find F when C is **i** 50° **ii** 80°

5 Draw a function machine to show how to find:
   a V using $V = l \times b \times h$ starting with l
   b S using $S = 2\pi rh$ starting with r
   c v using $v = u + at$ starting with t

6 Use the reverse machine for each part of question **5** to rewrite the formula. Make the letter you originally started with the subject.

7 Rearrange each formula, making r the subject.
   a $A = rh$ **b** $A = 2\pi rh$

8 **a** Draw a function machine to show how to find y; starting with x.
      i $y = px + q$
      ii $y = k(x - l)$
   b Use the reverse machine to rearrange the formula, making x the subject.

9 **a** Derive the formula for the perimeter, P, of this triangle.

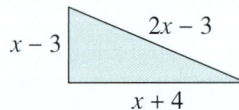

   b Rearrange this formula to make x the subject.
   c Use this rearranged formula to find x when the perimeter, P, is 30.
   d What are each of the side lengths of the triangle in part **a**?

The balance method

It is not always possible to use a function machine to change the subject of a formula, particularly when the chosen letter appears more than once (you will learn about this after Year 9).

Remember that a formula stays balanced like an equation. By keeping it balanced you can rearrange the formula to change the subject.

Example 11

Make R_1 the subject of the formula
$V = I(R_1 + R_2)$

....

Step 1: Divide both sides of the equation by I:

$$\frac{V}{I} = R_1 + R_2$$

Step 2: Subtract R_2 from both sides:

$$\frac{V}{I} - R_2 = R_1$$

Step 3: Turn the equation around:

$$R_1 = \frac{V}{I} - R_2$$

Note that step 3 is not an essential step. If R_1 is on its own, it is now the subject, even if it is on the right-hand side of the equals sign.

Exercise 7H

1 Using the balance method, make y the subject of each formula.

    a $C = x + y$ **b** $T = 3x + 2y$

    c $P = \dfrac{3y}{4}$ **d** $S = \dfrac{2}{3}y + 4$

2 Rearrange the formula $V = I(R_1 + R_2)$ so that the subject is:

    a R_2 **b** I

3 Rearrange the formula $I = \dfrac{PRT}{100}$ so that the subject is:

    a R **b** P

4 If $P = P_0(1 + \alpha t)$ rearrange the formula so that the subject is P_0.

5 Sort these formulae into two groups of equivalent formulae.

    a $2t = 8 - 4n$

    b $2t + 8 = {}^-4n$

    c $t = {}^-4 - 2n$

    d $4n = 8 - 2t$

    e $4n + 2t = {}^-8$

    f $n = 2 - \dfrac{1}{2}t$

    g $t = 4 - 2n$

    h $2t + 4n = 8$

    i $2t = {}^-4n - 8$

    j $2t + 4n - 8 = 0$

    k $n = {}^-2 - \dfrac{1}{2}t$

    l $4n = {}^-8 - 2t$

6 Rearrange the formula and make the given letter the subject.

    a l in $A = l \times b$

    b r in $C = 2\pi r$

    c I in $V = IR$

    d L in $S = \pi rL$

    e f in $v = u + ft$

    f x in $y = mx + c$

    g T in $P = \dfrac{RT}{V}$

    h R in $I = \dfrac{E}{R}$

7 Write down three formulae that can be rearranged to give $x = 4p - 5$.

7.4 Inequalities

In Year 7 you learned that:

$>$ means **greater than**

$<$ means **less than**.

The greatest value is at the big end of the inequality symbol.

greatest value BIG > small smallest value

smallest value small < BIG greatest value

You now need to know two more symbols.

$\geq$ means **greater than or equal to**.

$\leq$ means **less than or equal to**.

You can represent the inequality $x \geq 60$ on a number line like this:

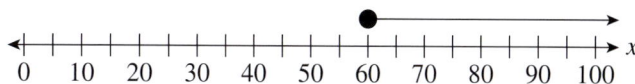

The shaded circle at 60 shows that 60 belongs to the interval and the arrow shows that the interval is **open**, which means that it carries on forever. If the circle was not shaded in, then 60 would not be part of the interval.

x can be **any** number greater than or equal to 60, e.g. 60, 64, 60.12, 74, 180.5 (the list goes on for ever).

Example 12

a Write down the inequality represented by the number line.
b Write down the largest two integers x could be.

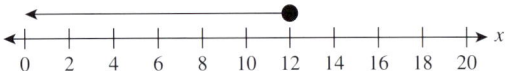

a All the values on the arrow are smaller than 12 and 12 is included in the interval as the circle is shaded at 12. So the inequality is $x \leq 12$.
b The largest two integers x could be are 11 and 12.

You also need to know about **closed** intervals.

Example 13

a Write down the inequality represented by the number line.
b Write down all the possible integer values x could be.

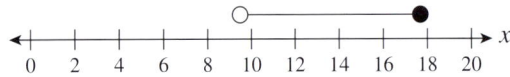

a All the values on the line are greater than 10 but do not include 10 as it is not a shaded circle. They are also less than or equal to 18 as the circle at 18 is shaded so the inequality is $10 < x \leq 18$
b The possible integers x could be are 11, 12, 13 14, 15, 16, 17 or 18.

Equivalent inequalities

You can also find inequalities that are equivalent inequalities.

Here is a diagram representing $x < 10$. The bar for x has less length than the bar for 10.

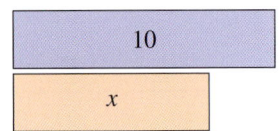

Now imagine adding 5 to the length of each bar.

Here is a new diagram to represent this. The bar for $x + 5$ has less length than the bar for $10 + 5$.

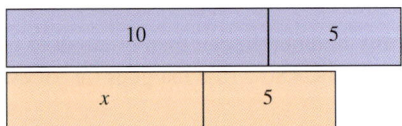

The equivalent inequality to $x < 10$ is:

$x + 5 < 15$

The inequality is equivalent because the same number has been added to both sides. Notice that the difference in length between the two bars is unchanged from the first diagram to the second.

Example 14

If $x \geq 4$, complete these equivalent inequalities.
a $x + 2 \geq \ldots$
b $2x \geq \ldots$
c $\ldots - \ldots \geq 1$

a $x + 2 \geq 6$
b $2x \geq 8$
c $x - 3 \geq 1$

Exercise 7I

1 Represent these inequalities on a number line.
    a $x \geq 4$ **b** $x < {}^{-}2$
    c $x \leq {}^{-}3$ **d** $2 \leq x < 10$
    e $4 < x \leq 8$

2 Write down the inequalities shown on these number lines.

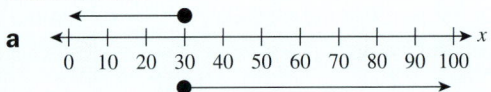

3 a What is the largest integer that x could be in question **2a**?

    b What is the smallest integer x could be in question **2b**?

    c What is the largest integer x could be in question **2c**?

    d What is the smallest integer x could be in question **2d**?

e What is the largest integer x could be in question **2e**?

f What is the smallest integer x could be in question **2f**?

4 Write down all the integers that satisfy the inequality $3 < x \le 7$

5 What is the largest negative integer that satisfies the inequality $x \le 0$?

6 Ricardo has a piece of string m cm long. Write the closed interval inequality, in terms of m, to show the following information.
The string is greater than or equal to 15 cm and the string is less than 45 cm.

7 Lukas thinks $x \le 7$ is equivalent to $x < 8$. Is he correct? Explain your answer.

8 If $x \le 9$, complete these equivalent inequalities.
a $x + 1 \le \dots$
b $3x \le \dots$
c $\dots - \dots \le 6$
d $\dfrac{x}{2} \le \dots$

9 If $2 \le x < 5$, complete these equivalent inequalities.
a $7 \le x + \dots < 10$
b $\dots \le 2x < \dots$
c $1 \le x \dots < 4$

10 Write down five inequalities that are equivalent to $x \le 10$.

11 The graph shows the line $y = 3$ and four points A, B, C and D.

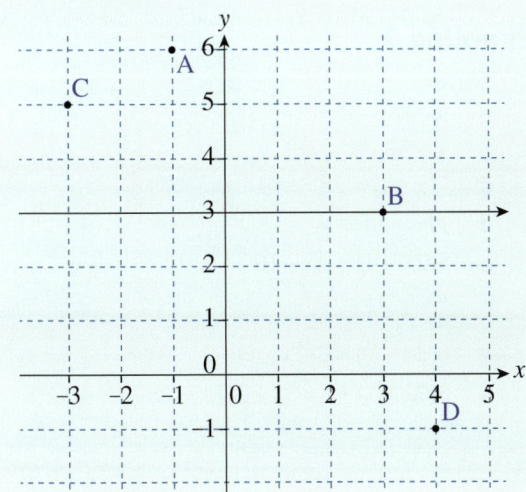

a Valentina picks a point from the graph. The y-coordinate of his point is greater than or equal to 3, which is written $y \ge 3$. Which of the points could he have picked?

b Mathias picks a point such that $x \ge 4$. Which point did he pick?

12 In this Venn diagram, x is a **single digit** positive integer.
Complete the Venn diagram showing all possible values.

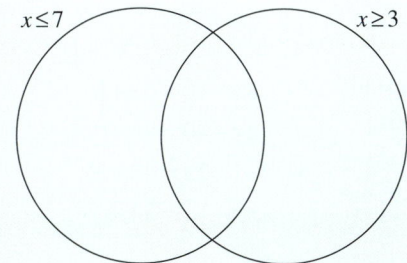

13 If $x > 7$, which of these statements are true?

$7 < x \qquad -x > -7 \qquad x + 1 \ge 8 \qquad 7 > x$

$14 < 2x \qquad x - 2 < 5$

14 a Explain why you cannot draw a triangle with side lengths 3 cm, 3 cm and 10 cm.

b Here is an isosceles triangle with all sides marked in centimetres.
Write down an inequality, in terms of x, to show the possible values x could be.

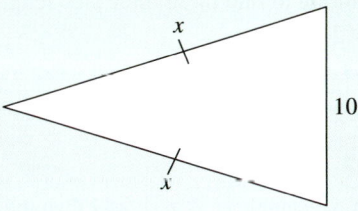

Consolidation

Example 1

Solve:

a $\dfrac{x}{3} - 4 = 7$ **b** $8(x-5) = 2(3x+4)$

a

$$\dfrac{x}{3} - 4 = 7$$

$[+4]$ $\dfrac{x}{3} = 11$

$[\times 3]$ $x = 33$

b $8(x-5) = 2(3x+4)$

$[\text{Expand}]$ $8x - 40 = 6x + 8$

$[-6x]$ $2x - 40 = 8$

$[+40]$ $2x = 48$

$[\div 2]$ $x = 24$

Example 2

The perimeter of this triangle is 21 cm. Find the side lengths.

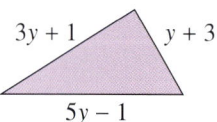

$[\text{Derive a formula}]$ $P = 3y + 1 + y + 3 + 5y - 1$

$[\text{Simplify}]$ $P = 9y + 3$

$[\text{Substitute 21 for } P]$ $21 = 9y + 3$

$[-3]$ $18 = 9y$

$[\div 9]$ $2 = y$

Use substitution to find these three side lengths.

$3y + 1 = 3 \times 2 + 1 = 7\,\text{cm}$

$y + 3 = 2 + 3 = 5\,\text{cm}$

$5y - 1 = 5 \times 2 - 1 = 9\,\text{cm}$

Example 3

One side of a rectangle is 3 cm shorter than the other side. The perimeter of the rectangle is 26 cm. Construct and solve an equation for the rectangle. Find the length of each side.

Let the longer side be x. The shorter side is $x - 3$.

The perimeter of the rectangle is 26, so
$26 = x + (x - 3) + x + (x - 3)$

Simplify:

$26 = 4x - 6$

$[+6]$ $32 = 4x$

$[\div 4]$ $8 = x$

So the longer side is 8 cm and the shorter side is 5 cm.

Example 4

Make x the subject of this formula.

$$y = 3(t + x)$$

$\dfrac{y}{3} = t + x$ [Divide both sides by 3]

$\dfrac{y}{3} - t = x$ [Subtract t from both sides]

Example 5

a Write down the inequality represented by the number line.

b Write down the smallest two integers x could be.

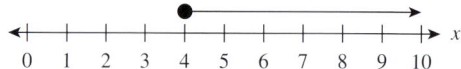

a $x \geq 4$ **b** 4 and 5

Example 6

a Draw the number line to represent the inequality $14 \leq x < 19$

b Write down all the possible integer values x could be.

a

b 14, 15, 16, 17 or 18

Exercise 7

1 Solve:

  a $5x - 8 = 32$

  b $x + 8 = 4x - 4$

  c $7(x + 1) = 28$

  d $11y - 24 = 3y$

  e $5(2x - 4) = 4x + 10$

  f $\dfrac{d}{4} - 1 = 1$

  g $6(2x - 7) = 2(4x - 1)$

  h $3(2x + 7) + 1 = 5x + 2(7x - 2)$

2 Represent these inequalities on a number line.
  a $x \geq 6$
  b $x \leq {}^{-}1$
  c $^{-}4 \leq x < 2$
  d $3 < x \leq 9$

3 Rearrange each formula to make the letter in brackets the subject.
  a $b = 2v + p$ (v)
  b $T = y - h$ (h)
  c $x = \frac{m}{3} + y$ (m)
  d $Y = 2(r + P)$ (r)
  e $\frac{b}{n} = j$ (n)
  f $h = \frac{d + s}{y}$ (s)

4 Write down the inequalities shown on these number lines.

  a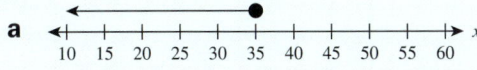

  b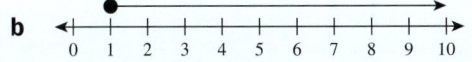

  c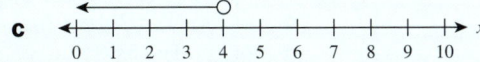

  d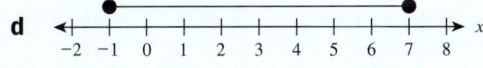

  e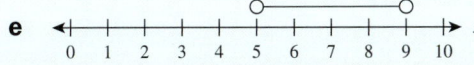

  f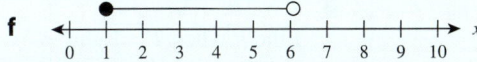

5 José bought n apples and some oranges. He had 4 times as many oranges as he had apples.
  a If José bought 3 apples how many oranges did he buy?
  b Write an expression showing how many oranges José bought.
  c If José bought 3 apples how many pieces of fruit did he buy altogether?
  d Derive a formula showing t, the total number of pieces of fruit José bought.
  e If José bought 30 pieces of fruit in total how many were oranges?

6 Write down all the integers that satisfy the inequality $^{-}2 \leq x < 4$

7 **a** Derive a formula for the perimeter, P, of this rectangle.

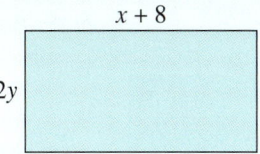

  b Derive a formula for the area, A, of this rectangle.

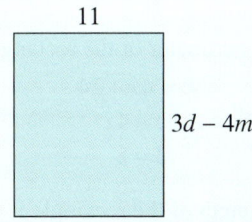

  c Use your answer to part **a** to find the value of y when $P = 76$ and $x = 20$
  d Use your answer to part **b** to find the value of d when $A = 110$ and $m = 2$

8 If $x \leq 3$ complete these equivalent inequalities.
  a $x + 2 \leq \dots$
  b $5x \leq \dots$
  c $\dots \leq {}^{-}1$

9 I started with a number, n. I subtracted 10 from this number and then multiplied the result by 7, and got the same number as I do when I add 18 to n and multiply the result by 3. What is n?

10 If $0 \leq x < 4$ complete these equivalent inequalities.
  a $7 \leq x + \dots < 11$
  b $\dots \leq 2x < \dots$
  c $^{-}3 \leq x \dots < 1$

Summary

You should know ...

1 How to construct and use formulae.
For example:

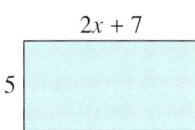

2x + 7

5

Write a formula for the perimeter, P, of the rectangle.
$P = 5 + 2x + 7 + 5 + 2x + 7$
$P = 4x + 24$
If the perimeter of the rectangle is 36, find its length.

$$36 = 4x + 24$$
$[-24] \qquad 12 = 4x$
$[\div 4] \qquad \quad 3 = x$

The length of the rectangle is $2x + 7$, or
$2 \times 3 + 7 = 13$

2 How to solve equations.
For example:

$$3(4d - 6) = 2(5d + 1)$$

[Expand] $\qquad 12d - 18 = 10d + 2$

$[-10d] \qquad \quad 2d - 18 = 2$

$[+18] \qquad \qquad 2d = 20$

$[\div 2] \qquad \qquad \; d = 10$

3 How to substitute integers into formulae.
For example:
$D = x^2 + 3p - 7$

Find p when $D = 12$ and $x = {}^-2$
$\qquad \qquad \qquad 12 = ({}^-2)^2 + 3p - 7$
[Simplify] $\qquad 12 = 3p - 3$
$[+3] \qquad \qquad 15 = 3p$
$[\div 3] \qquad \qquad 5 = p$

Check out

1 a Write a formula for the area, A, of this rectangle.

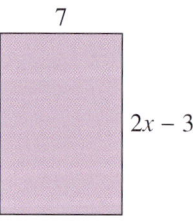

7

2x − 3

b If the area of this rectangle is 63, work out x.

2 Solve:
  a $7x - 3 = 2x + 42$
  b $5y + 8 = 8(y - 2)$
  c $4(3m - 1) = 2(4m + 6)$
  d $6(2x - 1) - 5x$
    $= 2(5x - 4) - 7$

3 $T = 2B^2 + 3n - 38$
Find n when $T = 1$ and
$B = {}^-3$

4 How to form equations to solve problems.
For example:
In four years' time James will be five times as old as he is now. What is his present age?

Let his present age be x years.
In four years' his age will be $x + 4$
Five times as old as his present age is $5x$

so	$x + 4 = 5x$
$[-x]$	$4 = 4x$
$[\div 4]$	$1 = x$

James is 1 year old now.

4 The width of a rectangle is w cm. Its length is 4 cm more than its width. If the perimeter of the rectangle is 28 cm, find its length and width.

5 How to change the subject of a formula.
For example:
Make x the subject of the formula $p = 7x - 2$

	$p = 7x - 2$
$[+2]$	$p + 2 = 7x$
$[\div 7]$	$\dfrac{p + 2}{7} = x$

5 Make x the subject of these formulae.
  a $t = 5x + m$
  b $k = \dfrac{x + 1}{3}$
  c $y = 4(x - r)$

6 How to find equivalent inequalities and represent inequalities on a number line.
For example:
This number line represents the inequality $^-1 \le x < 4$

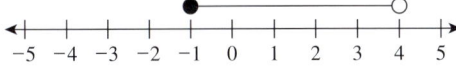

An equivalent inequality is $0 \le x + 1 < 5$ (adding 1 to all three parts).

6 Represent these inequalities on a number line.
  a $x \ge ^-1$
  b $x \le 20$
  c $^-9 \le x < ^-3$
  d $2 < x \le 7$

Geometry

Objectives

In this chapter you will learn about:

- the exterior angle of a triangle
- angles in parallel lines and intersecting lines
- finding the midpoint of a line segment.

What's the point?

Cycling is a popular sport across the world. Top cyclists practise on special cycle tracks called velodromes. The velodromes are usually banked at an angle of 42° to prevent cyclists from falling off at the bends.

Before you start

You should know ...

1 How to use letters to name an angle.

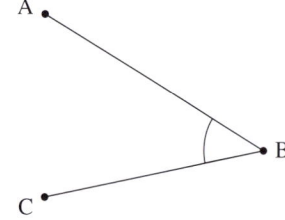

this is angle ABC.

Check in

1 Name the angles marked.

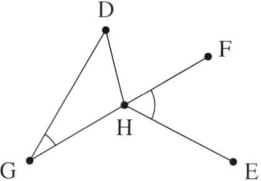

2 Angles in a triangle sum to 180°.

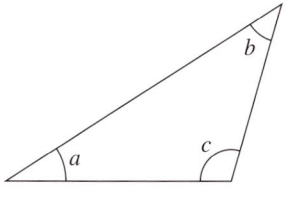

$$a + b + c = 180°$$

Angles on a straight line sum to 180°.

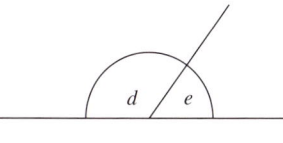

$$d + e = 180°$$

A complete turn is 360°.

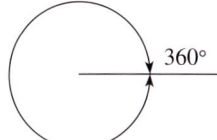

That is, angles at a point add up to 360°.

Angle sum in a quadrilateral is 360°.

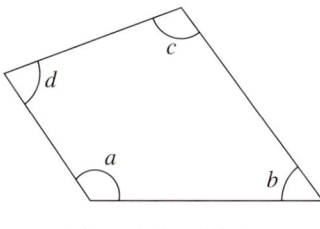

$$a + b + c + d = 360°$$

Angles in an equilateral triangle are all equal so must be 60°.

An isosceles triangle has two equal angles.

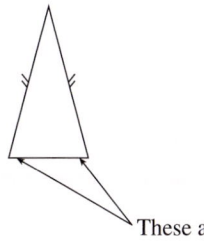

These are the two equal angles

2 Work out these missing angles.

a

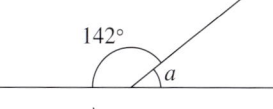

142° a

b

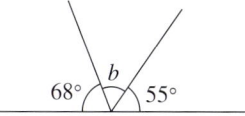

68° b 55°

c

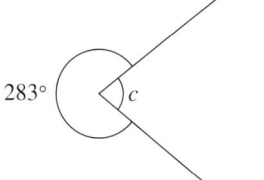

283° c

d

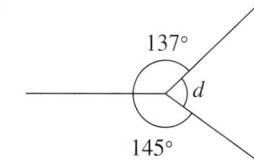

137° d 145°

e

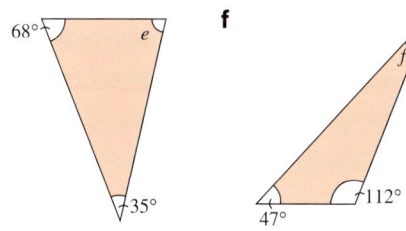

68° e 35°

f

47° 112°

g

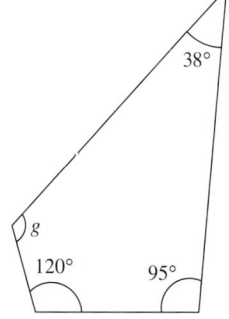

38° g 120° 95°

h

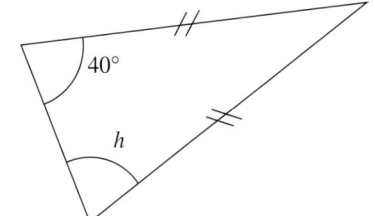

40° h

8.1 Exterior angle of a triangle

An **exterior angle** of a triangle is found by extending one of the sides of the triangle. The exterior angle x is marked in this diagram.

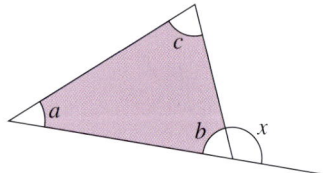

The exterior angle is related to the interior angles as shown in the proof.

$a + b + c = 180°$ because the angles in a triangle add up to 180°.
$b + x = 180°$ because angles on a straight line add up to 180°.

$a + b + c = b + x$ because they both equal 180°, subtracting b from both sides gives us $a + c = x$

Therefore the exterior angle of a triangle is equal to the sum of the two interior opposite angles.

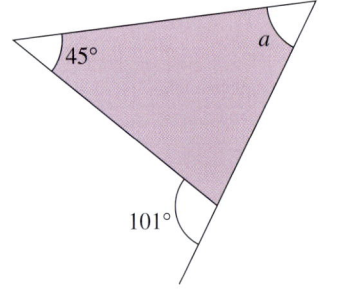

Example 1

Find the size of angle a.

The exterior angle of a triangle is equal to the sum of the two interior opposite angles
so $101° = 45° + a$

Subtracting 45° from both sides:
$a = 56°$

Exercise 8A

1 Find the missing angles.

a
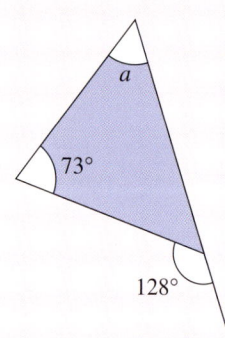

b
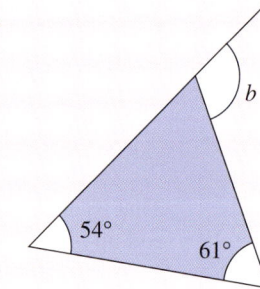

c
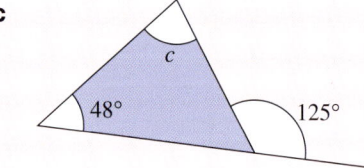

d
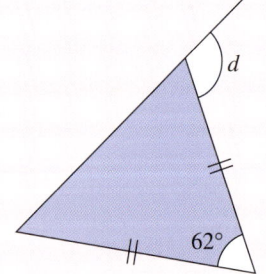

e
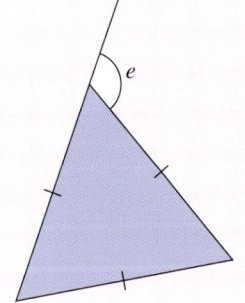

f

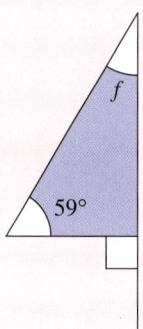

2 Which of these triangles is an impossible triangle?

a

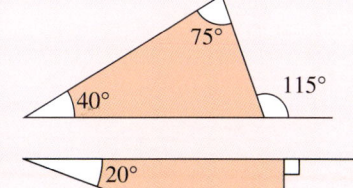

b

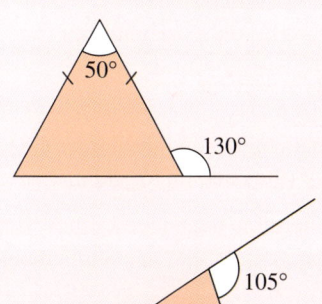

c

d

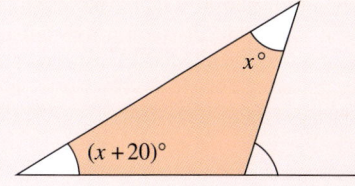

3 Find and simplify an expression for the marked exterior angle of this triangle.

④ Find the missing angles in these diagrams.

a

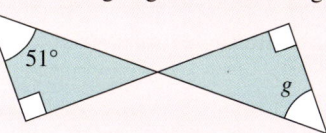

b

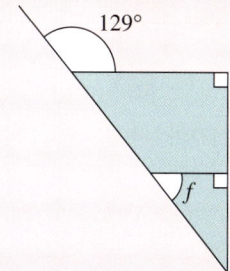

5 Write an equation in x and solve it to find the value of x.

a

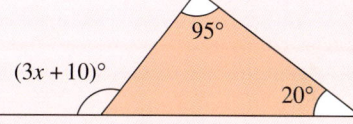

b

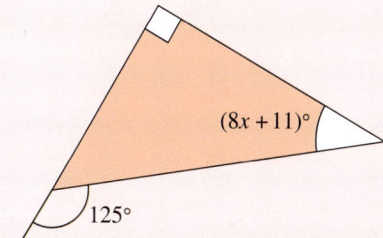

6 Find x.

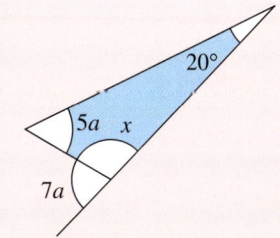

8.2 Angles in parallel lines and intersecting lines

Vertically opposite angles are equal.

These angles are vertically opposite.

Two pairs of vertically opposite angles are formed when two straight lines intersect.

When a line crosses two parallel lines you get other properties.

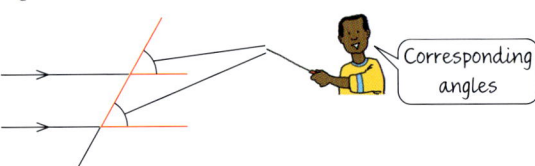

Corresponding angles

Corresponding angles are equal (look for an F shape).

Alternate angles

Alternate angles are equal (look for a Z shape).

You can use the ideas of corresponding and alternate angles to solve more complex problems involving missing angles.

Example 2

Find the missing angles a, b and c.

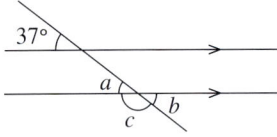

Write the reason for the answer.

$a = 37°$ (corresponding angles)
$a = b = 37°$ (vertically opposite angles)
$b + c = 180°$ (angles on a straight line)
So $37° + c = 180°$
hence $c = 180° - 37° = 143°$

Example 3

Find the missing angles in this parallelogram.

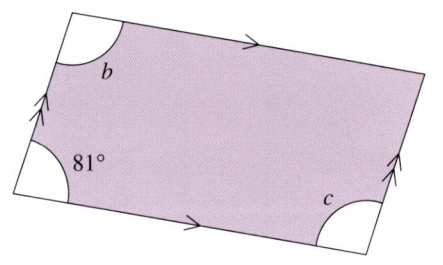

To find angle b, extend one side, as shown:

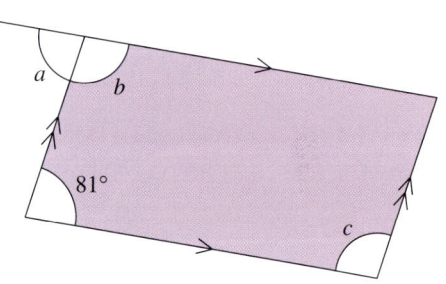

Angle $a = 81°$ (alternate angles)
$a + b = 180°$ (angles on a straight line)
$81° + b = 180°$
So $b = 99°$

This shows that angle b and the angle of $81°$ are supplementary (they add up to $180°$).

You can use the same technique to show that angle c and the angle of $81°$ are supplementary, so $c = 99°$ as well.

Remember to use common sense. In most questions to do with finding angles, diagrams will not be drawn to scale, so you cannot measure them. It is worth remembering that acute angles should be less than 90° and obtuse angles should be between 90° and 180° when you work them out: do a common-sense check of your answer.

Exercise 8B

1 Find the missing angles *a–i*. Give reasons for your answers.

a

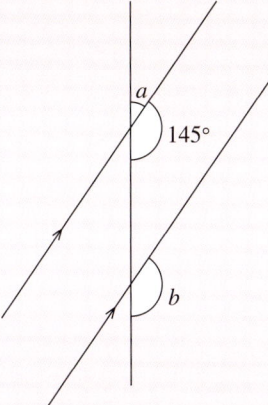

b

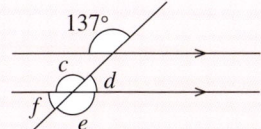

c

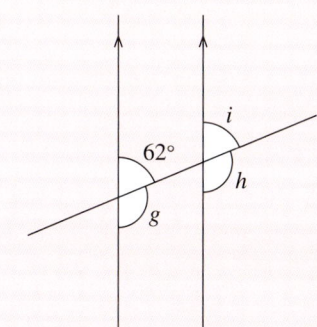

2

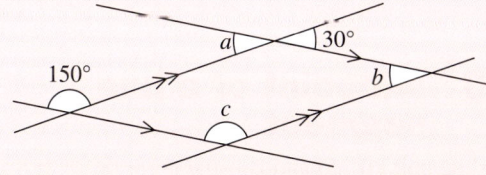

Copy and complete these sentences

a Angle *a* is . . .° because . . . .

b Angle *b* is . . .° because . . . .

c Angle *c* is . . .° because . . . .

3 Calculate angles *a–l*. Give reasons for your answers.

a

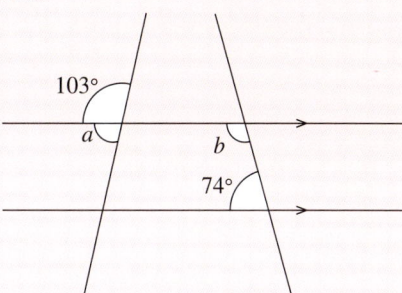

b

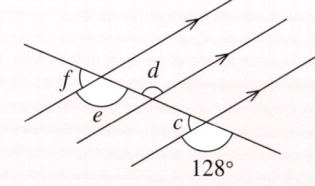

c

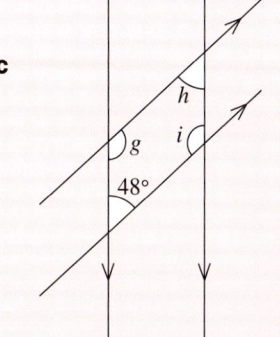

d

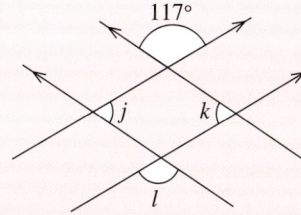

4 Find angles *a–f*.

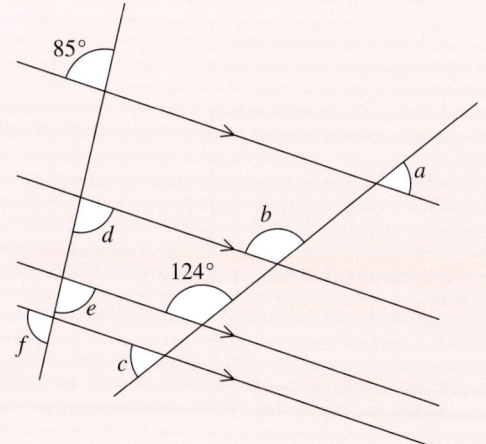

5 Without measuring any angles, say which diagrams below are drawn correctly. Give a reason for your answer.

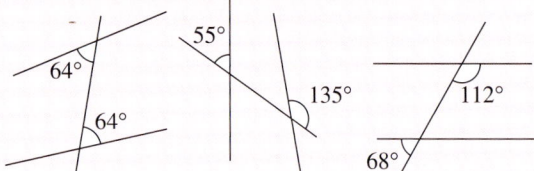

6 Find the marked angles *a–i* in these parallelograms.

a

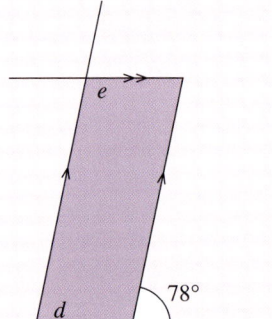

b

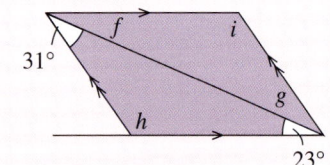

c

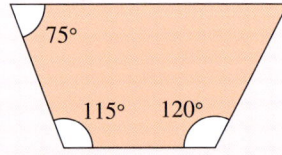

7 Here is a quadrilateral.

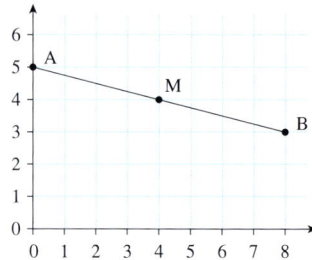

Samira thinks this quadrilateral is a trapezium.
Jasper thinks there is not enough information to be able to say.
Haib thinks it is not a trapezium.
Who is correct? Explain your answer.

8 Is AB parallel to CD? Justify your answer.

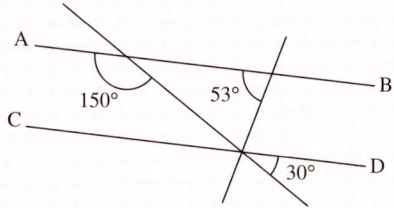

9 One interior angle of a parallelogram is 142°. What are the other angles?

10 a Find the values of *x* and *y*.

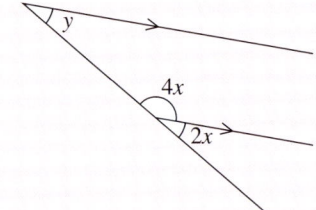

b Find the values of *m, n* and *p*.

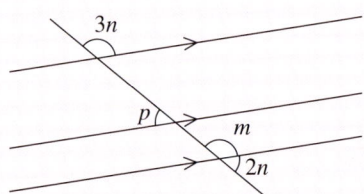

c Find the values of *t, u, v* and *w*.

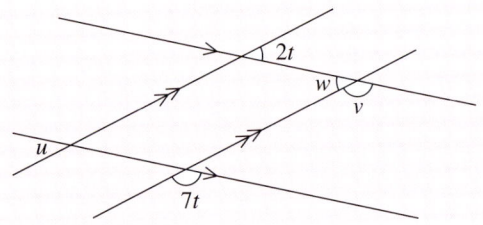

8.3 Midpoint of a line segment

A is the point (0, 5) and B is the point (8, 3). It is easy to find the midpoint of the line segment AB using a diagram. You can see by counting squares that the midpoint, M, is at (4, 4).

You don't have to draw the line segment to find the midpoint – there is another way.

The coordinates of the midpoint of a line segment AB, where A is the point (x_1, y_1) and B is the point (x_2, y_2), are

$$\left(\frac{x_1 + x_2}{2}, \frac{y_1 + y_2}{2} \right)$$

Can you see why this is the case?

Example 4

Find the midpoint of the line segment XY where X is the point $(^-3, ^-2)$ and Y is the point $(4, ^-4)$

....

Midpoint $= \left(\dfrac{^-3 + 4}{2}, \dfrac{^-2 + ^-4}{2} \right)$

$= \left(\dfrac{1}{2}, \dfrac{^-6}{2} \right)$

$= \left(\dfrac{1}{2}, ^-3 \right)$

Exercise 8C

1   **a**   Find the midpoint of the line segment AB where A = $(^-2, 4)$ and B = $(10, 2)$.
    b   Find the midpoint of the line segment CD where C = $(^-2, 0)$ and D = $(10, 6)$.
    c   Draw the line segments AB and CD from parts **a** and **b** on the same axes. Mark the point of intersection as point M. What are the coordinates of M? What do you notice?
    d   What sort of quadrilateral is ADBC?
    e   Measure angle AMC.
    f   Without measuring, what are angles BMD, AMD and BMC?

2   Find the midpoints of the line segments with end points:
    a   A(4, 3), B(2, 7)   **b**   C(1, $^-$2), D($^-$1, 6)
    c   E(2, $^-$5), F($^-$2, 3)   **d**   G(0, $^-$6), H($^-$3, 3)

3   A is the midpoint of LM.
    B is the midpoint of LN.
    X is the midpoint of AB.
    Find the coordinates of X.

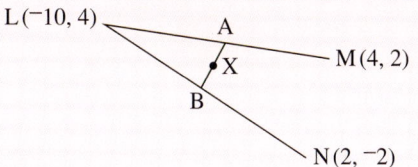

④ X is the midpoint of the line segment AB. The coordinates of X are (2, 4) and of A are ($^-$1, 3). Find the coordinates of B.

⑤ The midpoint of a line segment AB is (6, 2). Write down some possible coordinates for A and B.

⑥ This diagram shows a rectangle ABCD. What are the coordinates of vertices A, B and C?

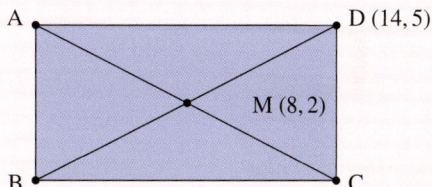

⑦ This diagram shows a square ABCD. What are the coordinates of the vertices B, C and D?

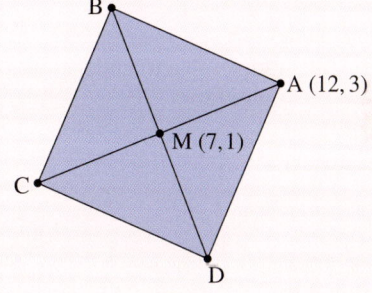

Consolidation

Example 1

Find angle a.

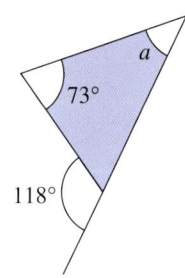

The exterior angle of a triangle is equal to the sum of the two interior opposite angles.
$118° = 73° + a$
So $a = 45°$

Example 2

Find the sizes of the missing angles a, b and c.

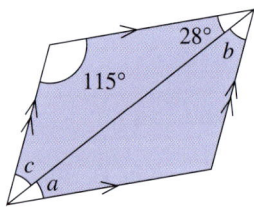

$a = 28°$ (alternate angles)
$28° + 115° + c = 180°$ (angles in a triangle add
                     up to $180°$)
So $c = 37°$
$b = c$ (alternate angles)
$b = 37°$

Example 3

Find the midpoint of the line segment AB if A = (7, 2) and B = (5, ⁻4).

Midpoint is at $\left(\dfrac{x_1 + x_2}{2}, \dfrac{y_1 + y_2}{2} \right)$

$= \left(\dfrac{7 + 5}{2}, \dfrac{2 + {}^-4}{2} \right) = \left(\dfrac{12}{2}, \dfrac{{}^-2}{2} \right) = (6, {}^-1)$

Exercise 8

1 Calculate angles a–d.

a

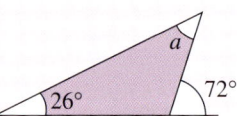

b

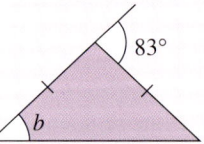

c

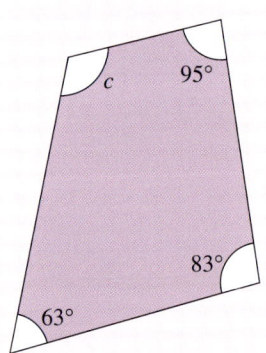

d

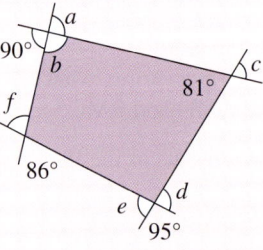

2 Without measuring, find the value of each angle, a–f.

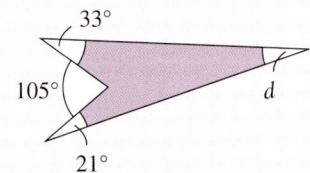

3

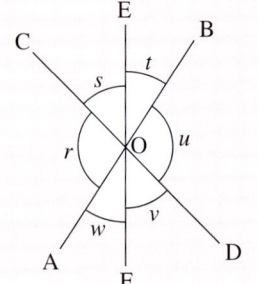

AB, CD and EF are straight lines that cross at O.

a If $s = 35°$ find v.

b If $r = 60°$ and $t = 65°$ find v.

c If $r + s = 120°$ and $v = 50°$ find u.

4

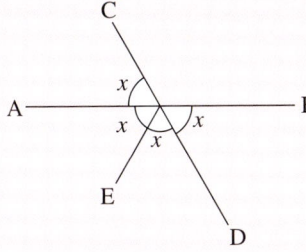

Find the size of x.

5 Calculate angles a–f. Give reasons for your answers.

a

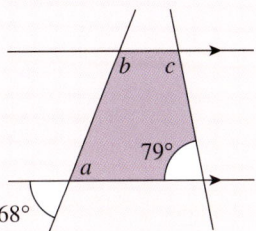

b

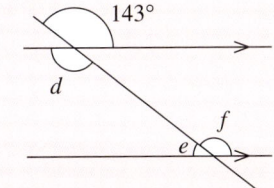

6 Calculate the sizes of angles a–e.

a

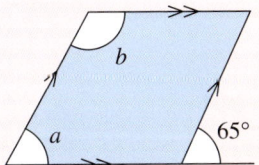

b

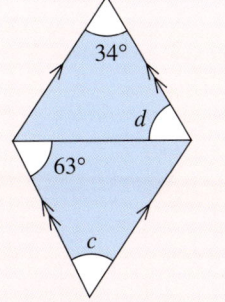

c

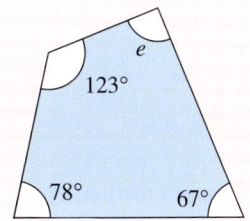

7 Copy and complete this proof.

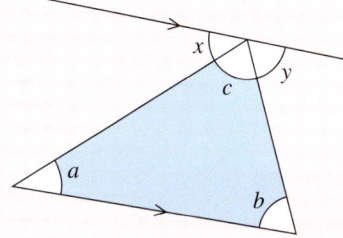

$x + y + c = 180°$ because . . . .

. . . . because alternate angles are equal.

$y = b$ because . . . .

So $x + y + c = a + b + c = . . .°$.

Therefore the angles of a triangle always add up to . . .°.

8 Write down the size of the angles marked by letters. Give reasons for your answers.

a

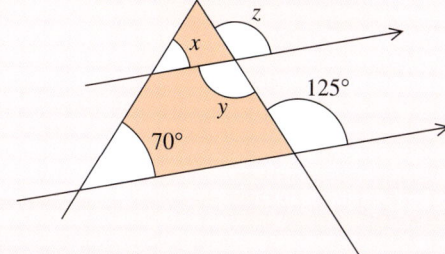

b

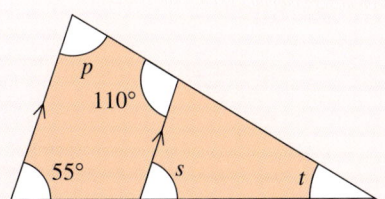

c

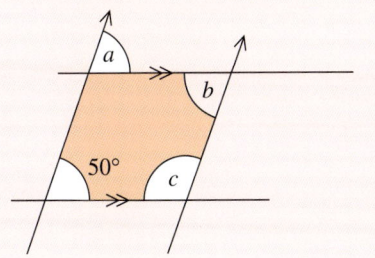

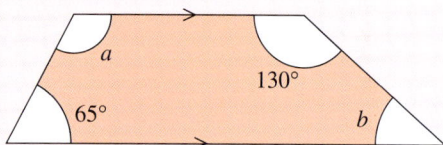

d

9 Find the missing angles marked by letters in these parallelograms.

a

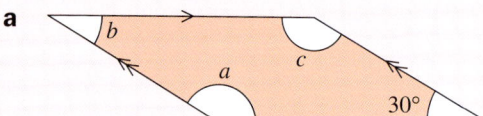

b

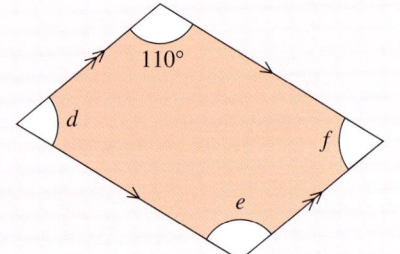

c

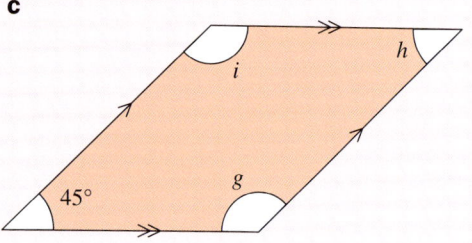

10 a Find the midpoint of the line segment CD if C = (3, 4) and D = (⁻1, 8).
b Find the midpoint of the line segment EF if E = (6, 9) and F = (3, ⁻5).

11 a Find the midpoint of the line segment AC where A = (⁻4, 2) and C = (4, 0).
b Find the midpoint of the line segment BD where B = (1, 5) and D = (⁻1, ⁻3).
c Plot points A, B, C and D from parts **a** and **b** on the same axes. What sort of quadrilateral is ABCD?
d Draw in the diagonals of this quadrilateral. Where do they intersect? What do you notice?

12 Find all of the angles inside this quadrilateral.

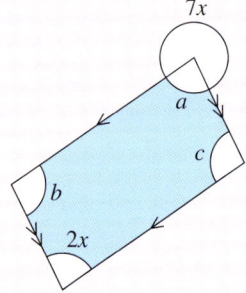

13 AB is a line segment with A(4, 10). The midpoint of AB is (7, ⁻1). Find the coordinates of B.

Summary

You should know …

1 Vertically opposite angles are equal.

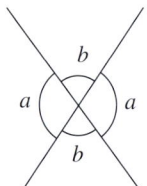

Check out

1 Calculate the size of angles *a* and *b*.

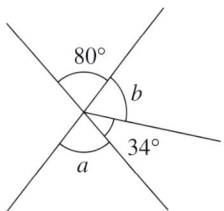

2 a Alternate and corresponding angles are equal.

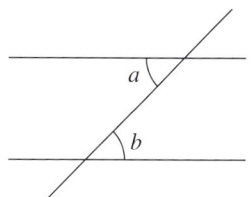

a and *b* are alternate angles so $a = b$

b

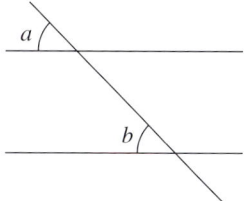

a and *b* are corresponding angles so $a = b$

3 The exterior angle of a triangle is equal to the sum of the two interior opposite angles.

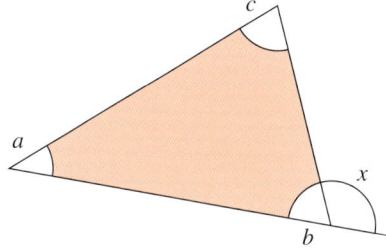

$a + c = x$

4 How to solve geometric problems using properties of angles. *For example:*

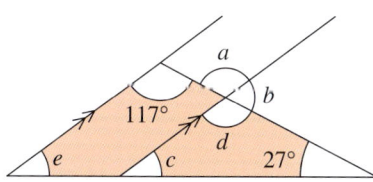

Angle $a = 117°$ (alternate angles)
Angle $b = 63°$ (angles on a straight line add up to $180°$)
Angle $d = 117°$ (vertically opposite angles)
Angle $c = 36°$ (angles in a triangle add up to $180°$)
Angle $e = 36°$ (corresponding angles)

5 How to find the midpoint of a line segment. *For example:*
$F = (3, {}^-2)$ and $G = (8, 10)$

Midpoint is at $\left(\frac{x_1 + x_2}{2}, \frac{y_1 + y_2}{2}\right)$

$= \left(\frac{3+8}{2}, \frac{{}^-2+10}{2}\right) = \left(\frac{11}{2}, \frac{8}{2}\right) = (5.5, 4)$

2 Find the size of angles *a*–*d*.

a

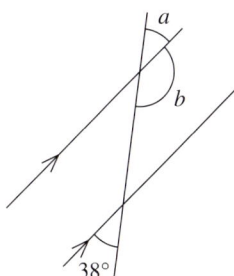

b

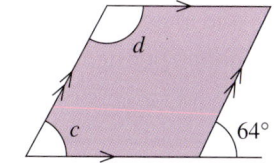

3 Complete this proof for the triangle shown on the left.

$a + b + c = 180°$ because . . . .

. . . because angles on a straight line add up to $180°$.

$a + b + c = b + x$ because they both equal $180°$.

Subtracting *b* from both sides gives us . . . .

4 Find the missing angles *a*–*e*, giving reasons for your answers.

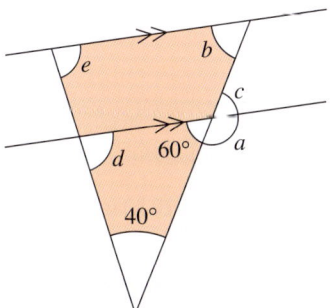

5 Find the midpoint of line segment XY if:
a $X = (7, 7)$ and $Y = ({}^-1, 3)$
b $X = (2, {}^-4)$ and $Y = (7, 10)$

Decimals, percentages and fractions

Objectives

In this chapter you will learn about:

- multiplying decimals by decimals and integers
- dividing by decimals
- increasing and decreasing by a percentage
- simplifying calculations with fractions and decimals.

What's the point?

Percentages are everywhere. Your teachers probably give you test results as percentages. Percentages are regularly used in the news, for example: 'Unemployment has fallen by 2%'. Percentages are used in the economy – for example, you will pay a percentage of your earnings in tax.

Before you start

You should know ...

1 How to convert a fraction to a decimal by using equivalent fractions or dividing.
   For example:

   $$\frac{2}{5} = \frac{4}{10} = 4 \div 10 = 0.4$$

   $$\frac{3}{8} = \frac{0.\ 3\ 7\ 5}{8) \overline{3.^30^60^40}}$$

Check in

1 Write these fractions as decimals.

   a $\frac{3}{10}$ b $\frac{4}{5}$

   c $\frac{5}{8}$ d $\frac{7}{20}$

2 To add or subtract fractions their denominators must be the same.

For example:

$$\frac{3}{4} + \frac{2}{5}$$

$$= \frac{15}{20} + \frac{8}{20}$$

$$= \frac{23}{20}$$

$$= 1\frac{3}{20}$$

You can't add 3 fourths to 2 fifths

but you can add 15 twentieths to 8 twentieths.

3 How to do long multiplication.

For example:

```
        3   7
    ×   2   5
    ─────────
    1   8   5      ← This row is 37 × 5

+   7   4   0      ← This row is 37 × 20
    ─────────
    9   2   5      Add the 2 rows above
```

2 a Work out:

i $\frac{1}{2} + \frac{1}{4}$ **ii** $\frac{2}{3} + \frac{1}{4}$

iii $3\frac{1}{2} + \frac{2}{3}$ **iv** $1\frac{4}{7} + \frac{2}{9}$

v $\frac{2}{3} - \frac{1}{2}$ **vi** $3\frac{3}{4} - 1\frac{7}{8}$

b Beatrice drinks $\frac{2}{3}$ litre of milk from a $1\frac{1}{2}$ litre bottle. How much milk is left in the bottle?

3 Work out:
   a 35×24
   b 49×23
   c 78×52
   d 124×67
   e A school has 48 classes of 26 students. How many students are there altogether?

9.1 Multiplying by decimals

Multiplying decimals is like multiplying whole numbers. The only difference is you have to put the decimal point in the right place. You can do this from an estimate of the answer.

Example 1

Work out 7.4×8

The estimate is $7 \times 8 = 56$

```
    7.4
  ×   8
  ─────
  59₃.2
```
The digits in the answer are 592
The estimate is 56
Place the decimal point to give 59.2

That is, $7.4 \times 8 = 59.2$

The method is the same even if both numbers are decimals.

Example 2

Work out 2.3×0.6

The estimate is $2 \times 1 = 2$

```
    2.3
  × 0.6
  ─────
  1.3₁8
```
The digits are 138
The estimate is 2
Place the decimal point to give 1.38

That is, $2.3 \times 0.6 = 1.38$

Example 2 can be thought of as follows.
Do the sum 2.3×0.6 as if there are no decimal places: as 23×6.

Notice that

$$23 \times 6 = 138$$
$$\downarrow \quad \downarrow \quad \downarrow$$
$$2.3 \times 0.6 = 1.38$$

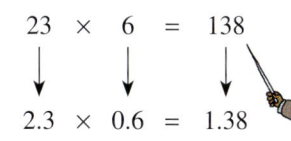

2.3 is 10 times smaller than 23
0.6 is 10 times smaller than 6
so 1.38 is 100 times smaller than 138.

If there are two decimal places in the product, there must be two decimal places in the answer.

The product and the answer both have two decimal places.

$$2.\underline{3} \times 0.\underline{6} = 1.\underline{38}$$

You can also look at patterns to see where to place the decimal point.

$0.3 \times 200 = 60$

$0.3 \times 20 = 6$

$0.3 \times 2 = 0.6$

$0.3 \times 0.2 = 0.06$

$0.3 \times 0.02 = 0.006$

$0.\underline{3} \times 0.\underline{002} = 0.\underline{0006}$

Notice there are four decimal places in the question (underlined) so there are four decimal places in the answer.

You can also multiply by negative integers.

Example 3

$^-23 \times 0.46$

Work out the answer to 23×46 ignoring the negative sign and the decimal place.

$$\begin{array}{r} 23 \\ \times\ 46 \\ \hline 13\,8 \\ 9\,20 \\ \hline 1058 \end{array}$$

There are two decimal places in the question so there should be two decimal places in the answer.

$^-23 \times 0.46 = ^-10.58$

A positive multiplied by a negative makes a negative.

Exercise 9A

1 Copy and complete this table.

Calculation	Estimate	Answer
3.2×4	12	
6.9×9		
32.5×8		
1.68×3		
12.4×13		

2 Work out:
  a 0.4×8 **b** $0.9 \times ^-6$
  c 2.3×5 **d** 4.7×8
  e $^-6 \times 1.4$ **f** $^-12 \times 0.6$
  g 11×0.42 **h** $^-4 \times 0.65$
  i $3.24 \times ^-8$ **j** 4.85×12

3 Find the cost of 12 pencils each selling for $0.45.

4 If a packet of rice holds $1.4\,\text{kg}$, what is the mass of 6 packets?

5 Given that $45 \times 19 = 855$, find:
  a 45×1.9 **b** $4.5 \times ^-19$
  c 45×0.19 **d** 4.5×1.9
  e 0.45×19 **f** 0.45×1.9

6 Explain why 23.4×0.06 and 2.34×0.6 have the same answer.

7 Work out:
  a 3.4×0.5 **b** 7.12×0.3
  c 0.4^2 **d** 7.15×0.7
  e 8.2×0.02 **f** 4.8×0.06
  g 62.3×0.4 **h** 0.34×0.08
  i 0.7×4.52 **j** 0.33×6.4
  k 0.09×53.1 **l** 0.06^2
  m 0.23×1.4 **n** 0.38×2.6
  o 2.3^2 **p** $^-0.7 \times 3.7$
  q $^-2.61 \times ^-0.04$ **r** $(^-0.2)^2$

8 What is the total length of 20 pieces of pipe each $3.25\,\text{m}$ long?

9 A tyre costs $153.94.

  What would be the cost of 6 tyres?

10 Another way to multiply decimals is to consider equivalent calculations. Look at the example below.
$$4.23 \times 0.4 = (4.23 \times 4) \div 10$$
$$= 16.92 \div 10 = 1.692$$

Use this method to copy and complete:

a $3.17 \times 0.3 = (\square \times \square) \div 10$

$\qquad = \square \div 10 = \square$

b $21.8 \times 0.04 = (\square \times \square) \div 100$

$\qquad = \square \div 100 = \square$

c $9.87 \times 0.8 = (\square \times \square) \div \square$

$\qquad = \square \div \square = \square$

d $16.5 \times 0.03 = (\square \times \square) \div \square$

$\qquad = \square \div \square = \square$

e $458.3 \times 0.002 = (\square \times \square) \div \square$

$\qquad = \square \div \square = \square$

f $0.3 \times 17.4 = (\square \times \square) \div \square$

$\qquad = \square \div \square = \square$

g $0.6 \times 9.84 = (\square \times \square) \div \square$

$\qquad = \square \div \square = \square$

11 Complete this multiplication grid.

×	0.8	1.2
⁻5		
	0.4	
	0.2	
		0.15

12 A square has perimeter 22 cm. What is the area of this square?

13 Another way to multiply decimals is to change them to fractions. Look at the example below.

$0.6 \times 4.32 = \dfrac{6}{10} \times 4\dfrac{32}{100}$

$\qquad = \dfrac{6}{10} \times \dfrac{432}{100}$

$\qquad = \dfrac{2592}{1000}$

$\qquad = 2\dfrac{592}{1000}$

$\qquad = 2.592$

Use this method to work out:

a 1.3×1.6 **b** 4.6×0.7

c 0.3×0.4 **d** 0.32×0.8

9.2 Dividing by decimals

When you divide a decimal by a decimal, turn the divisor into a whole number.

Example 4

Work out $3.66 \div 0.6$

Multiply by 10 to turn 0.6 into a whole number.

$3.66 \div 0.6 = \dfrac{3.66}{0.6}$

$\qquad = \dfrac{3.66 \times 10}{0.6 \times 10}$

$\qquad = \dfrac{36.6}{6}$

$\qquad = 6.1$

$6\overline{)36.6} \quad 6.1$

In Example 4, both the numerator and denominator were multiplied by the same number. You had to do this to keep the answer the same, using an equivalent fraction method.

Exercise 9B

1 Look at this division.

$72 \div 0.9 = \dfrac{72}{0.9}$

    a Is it easier to divide by 0.9 or by 9?

    b By what must you multiply 0.9 to give 9?

    c So by what must you multiply the 72?

    d Copy and complete:

$\dfrac{72}{0.9} = \dfrac{72 \times 10}{0.9 \times 10} = \dfrac{\square}{\square} = \square$

2 Look at this division.

$^-6 \div 0.2 = \dfrac{^-6}{0.2}$

Copy and complete this equivalent calculation.

$\dfrac{^-6}{0.2} = \dfrac{^-6\square 10}{0.2 \times 10} = \dfrac{\square}{\square} = {^-\square}$

3 Work out:

    a   **i** $64 \div 8$ **ii** $6.4 \div 0.8$

    b   **i** $35 \div 5$ **ii** $3.5 \div 0.5$

    c   **i** $^-40 \div 5$ **ii** $^-4 \div 0.5$

    d   **i** $48 \div 3$ **ii** $4.8 \div 0.3$

    e   **i** $36.4 \div 4$ **ii** $3.64 \div 0.4$

Compare your answers to parts **i** and **ii** in each case. What do you notice?

4 Work out:

    a $5.6 \div 0.4$ **b** $^-64 \div 0.4$

    c $^-57 \div 0.3$ **d** $1.25 \div 0.5$

    e $69 \div 0.3$ **f** $14.4 \div 1.2$

    g $1.38 \div 0.6$

5 Another way to divide decimals is to consider equivalent calculations.

$2.36 ÷ 0.4 = (2.36 × 10) ÷ 4$
$= 23.6 ÷ 4 = 5.9$

Use this method to copy and complete:

a $7.35 ÷ 0.3 = (\square × 10) ÷ \square$
$= 73.5 ÷ \square = \square$

b $6.88 ÷ 0.4 = (\square × 10) ÷ \square$
$= \square ÷ \square = \square$

c $10.08 ÷ 0.8 = (\square × \square) ÷ \square$
$= \square ÷ \square = \square$

(d) $30.1 ÷ 0.07 = (\square × \square) ÷ \square$
$= \square ÷ \square = \square$

(e) $1.284 ÷ 0.006 = (\square × \square) ÷ \square$
$= \square ÷ \square = \square$

6 The area of this rectangle is $5.4\,cm^2$. Work out the missing side length.

0.9 cm

? cm

7

A centipede is timed as travelling 4.65 m in 7.5 s. What is its speed in metres per second?

8

1.3 kg of Ultra Tide soap powder costs $10.66. How much does:
a 1 kg of powder cost
b 0.75 kg of powder cost?

9.3 Percentage increase and decrease

A teacher earning $30 000 has a 5% increase in annual salary. There are two ways of working out the teacher's new salary.

The slow way
Work out 5% of 30 000:
$0.05 × 30 000 = 1500$
$30 000 + 1500 = 31 500$

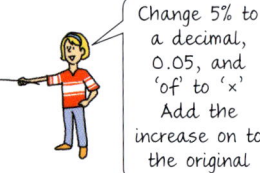

Change 5% to a decimal, 0.05, and 'of' to '×' Add the increase on to the original

The faster way
This can be done in a single calculation. An increase of 5% means that you have 105% of the original.

Work out 105% of 30 000:
$1.05 × 30 000 = 31 500$

Change 105% to a decimal, 1.05, and 'of' to '×'.

Both methods show that the teacher's new salary is $31 500.

Example 5

In a sale, a coat costing $350 was reduced by 20%. What was its sale price?

....

$100\% - 20\% = 80\%$ so the coat now costs 80% of its original price.

80% of $350 = 0.8 × 350 = 280$

The sale price is $280

Example 6

Increase 140 by 300%.

....

$100\% + 300\% = 400\%$ of original.

400% of $140 = 4 × 140 = 560$

Absolute change

In Example 5, when $350 was reduced by 20% the absolute change was ⁻$70.

In Example 6, when 140 was increased by 300% the absolute change was $140 + 140 + 140 = 420$.

So for percentage decreases the absolute change is negative.

If 140 is decreased by 200% the absolute change is $140 - 140 - 140 = {}^{-}140$.

Exercise 9C

1 Match the boxes. The first is done for you.

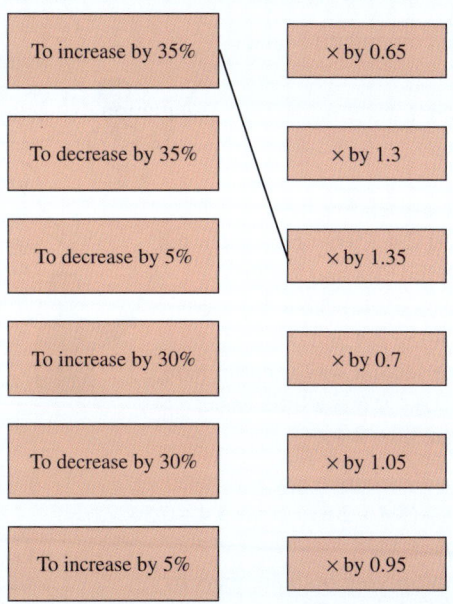

To increase by 35%	× by 0.65
To decrease by 35%	× by 1.3
To decrease by 5%	× by 1.35
To increase by 30%	× by 0.7
To decrease by 30%	× by 1.05
To increase by 5%	× by 0.95

2 Kulwinder and Lee were working out the price of a car originally costing $24 000 after a tax of 8% was added on. Here is their working.

Kulwinder
24000 × 1.8 = 432000
New cost $432 000

Lee
24000 × 0.92 = 22080
New cost $22 080

Both have made mistakes. Describe their mistakes and write down what the correct working should be.

3 To increase something by 15% the multiplier you should use is 1.15. What multiplier should you use for:

   a an increase of 12%
   b a decrease of 17%
   c a decrease of 60%
   d an increase of 24%
   e a decrease of 4%
   f an increase of 8%
   g a decrease of 18%
   h an increase of 80%
   i an increase of 12.5%
   j a decrease of 5.5%
   k an increase of 400%
   l a decrease of 300%?

A multiplier is the number you multiply by.

4 A farmer produced 7000 tonnes of crop last year. This year the mass of his crop was down 23%. What was the mass of crop he produced this year?

5 Work out the new values.
   a Decrease 200 km by 15%
   b Increase $750 by 50%
   c Increase 425 ml by 8%
   d Decrease 5750 cm by 24%
   e Increase 43 kg by 200%
   f Increase 300 people by 350%

6 A radio is priced at $120. What is the selling price if 15% value added tax is charged?

7 Factory workers want a 12% increase in pay. How much would a worker get if they currently receive:
   a $225 a week
   b $350 a week?

8 In a restaurant a service charge of 10% is added to the price of the meal. What will be the total bill, when service charge is added, for a meal costing $78?

9 After harvesting, the mass of honey in a row of beehives went down by 12%. If there was originally a mass of 85 kg, how much honey was left in the beehives?

10 Last year at Southfield High School, 150 students passed their maths paper. This year there was a 4% increase in passes. How many students passed their maths paper this year?

11 The insurance premium on Mr Masood's car normally costs $380. With a no-claims discount the premium is reduced by 25%. What is his reduced premium?

12 A runner has a mass of 60 kg at the start of a marathon. During the race his body mass is reduced by 4%. What is his body mass at the end of the race?

13 A machine produces car parts at a rate of 260 per hour. How many car parts per hour does the machine produce if:
   a it develops a fault and the rate is reduced by 15%
   b an improvement is made to the machine and the rate is increased by 20%?

14 If 150 is decreased by 200% which of these represents the absolute change?
   150   ⁻150   300   ⁻300   200   ⁻200

15 For your answers to question **5** work out the absolute change.

16 Anna earns \$24 000 a year. Her boss gives her a choice of a 4% pay rise or \$75 extra each month.
  a Which new salary is highest?
  b By how much is it higher?

17 Which of these multipliers increases something by 800%?
  800   8   9   0.8   1.8   900   1.9

18 A car bought for \$50 000 depreciates in value by 20% each year.

  (**Note:** 'depreciates' means 'goes down in value'.)
  a What is its value after one year?
  b What is its value after two years?

19 a Find the value of the car in question **18** after five years.
  b What would be its value after five years if the depreciation rate was only 10%?

Ask your teacher for a copy of the Percentages game from Teacher Handbook 8. Play to see who can win.

9.4 Simplifying calculations with fractions and decimals

You have already learned about using BIDMAS, the order of operations to do calculations.

You are now going to apply this to questions with fractions and decimals.

Fractions

Brackets first

then **I**ndices

then **D**ivision and **M**ultiplication

then **A**ddition and **S**ubtraction.

Example 7

Work out: **a** $7 \times \left(\frac{3}{5} - \frac{1}{4} \right)$ **b** $\frac{1}{3} + 2 \times \frac{3}{11}$

a $7 \times \left(\frac{3}{5} - \frac{1}{4} \right)$

$= 7 \times \left(\frac{12}{20} - \frac{5}{20} \right)$

$= 7 \times \frac{7}{20} = \frac{49}{20} = 2\frac{9}{20}$

Brackets first

Multiplication before addition

b $\frac{1}{3} + 2 \times \frac{3}{11}$

$= \frac{1}{3} + \frac{6}{11} = \frac{11}{33} + \frac{18}{33} = \frac{29}{33}$

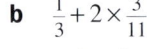

Laws of arithmetic

The laws of arithmetic can make calculations with fractions and decimals easier.

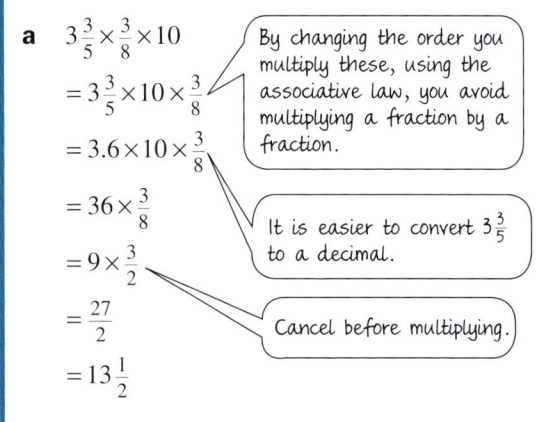

- The commutative law: $2 \times \frac{2}{5} = \frac{2}{5} \times 2 = \frac{4}{5}$

  Multiplications can be done in any order.

  When adding three numbers, you can add any pair of numbers first.

- The associative law:
  $3.8 + 2.4 + 0.6 = 6.2 + 0.6 = 6.8$
  (adding 3.8 and 2.4 first)
  or $3.8 + 2.4 + 0.6 = 3.8 + 3 = 6.8$
  (adding 2.4 and 0.6 first).
  The second way is easier here, as $2.4 + 0.6$ is a little easier than $3.8 + 2.4$.

- The distributive law:
  $0.5 \times (3.4 + 2.6) = 0.5 \times 3.4 + 0.5 \times 2.6 =$
  $1.7 + 1.3 = 3$
  (multiplying both numbers by 0.5 before adding)
  or $0.5 \times (3.4 + 2.6) = 0.5 \times 6 = 3$
  (doing $3.4 + 2.6$ first is easier here).

  When multiplying a sum by a number, each number in the sum can be multiplied by the number before adding.

Example 8

Calculate:

a $3\frac{3}{5} \times \frac{3}{8} \times 10$ **b** $14.83 \times 42 + 5.17 \times 42$

a $3\frac{3}{5} \times \frac{3}{8} \times 10$

$= 3\frac{3}{5} \times 10 \times \frac{3}{8}$

$= 3.6 \times 10 \times \frac{3}{8}$

$= 36 \times \frac{3}{8}$

$= 9 \times \frac{3}{2}$

$= \frac{27}{2}$

$= 13\frac{1}{2}$

By changing the order you multiply these, using the associative law, you avoid multiplying a fraction by a fraction.

It is easier to convert $3\frac{3}{5}$ to a decimal.

Cancel before multiplying.

b $14.83 \times 42 + 5.17 \times 42$
  $= (14.83 + 5.17) \times 42$
  $= 20 \times 42$
  $= 840$

Using the distributive law add the decimals first; this makes the calculation much easier.

Exercise 9D

1 Here is Masadi's homework.

$$3\frac{2}{9}+2\frac{1}{8}-2\frac{2}{9}$$

$$=3\frac{16}{72}+2\frac{9}{72}-2\frac{16}{72}$$

$$=5\frac{25}{72}-2\frac{16}{72}$$

$$=3\frac{9}{72}$$

$$=3\frac{1}{8}$$

Masadi said, 'I can work this out in my head in a much easier way.' Explain a method Masadi could have used.

2 Here is Jane's method for making a calculation easier.

$$4.86\times9.9$$

$$=4.86\times(10-0.1)$$

$$=4.86\times10-4.86\times0.1$$

$$=48.6-0.486$$

$$=48.114$$

Use Jane's method to work out:

a 58.23×9.9 **b** 9.9×28.71

c 0.82×0.99 **d** 23.6×1.1

3 Use laws of arithmetic to make these easier to work out.

a $4\frac{7}{8}\times63+5\frac{1}{8}\times63$

b $2.12\times0.04\times100$

c $5\frac{2}{3}+2\frac{3}{5}-2\frac{2}{3}$

d $23\times10\frac{7}{15}-23\times7\frac{7}{15}$

4 Work out:

a $2\times\left(\frac{3}{4}+\frac{1}{3}\right)$

b $\left(\frac{5}{8}-\frac{1}{4}\right)\times6$

c $3\times\left(\frac{7}{8}+\frac{2}{5}\right)$

5 Work out:

a $\frac{7}{8}-5\times\frac{1}{6}$

b $\frac{2}{3}+3\times\frac{3}{10}$

c $\frac{9}{10}-2\times\frac{3}{8}$

6 Work out:

a $\frac{7}{8}+3\times\frac{1}{4}+\frac{1}{2}$

b $\frac{3}{4}+\frac{1}{5}\times4-\frac{1}{2}$

c $\left(\frac{3}{5}-\frac{1}{20}\right)\times\left(\frac{2}{3}+\frac{4}{5}+\frac{8}{15}\right)$

Consolidation

Example 1

What is 3.2×0.06?

$$
\begin{array}{r}
32 \\
\times 6 \\
\hline
192
\end{array}
$$

The digits are 192.

There are 3 decimal places in $3.\underline{2}\times0.0\underline{6}$ so there are 3 decimal places in the answer.

So $3.2\times0.06=0.192$

Example 2

$17.37\div0.3$

$$\frac{17.37}{0.3}=\frac{17.37\times10}{0.3\times10}$$

$$=\frac{173.7}{3}$$

$$
\begin{array}{r}
57.9 \\
3\overline{)173.7}
\end{array}
$$

So $17.37\div0.3=57.9$

Example 3

A newspaper headline reads:

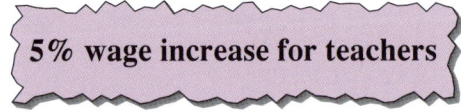

5% wage increase for teachers

A teacher currently earns $2500 a month. What will their new pay be?

Here is how to calculate the teacher's new pay.

$100\% + 5\%$ is 105%

105% of $2500

$= 1.05 \times \$2500$

$= \$2625$

Example 4

An apple tree had 475 apples on it last year. This year there were 24% fewer apples on the tree. How many apples were on the tree this year?

To decrease 475 by 24%, find 76% of 475:

$0.76 \times 475 = 361$

$100\% - 24\%$ is 76%

There were 361 apples on the tree this year.

Example 5

Work out $78 \div \left(2\frac{3}{5} - 1\frac{5}{8}\right)$

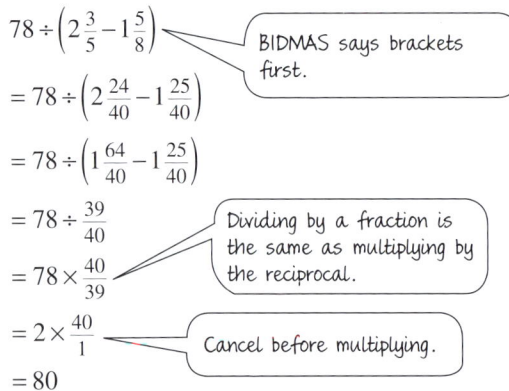

$78 \div \left(2\frac{3}{5} - 1\frac{5}{8}\right)$

BIDMAS says brackets first.

$= 78 \div \left(2\frac{24}{40} - 1\frac{25}{40}\right)$

$= 78 \div \left(1\frac{64}{40} - 1\frac{25}{40}\right)$

$= 78 \div \frac{39}{40}$

$= 78 \times \frac{40}{39}$

Dividing by a fraction is the same as multiplying by the reciprocal.

$= 2 \times \frac{40}{1}$

Cancel before multiplying.

$= 80$

Exercise 9

1 Work out:

a $^-3 \times 4.7$ **b** $^-2 \times 0.24$

c $1.28 \times ^-5$ **d** $0.044 \times ^-200$

2 $234 \times 29 = 6786$

Use this fact to work out:

a 23.4×2.9 **b** $2.34 \times ^-29$

c 0.0234×290 **d** 2.34×0.029

e $6786 \div 2.9$ **f** $6786 \div 23.4$

3 Work out:

a 4.6×0.3 **b** 0.38×0.04

c 13.68×0.7 **d** 41.6×0.06

e 1.3×2.4

4 Work out:

a $4 \div 0.2$ **b** $36 \div 0.9$

c $4.78 \div 0.2$ **d** $57.6 \div 0.4$

e $7.11 \div 0.9$

5 The cost of 9 kg of pistachio nuts is $111.24.

a What is the cost of 10 kg?

b What is the cost of 10.6 kg?

6 What is the area of this rectangle?

5.4 cm

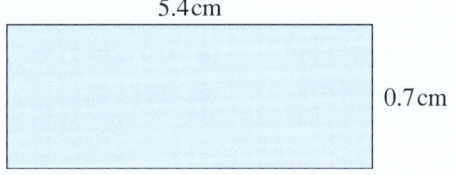

0.7 cm

7 A square garden plot requires 63.4 m of fence to enclose it.

a What is the area of the plot?

b If the plot's area is increased by 10 m² but it still remains square, how much fencing is required to enclose it?

Give your answers to 1 d.p.

8 To decrease something by 32%, the multiplier you should use is 0.68. What multiplier should you use for:

a an increase of 2%

b a decrease of 27%

c a decrease of 53%

d an increase of 35%

e an increase of 40%?

9 A sales tax of 3% is placed on goods. What is the selling price of a bicycle priced at $950 before tax?

10 A manufacturer produced 9000 cars last month. This month production was down by 12% from last month. How many cars were produced this month?

11 Find how much you would have to pay on a bill of $65 if a discount was given of either:

    a 7%   **b** 9%

12 Work out:

    a $12 \div \left(1\frac{2}{3} - 1\frac{2}{5}\right)$

    b $16\frac{7}{15} \times 17 + 23\frac{8}{15} \times 17$

    c $\left(2\frac{7}{8} - 1\frac{1}{4}\right) \times 10$

Summary

You should know …

1 You can multiply decimals by decimals.

For example: 2.8×0.3

$$\begin{array}{r} 28 \\ \times\ 3 \\ \hline 84 \end{array}$$

The digits in the answer are 84.

There are 2 decimal places in 2.8×0.3 so there are 2 decimal places in the answer.

So $2.8 \times 0.3 = 0.84$

2 To divide a decimal by a decimal you turn the divisor into a whole number.

For example: $3.8 \div 0.4 = \dfrac{3.8}{0.4} = \dfrac{3.8 \times 10}{0.4 \times 10} = \dfrac{38}{4} = 9.5$

3 How to increase and decrease by a given percentage.

For example:

To increase 325 by 28% find 128% of 325:
$1.28 \times 325 = 416$

To decrease 3800 by 15% find 85% of 3800:
$0.85 \times 3800 = 3230$

100% + 28% is 128%

100% - 15% is 85%

Check out

1 Calculate:

    a 3.2×0.5

    b 0.4×7.4

    c 12.2×0.6

    d 8.4×0.03

    e 9.42×0.8

    f 1.36×0.42

2 Work out:

    a $8 \div 0.5$

    b $4 \div 0.2$

    c $3.6 \div 0.3$

    d $5.8 \div 0.2$

    e $12.1 \div 1.1$

3 Work out the new values.

    a Increase $80 by 5%

    b Decrease 800 km by 45%

    c Increase 625 g by 16%

    d Decrease 32 ℓ by 12.5%

4 How to apply the order of operations to calculations involving decimals or fractions.

For example:

Work out $\left(4\frac{2}{5} - 2\frac{2}{3}\right) \times 6$

$= \left(4\frac{6}{15} - 2\frac{10}{15}\right) \times 6$ ← BIDMAS says brackets first.

$= \left(3\frac{21}{15} - 2\frac{10}{15}\right) \times 6$

$= 1\frac{11}{15} \times 6$

$= \frac{26}{15} \times 6$ ← Cancel before multiplying.

$= \frac{26}{5} \times 2$

$= \frac{52}{5}$

$= 10\frac{2}{5}$

4 a $74 \div \left(2\frac{1}{9} - 1\frac{7}{10}\right)$

b $23 - 4.8 \times 2.4$

Presenting data and interpreting results

What's the point?

Statistics are used all the time in the real world, in many different areas. For example, a good way to see the profits or losses of a company is by drawing a graph that shows visually whether the general trend is up or down, and can be read far more quickly and easily than a page of numbers.

Before you start

You should know ...

1 How to calculate the mean, mode, median and range of a given set of data.

For example:

Carlene scored a total of 520 marks in 8 subjects in her last exam. Her mean score was $\frac{520}{8} = 65$.

A reminder:

Mean = total of values ÷ number of values

Mode = most common value

Median = middle value when all values placed in numerical order

Range = highest value subtract lowest value

Check in

1 Find the mode, median, mean and range of these numbers.

15, 13, 13, 12, 11, 11, 10
10, 10, 10, 9, 7, 8, 8, 7

2 How to read information from graphs.
For example:

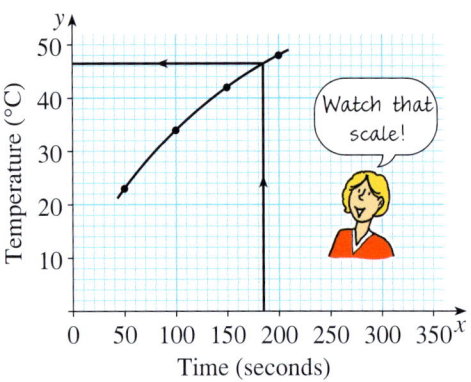

The temperature at 185 seconds is 46.5°C.

2 A baby's temperature was taken every 30 minutes starting at 5 o'clock.

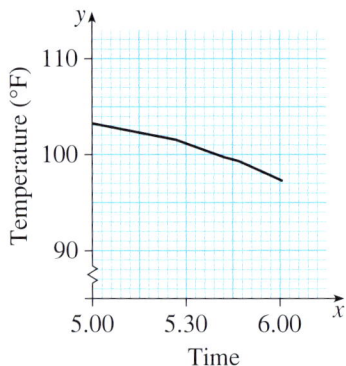

a At what time was the baby's temperature 100°F?

b What was the baby's temperature at 5.18?

10.1 Time series graphs

A **time series graph** is used to show a trend over time.

Example 1

a Draw a time series graph to show Bently's height on 1 January each year, from his birth in 2017.

Year	2017	2018	2019	2020	2021
Height (cm)	40	65	80	90	100

b Estimate his height in July 2019.

....

a

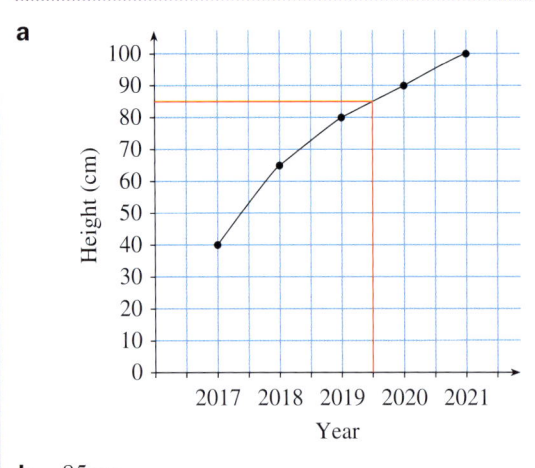

b 85 cm

It makes sense to join all the plotted points with straight lines as you assume that growth is continuing at a steady rate between the times Bently's height is measured.

Notice that in Example 1 the scale chosen is important. A change of scale can make the graph look different:

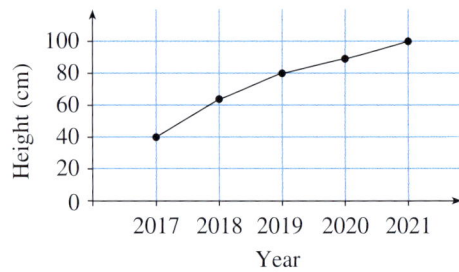

You need to choose scales and axes carefully when drawing charts. If you don't, your graphs could mislead!

Exercise 10A

1 a Draw a line graph to show the change of temperature over the course of a day. Start the temperature axis at 20°C.

Time	6 a.m.	8 a.m.	10 a.m.	12 noon	2 p.m.	4 p.m.
Temperature (°C)	24	25	27	29	29	26

b What do you think the temperature was at
   i 7 a.m.   **ii** 3 p.m.?
c At what times do you think the
   temperature was 28°C?
d Why does it make sense to join the points?
e Redraw the graph with the temperature
   axis beginning at 0°C. Which graph is
   easier to read?

2 Kamil was conducting a science experiment
   to see how boiling water in a mug cools in
   10 minutes. He took 9 measurements of the
   temperature, as shown in this table.

Time	Temperature (°C)
10:00	100
10:01	96
10:02	91
10:03	87
10:04	
10:05	83
10:06	80
10:07	81
10:08	77
10:09	75
10:10	74

a Draw a line graph to show the
   temperature of water in a mug as it
   cools over 10 minutes.
b Maria says Kamil must have made a
   mistake in recording of one of the
   temperatures.
   i Which temperature is she talking about?
   ii How does she know a mistake has
      been made?
c Kamil forgot to take the temperature at
   10:04. What do you think the temperature
   of the water was at this time?

3 The graph shows temperatures during a day
   in July in Rome, Italy.

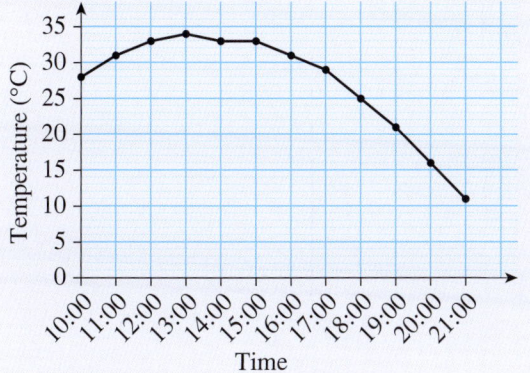

a At what time was the highest temperature
   recorded?
b Estimate the temperature at 17:30.
c The temperature stayed the same for an
   hour. When was this?
d Which hour showed the greatest rise in
   temperature?
e Which hour showed the greatest fall in
   temperature?
f Estimate the time when the temperature
   was 23°C.

4 A football match was due to start at 15:00.
   The sports stadium gates were opened at
   14:00. The number of people inside the
   stadium was recorded every 10 minutes.
   The results are shown in this table.

Time	Number of people inside stadium
14:00	0
14:10	4000
14:20	15000
14:30	23000
14:40	30000
14:50	34000
15:00	35000

a Draw a line graph to show the number of
   people in the stadium between 14:00
   and 15:00.
b Use your graph to estimate how many
   people were in the stadium at 14:25.
c In which 10-minute period did the most
   people enter the stadium?
d What have you assumed when answering
   part **b**?

10.2 Stem-and-leaf diagrams

A stem-and-leaf diagram is similar to a frequency
diagram in that the shape of the diagram gives an
overall picture of the trend. The advantage of a stem-
and-leaf diagram over a frequency diagram is that you
can still see all the data values. Bars in frequency
diagrams are often drawn vertically, while in stem-and-
leaf diagrams numbers are usually written horizontally.

A stem-and-leaf diagram has four main features.

• The key – this explains what the stem and leaves
   are worth.

151

- The stem – this is often the tens column (but not always – check the key to find out).
- The leaves – these are often the units column (but not always – check the key to find out). Note the leaves *must be in numerical order*.
- The spacing – numbers *must be evenly spaced* to maintain the shape clearly.

Example 2

Draw a stem-and-leaf diagram for the data below, which shows the height, in centimetres, of 15 plants.

14, 34, 21, 20, 36, 45, 56, 52, 33, 44, 63, 49, 47, 49, 58

Look at the numbers to see what the stems should be. The tens columns for these numbers are 1, 2, 3, 4, 5 and 6. Write these in a vertical list with a vertical line to the right of them. Write the first two values, 14 cm and 34 cm, in the stem-and-leaf diagram and include a key. Make sure you label the diagram 'Unordered' so that you remember to put it in order later.

Unordered

Stem	Leaves
1	4
2	
3	4
4	
5	
6	

Key

1 | 4 means 14 cm

Beginning with an *unordered* stem-and-leaf diagram is the best way to make sure that you don't miss any values out, particularly when there are a lot of values in a list. Starting from the beginning and working your way through the data, write in the other values. Don't worry too much about spacing at this point as this is just rough work.

Your stem-and-leaf diagram should look now look like this.

Unordered

Stem	Leaves
1	4
2	1 0
3	4 6 3
4	5 4 9 7 9
5	6 2 8
6	3

Key

1 | 4 means 14 cm

Now place all the leaves in numerical order and make sure that they are evenly spaced. Label the new diagram 'Ordered'.

Ordered

Stem	Leaves
1	4
2	0 1
3	3 4 6
4	4 5 7 9 9
5	2 6 8
6	3

Key

1 | 4 means 14 cm

Notice that if a bar chart or frequency diagram was used for the data in Example 2, only the shape would be known, not the original values.

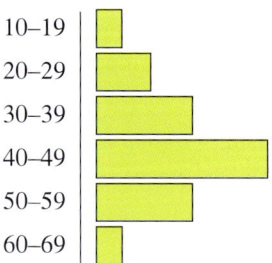

A useful feature of stem-and-leaf diagrams is that they are good for finding averages.

```
1 | 4
2 | 0 1
3 | 3 4 6
4 | 4 5 7 9 9
5 | 2 6 8
6 | 3
```

Key

1 | 4 means 14 cm

Using this stem-and-leaf diagram, you can see that the mode is 49 cm as that is in the only row with repeated digits. Because the numbers are in order you can easily find the middle number: the median is 45 cm. Finding the mean is also possible as you have all the original data. The mean for this data is:

$(14 + 20 + 21 + 33 + 34 + 36 + 44 + 45 + 47 + 49 + 49 + 52 + 56 + 58 + 63) \div 15 = 41.4$ cm to 1 d.p.

Exercise 10B

1 For a school project, Guntur surveyed 15 streets in his town to see how many houses there were in each street. These were the results.

17, 55, 48, 57, 61, 30, 26, 57, 46, 51, 40, 60, 58, 31, 42

His friends Kersen, Fatima and Eva drew these stem-and-leaf diagrams.

Kersen

```
1 | 7
2 | 6
3 | 0 1
4 | 0 2 8
5 | 1 5 7 7 8
6 | 0 1
```

Fatima

```
1 | 7
2 | 6
3 | 0 1
4 | 8 6 0
5 | 5 7 7 1 8
6 | 1 0
```

Key

1 | 7 means 17 houses

Eva

```
1 |   7
2 |   6
3 |   0 1
4 |   0 2 6 8
  5| 1 5 7 7 8
6 |   0 1
```

Key

1 | 7 means 17 houses

Explain what is wrong with each of these stem-and-leaf diagrams.

2 a Draw an ordered stem-and-leaf diagram from this unordered diagram about the masses of spare parts for a machine.

Unordered

```
2 | 7
3 | 3 1 9
4 | 6 8 8 1 5
5 | 3 5 2 5 0 5 9
6 | 5 8 6 1
7 | 2 3
```

Key

2 | 7 means 2.7 kg

b In the stem-and-leaf diagram in part **a**, which two numbers are represented by this row?

7 | 2 3

c What is the modal mass?
d What is the median mass?
e What is the smallest mass?
f What is the largest mass?

3 All of the students in a class were asked by their maths teacher to estimate the size of an angle that she drew on the board. These were the results.

45, 62, 38, 51, 44, 38, 36, 43, 50, 44, 40, 55, 48, 37, 32, 38, 29, 40, 41, 35, 46, 25, 42, 41, 39, 39, 45, 32, 43, 46

a Draw a stem-and-leaf diagram for this data.
b What is the modal guess?
c What is the median guess?
d Only one student guessed the angle correctly. Which angle do you think this was? Justify your answer.

4 The two data sets below show the heights of all the teachers in Yasmina's school, measured in centimetres.

Set A

162, 153, 157, 159, 165, 168, 190, 184, 181, 173, 174, 172, 166, 164, 155, 162, 171, 153, 164

Set B

192, 182, 176, 190, 168, 180, 179, 177, 174, 191, 185, 181, 183, 186, 167, 183, 177, 182, 158

a Copy and complete the two stem-and-leaf diagrams started below.

```
15 | 3
16 |
17 |
18 |
19 |
```

Key

15 | 3 means 153 cm

```
15 | 8
16 |
17 |
18 |
19 |
```

Key

15 | 8 means 158 cm

b One set of data is for the male teachers, the other set is for the female teachers. Which set, A or B, do you think represents the male teachers? Why?

c Is it easier to answer part **b** by looking at the stem-and-leaf diagrams or by looking at the lists of numbers?

5 Draw a line on a piece of paper. Ask your classmates to estimate the length of the line, in millimetres. Draw a stem-and-leaf diagram for this data.

6 This stem-and-leaf diagram shows the number of watches sold by a jeweller each day, for a period of 20 days.

```
0 | 4  6  7
1 | 1  3  3  3  6  8  9
2 | 1  2  4  7  7  7
3 | 1  2  4
4 | 1
```

Key

4 | 1 means 41

a Work out the mean, median, mode and range for this data.

The next day 27 watches are sold.

b Without doing any calculations, what effect does this extra value have on the mean, median, mode and range? Why?

10.3 Histograms

Histograms are very similar to bar charts. A histogram is usually used to show continuous data, while a bar chart is used to represent discrete data. Histograms are also often used for grouped data.

For example, look at these heights of 25 children measured to the nearest centimetre.

139, 141, 142, 142, 145, 146, 147, 147, 151, 151, 152, 152, 153, 153, 154, 156, 156, 157, 157, 158, 160, 161, 162, 162, 166

Drawing this information on a graph gives:

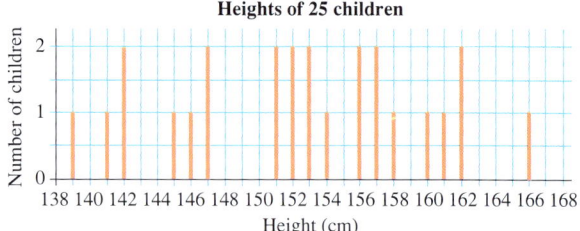

Such a graph is not very useful – explain why.

Instead it is better to construct a grouped frequency table. Using groups 135–139, 140–144, etc. this is the frequency table.

Class interval	Tally	Frequency
135–139	I	1
140–144	III	3
145–149	IIII	4
150–154	IИ II	7
155–159	IИ	5
160–164	IIII	4
165–169	I	1

The histogram of this data is:

Histogram of heights of 25 children

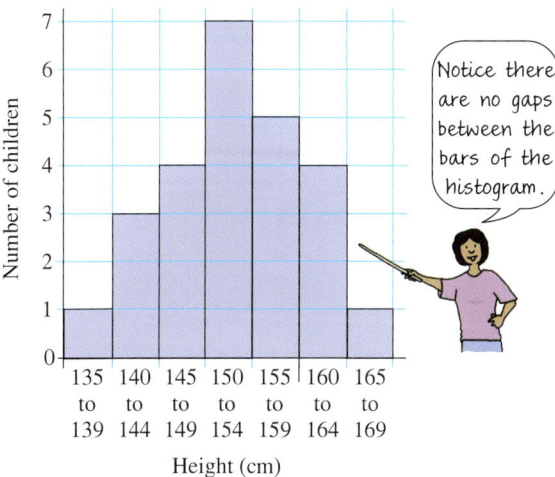

Notice there are no gaps between the bars of the histogram.

Note that it is more usual to label the horizontal axis as shown below, instead of using the '135 to 139' notation.

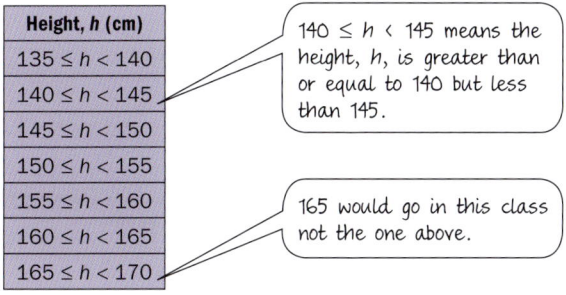

135 140 145 150 155 160 165 170
Height (cm)

It is useful to write the classes using inequalities when data is not rounded. For the data above this would be:

Height, h (cm)
$135 \le h < 140$
$140 \le h < 145$
$145 \le h < 150$
$150 \le h < 155$
$155 \le h < 160$
$160 \le h < 165$
$165 \le h < 170$

$140 \le h < 145$ means the height, h, is greater than or equal to 140 but less than 145.

165 would go in this class not the one above.

When data is grouped and if you don't have the original data you cannot say what the mode is. But you can identify the modal class. The **modal class** is the group with the highest frequency.

Example 3

Mass, m (kg)	Frequency
$40 \le h < 45$	7
$45 \le h < 50$	12
$50 \le h < 55$	6
$55 \le h < 60$	3

a Which class interval would 50 go in?
b What is the modal class?

....

a $50 \le h < 55$
b $45 \le h < 50$

Exercise 10C

1 a Draw a histogram for the heights of the 25 children using intervals of:
  i 10 cm   **ii** 20 cm
  b Look at all the charts for this data. Which one shows best how heights vary?

2 The masses, in kilograms, of 25 children are:

Mass, m (kg)	No. of children
$10 \le m < 20$	1
$20 \le m < 30$	4
$30 \le m < 40$	7
$40 \le m < 50$	12
$50 \le m < 60$	1

a Draw a histogram to show this information.
b What is the modal mass?

3 The heights of 30 plants, in centimetres, 6 weeks after planting were:

Height, h (cm)	Frequency
$5 \le h < 10$	5
$10 \le h < 15$	9
$15 \le h < 20$	8
$20 \le h < 25$	6
$25 \le h < 30$	2

Draw a histogram to show the data.

4 The heights in centimetres of 30 different plants were measured 6 weeks after planting. The results are shown in this histogram.

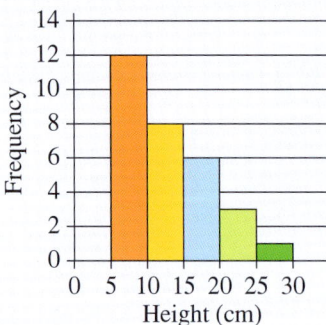

a Copy and complete this table.

Height, h (cm)	
$5 \leq h < 10$	
	8
$15 \leq h < 20$	
$20 \leq h < 25$	
	1

b Compare the two histograms from questions **3** and **4**. Do you think these plants were the same species? Justify your answer.

5 The waiting time, in minutes, for 25 patients to see a doctor was:

8, 35, 12, 45, 4, 15, 38, 28, 30, 23, 14, 38, 53, 26, 33, 32, 15, 18, 48, 37, 34, 34, 28, 51, 16

a Construct a frequency table using an interval of your choice to show the data.
b Draw a histogram to illustrate the data.
c How many patients waited less than 10 minutes?
d What percentage of patients waited half an hour or more?
e Was your interval choice the best?

6 The masses, in kilograms, of 30 items to be transported are shown below.

7, 13, 18, 24, 46, 29, 31, 34, 36, 37, 38, 38, 41, 43, 43, 44, 45, 46, 47, 48, 49, 52, 54, 55, 56, 59, 63, 72, 85, 97

a Jalad drew this histogram for the data.

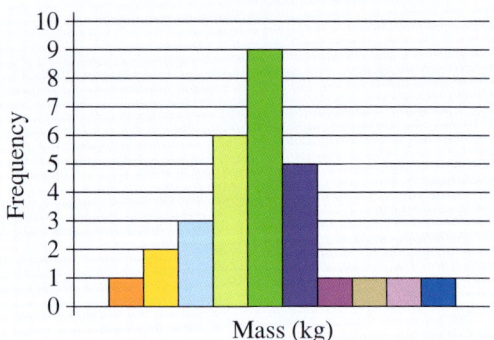

What class intervals did he use?

b Amy drew this histogram for the data.

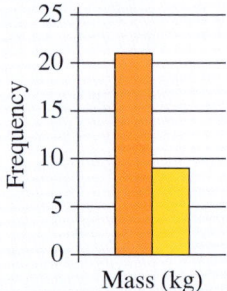

What class intervals did she use?

c Compare the two histograms. Which do you think is better? Why? Can you think of a reason why the other one might be used?

Activity

Measure the heights of the students in your class.
1 Construct a suitable frequency table to show this data.
2 Draw a histogram to illustrate the distribution of heights in your class.
3 Repeat the process, but this time draw separate tables and histograms for boys and girls.
4 Comment on your results.

10.4 Interpreting and comparing data and diagrams

Comparing distributions means looking for what is similar and what is different about them. In particular, we usually look for which values are larger or smaller, for example which distribution has the larger mean, median, mode or range.

Interpreting tables, graphs and diagrams means understanding what these show. For example, a larger range shows that data is more spread out or more varied.

Often when you have interpreted and compared data you will be able to write some sort of **conclusion**. In question **4** of Exercise 10B you interpreted the two stem-and-leaf diagrams and compared them to come to a conclusion based on female and male heights.

The next exercise is all about interpreting and comparing data.

Example 4

The holiday destinations of 300 students during 2018 are recorded in the first pie chart. The second pie chart shows the holiday destinations of 600 students during 2019.

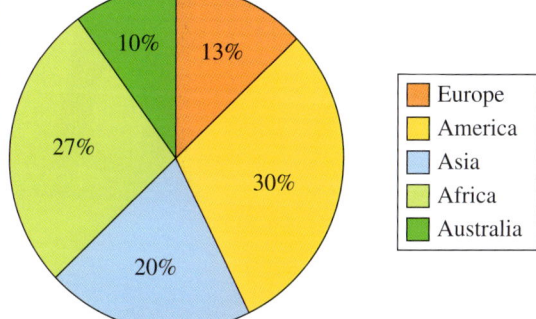

2018 holiday destinations

■	Europe
■	America
■	Asia
■	Africa
■	Australia

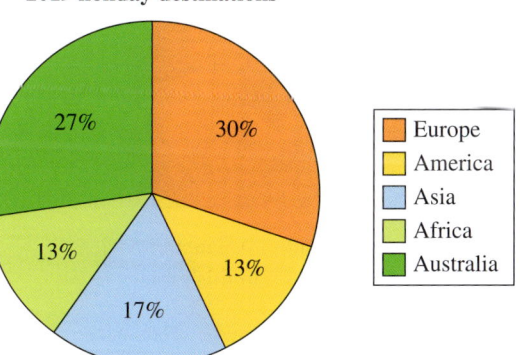

2019 holiday destinations

■	Europe
■	America
■	Asia
■	Africa
■	Australia

Compare the two pie charts by considering:

a   the most and least popular resorts
b   the number of students travelling to Asia each year

a   The most popular two holiday destinations in 2018 were America and Africa, but these were the least popular in 2019. The least popular destination in 2018 was Australia. However, this was one of the most popular destinations in 2019.

b   20% of students went to Asia in 2018. This was 60 students in total ($0.2 \times 300 = 60$). 17% of students went to Asia in 2019. This was 102 students in total ($0.17 \times 600 = 102$). So although a lower percentage of students went to Asia in 2019, this represented a higher number.

You have seen in Example 4 that it is important to be careful when pie charts represent different total numbers. You must make sure you talk about percentages and numbers with care.

Exercise 10D

1   This histogram shows the heights of some men.

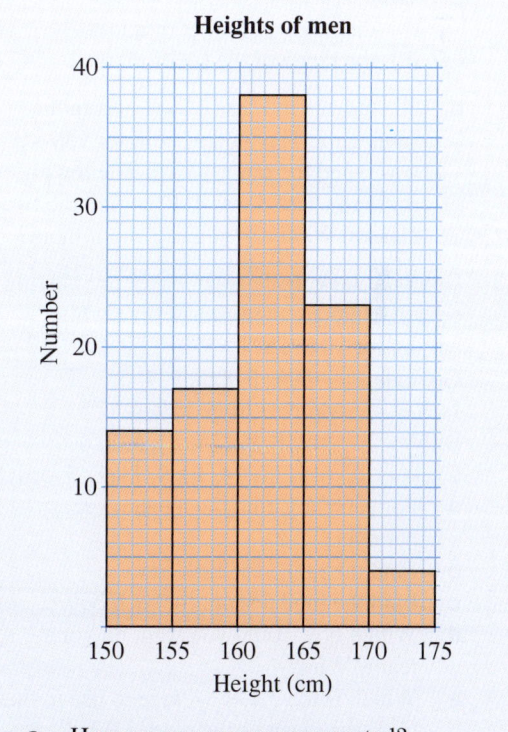

Heights of men

a   How many men are represented?
b   How many men are shorter than 160 cm?
c   How many men are 165 cm or taller?
d   Is it possible to say how tall the tallest man is? Explain your answer.

2 30 students from Class 2 took part in a quiz. This bar chart shows their scores.

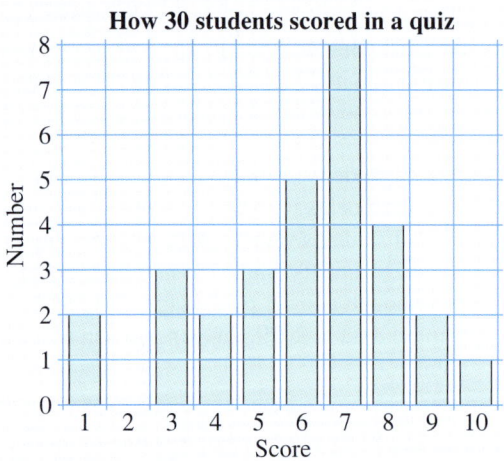

How 30 students scored in a quiz

a How many students scored:
   i the lowest mark
   ii the highest mark?

b What mark did most students score?

c What is:
   i the mode of the set of marks
   ii the range of the set of marks?

d 30 students in Class 3 took part in the same quiz. The modal score for Class 3 was 6 and the range was 5. The lowest score in Class 3 was 4. Compare the two classes. What can you say about them?

3 Govinda, Jack, Billy and Chakor are playing cricket. Here are their runs scored while batting for 10 overs.
Govinda   10, 8, 5, 8, 5, 9, 4, 5, 9, 7
Jack   6, 7, 1, 4, 5, 2, 2, 4, 4, 5
Billy   12, 8, 9, 1, 1, 54, 5, 3, 6, 1
Chakor   15, 11, 14, 10, 9, 11, 8, 5, 5, 12

a For each boy, find the mean score per over.

b Find the range for each boy.

c Find the modal score for each boy.

d Which boy would you want on your team? Why?

e Which is the better average to use in this case, the mean or the mode? Why?

4 a In the following graph the solid line represents the number of baboons. The dotted line represents the number of cheetahs. Compare the shapes of the two graphs. What conclusions can you reach?

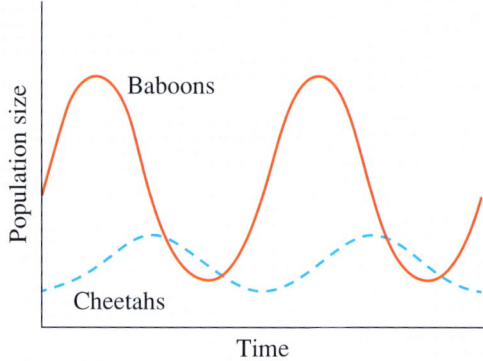

b Suggest a possible scale for the horizontal axis. Justify your choice.

5

The masses of adult Asian and African elephants are given in this table.

Mass (kg)	Frequency	
	African	Asian
3000–4000	2	8
4000–5000	4	23
5000–6000	19	7
6000–7000	10	1
7000–8000	6	0

a Show this data on two frequency graphs.

b What is the modal mass of:
   i Asian elephants
   ii African elephants?

c Three elephants each have a mass of:
   i 7250 kg
   ii 3245 kg
   iii 5050 kg
   Decide which species you think each of the elephants is.

d How confident can you be about your answers to part **c**?

6 An artist showed four paintings, A, B, C and D, to some adults and asked them which one they liked best. She then showed them to some children. The pie charts below show the adults' and children's choices.

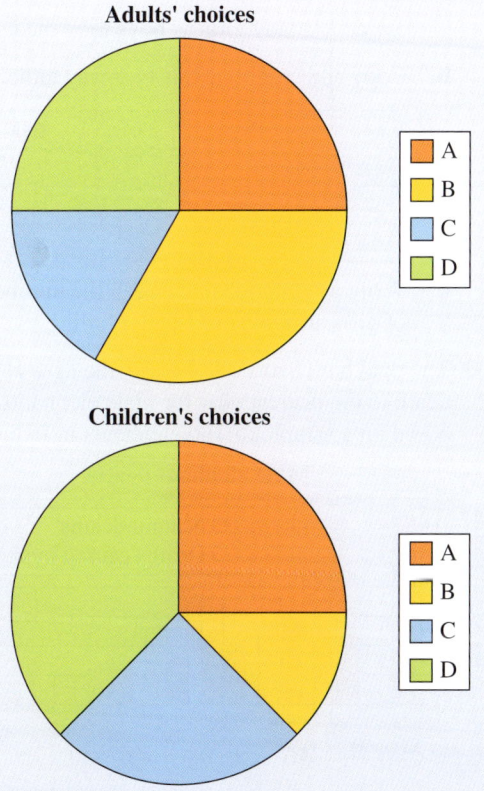

Which of the following statements are:
i true
ii false
iii impossible to say?
Give reasons for your answers.

a The same percentage of adults and children liked painting A.

b More adults than children liked painting B.

c More children liked painting D than painting A.

d A higher percentage of children than adults liked painting D.

e Painting B was the most popular among the adults.

f The same number of adults liked painting A and painting D.

g A lower percentage of adults than children liked painting C.

h Painting A was the most popular among the children.

7 Two samples of 80 people at a school were surveyed to see how far they lived from the school. The results are shown in the histograms below.

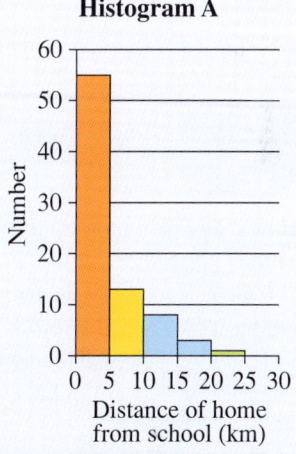

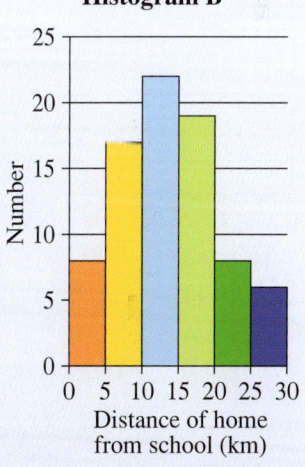

a In histogram A, what was the modal distance from school?

b In histogram B, what was the modal distance from school?

c Is this statement true, false or impossible to say: 'The range for histogram A was 24'? Give a reason for your answer.

d One histogram was for the students and one was for the teachers. Which do you think was for the teachers? Justify your answer.

e Write a sentence comparing the two histograms.

8 The table below shows how long the battery lasts on two different laptops, type A and type B.

Number of laptops still working after	Type A	Type B
0 h	32	38
1 h	30	37
2 h	29	35
3 h	26	28
4 h	21	15
5 h	11	7
6 h	3	2

a How many laptops were there in the sample altogether?

b Is it easy to see from the data above which laptop seems to last longer?

c Copy and complete this line graph.

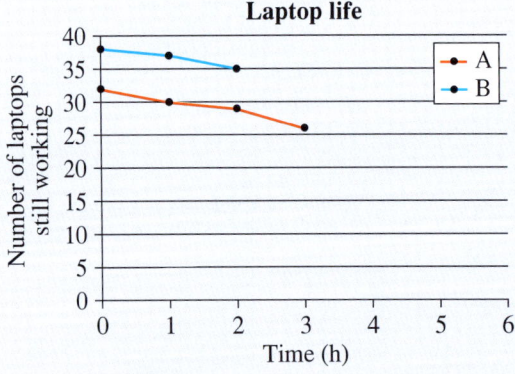

d After 3.5 hours how many laptops of:
  i type A
  ii type B
  are still working?

e What percentage of the original laptops are still working after 3.5 hours?

f Is it easy to see from the graph which laptop seems to last longer?

9 The pie charts show the number of students passing or failing their driving test on the first attempt. The sample is taken from one year group in Cransley High School.

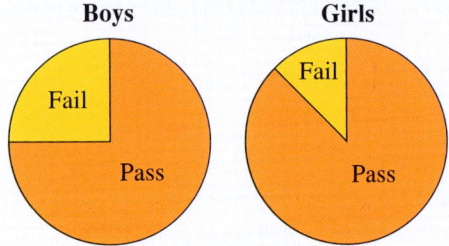

a Write a comparison between the percentage of girls and boys passing or failing.

The number of boys who failed the test first time is 20.

Twice as many girls as boys took the test.

b Copy and complete this two-way table.

	Boys	Girls	Total
Pass			
Fail	20		
Total			

c Write a comparison between the numbers of girls and boys passing or failing.

10 The dual bar chart below shows the ages at death to the nearest year for male elephants, based on a sample of 166 elephants in India.

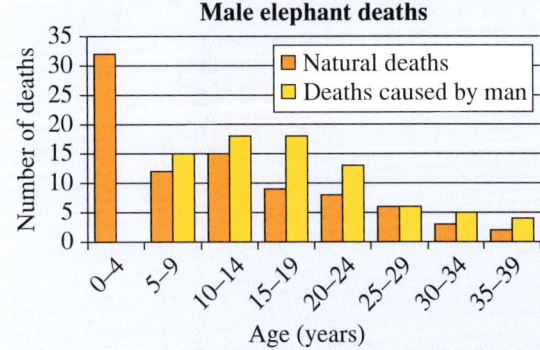

a How do the bars for the 0–4 range compare to all the other age ranges? Why do you think this is the case?

b What is the modal age of death of these male elephants for:
  i natural deaths
  ii deaths caused by man?

c Is it possible to work out the range for the ages of death? Give a reason for your answer.

d Can you suggest any possible problems with the data?

11 Here is some summary data showing information about the test results for 30 boys and 30 girls. The test was marked out of 100.

	Boys	Girls
Mean	72	65
Median	70	64
Mode	60	82
Range	27	35

Use one of these words, or a number, to complete each of these sentences.

mean median mode range boys girls

a The data that shows how consistent the students are is the . . . as it shows how close together the marks are.

b The . . . is not the best average to compare because it doesn't fit the trend of the other averages.

c The . . . and the . . . both show that the boys had higher marks on average.

d . . . were more consistent in their marks than . . ..

e The mean shows that boys had higher marks on average than girls by . . . marks.

f The best average to use for this particular situation is the . . ..

12 Jess thinks that the reaction times (in seconds) of children is faster than the reaction times of adults. She uses a computer simulation game to collect this reaction time data.

Children: 4, 5, 5, 8, 10, 11, 11, 11, 12

Adults: 5, 6, 7, 9, 10, 11, 12, 13, 19

a Explain why the mode is not a suitable average to use for this data, to compare children and adults.

b Using data to support your conclusions decide who has the fastest reaction times and is more consistent out of the children and adults.

c Explain why the average you chose is better than the other possible average you could have used.

13 Here is a time series graph showing the value of two different cars over a four-year period.

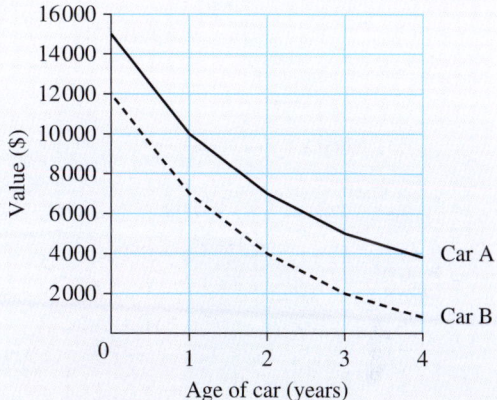

a During which year did each car depreciate most (lose the greatest value)? How does the time series graph show this?

b Describe the trend in the values of each car.

c Use the graph to estimate the value of car A after:
i 6 months
ii 18 months.

d How much had car B depreciated by after:
i the first year
ii four years?

e Henrik thinks not all cars depreciate in value. He says, 'Some rare and very old cars can increase in value.' Explain how the graph would show this if he plotted it on the axes above.

f Both cars were the same make and model. Annabel thinks you can tell from the graph which car is newer. Sasha thinks that the graph cannot necessarily show which car is newer. Give a reason why both could be correct.

14 These two pie charts show how students in Year 10 spend their free time.

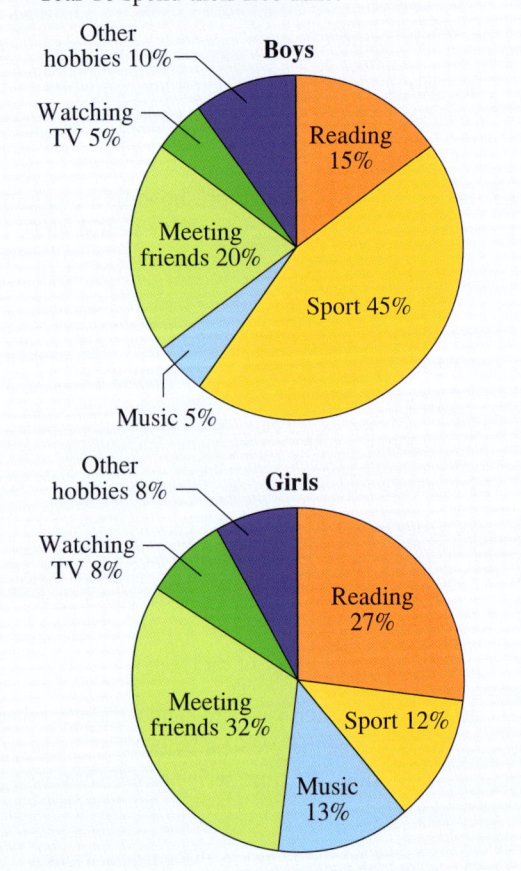

Boys

Other hobbies 10%
Watching TV 5%
Reading 15%
Meeting friends 20%
Sport 45%
Music 5%

Girls

Other hobbies 8%
Watching TV 8%
Reading 27%
Meeting friends 32%
Sport 12%
Music 13%

a Compare the percentages in the two pie charts.

b Boys spent 10% of their time on 'Other hobbies' while girls only spent 8% of their time in this way. There were 8 boys who spent their time on other hobbies. If there were 180 students surveyed in total, how many girls spent their time on other hobbies?

c Write a paragraph comparing the way Year 10 boys and girls spend their free time.

Consolidation

Example 1

a Draw a stem-and-leaf diagram for these marks out of 50 for 11 students in a recent chemistry test.

11, 47, 34, 21, 20, 36, 45, 48, 50, 34, 44

b What was the modal score in the test?

....

a

```
1 | 1
2 | 0 1
3 | 4 4 6
4 | 4 5 7 8
5 | 0
```

Key
1 | 1 means 11 marks out of 50

b The modal score in the test was 34.

Example 2

The heights of 25 children, measured to the nearest centimetre, are shown in the histogram.

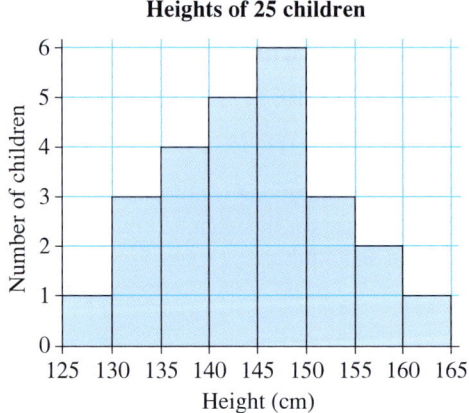

Heights of 25 children

Number of children (y-axis, 0 to 6)
Height (cm) (x-axis, 125 to 165)

a Which interval is the mode?
b How many children were 150cm or more in height?
c Which interval for height contained exactly five children?

....

a The mode is from 145 to 150.
b Six children were 150cm or more in height.
c There were exactly five children in the 140 to 145cm height interval.

Example 3

The two frequency diagrams below show the scores of 40 students in their biology and physics tests.

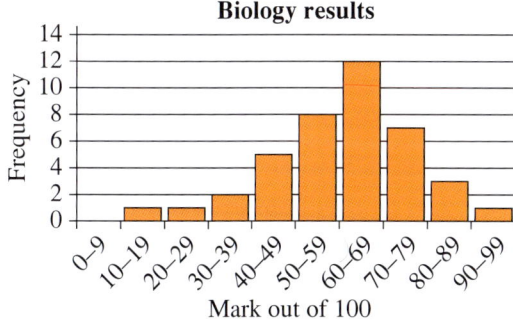

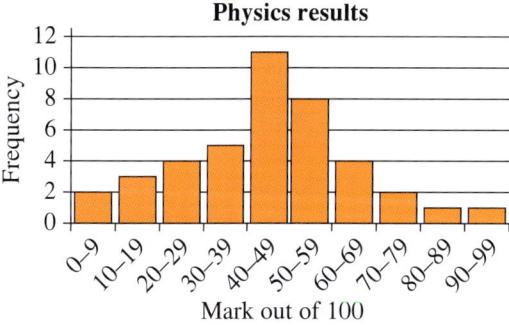

Which was the harder test? Justify your answer.

....

The modal score for biology is 60–69 and the modal score for physics is 40–49. More candidates scored lower in physics so it seems that physics was the harder test.

Exercise 10

1 The masses of 50 students in a school are shown in this table.

Mass, m (kg)	No. of children
$40 \leq m < 45$	8
$45 \leq m < 50$	12
$50 \leq m < 55$	17
$55 \leq m < 60$	9
$60 \leq m < 65$	4

a Draw a histogram of the data.
b What is the modal mass interval?
c How many students had a mass of 55kg or more?

A different group of students had the following masses.

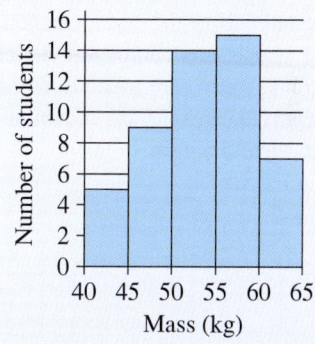

d Which group of students do you think was younger? Give a reason for your answer.

2 The mass of a baby during its first 24 months of life is shown in this graph.

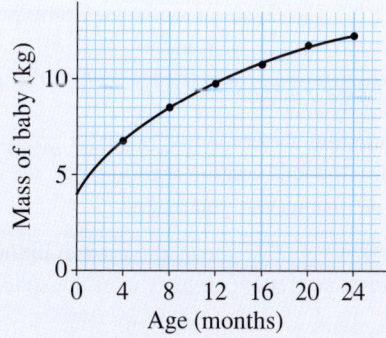

a What was the mass of the baby:
  i at birth
  ii after 6 months
  iii after 19 months?

b At what age was the baby's mass:
i 5 kg **ii** 9 kg **iii** 7.3 kg?

c Describe the trend in the baby's mass over the two years.

3 Half of a class of 32 students completed a psychological test and were timed while doing so.

a Draw a stem-and-leaf diagram for the following list of times taken, in seconds, to complete the test.

7.2, 8.9, 5.8, 7.4, 10.3, 6.1, 8.3, 6.2, 7.4, 6.7, 7.2, 9.1, 7.4, 8.3, 9.4, 5.7

b What is **i** the mean
ii the median
iii the modal time?

c What is the range of times?

The second half of the class watched while the first 16 students completed the test. Then the second group were asked to complete the same test. Their results are shown in this stem-and-leaf diagram.

```
 5 | 3
 6 | 0  4  7  8
 7 | 0  0  1  2  4  7
 8 | 3  4
 9 | 1  3
10 | 1
```

Key
5 | 3 means 5.3 seconds

d Work out all three averages and the range for the second set of students. Compare the two groups of students.

e Which average was the least helpful in part **d**? Why?

f Thinking about the two different groups (the first half and the second half), what conclusions can you make?

4 The following table shows the growth in the number of staff at a company over the space of 20 years. (For the purpose of this question, assume that no one ever leaves the company.)

Year	Number employed
1	7
4	25
8	40
12	55
16	60
20	65

a Draw a line graph for this time series.
b Estimate how many staff the company employed after 7 years.
c Estimate in which year the 50th person was employed by the company.
d Describe the trend shown in the graph.

5 This pie chart shows how a local government agency allocates a budget to different areas.

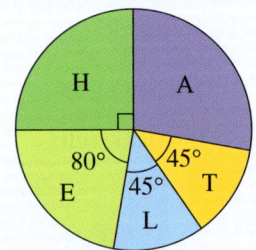

A: Agriculture
E: Education
H: Health
L: Labour
T: Transportation

The agency allocates $30 000 to Education.
Calculate:
a the total budget
b the amount allocated to
i Health
ii Labour
iii Agriculture
c A different region allocates its budget as shown in the pie chart.

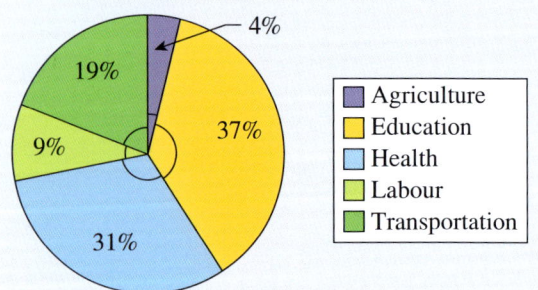

This region spends $6400 of its budget on Agriculture. Copy and complete this table.

Area	Budget allocation ($)
Agriculture	6400
Education	
Health	
Labour	
Transportation	
Total budget	

d Compare the two regions. What conclusions can you make?

e One of the regions is an inner-city region and the other is rural. Which do you think is in the city and why?

Summary

You should know …

1 How to draw and interpret line graphs for time series.

For example:

The number of tourists visiting a resort were recorded during the first 10 days in August. The total visitors to the resort are recorded on this time series graph. For example, there were 78 tourists on 1 August and 72 tourists on 2 August, so there were 150 tourists in total by 2 August.

During which day in August do you think there was a big tourist market?

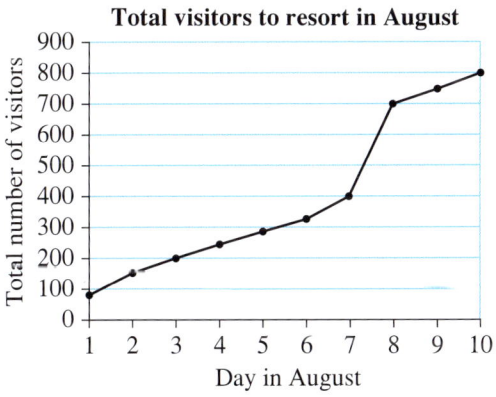

Total visitors to resort in August

There is a big jump on 8 August from the previous day, so 8 August is likely to be the day of the big tourist market.

2 How to draw and interpret stem-and-leaf diagrams.

For example:

Draw a stem-and-leaf diagram for the following prices, in cents, of snacks sold in a school canteen.

43, 99, 55, 74, 62, 24, 83, 29, 65, 89, 33, 34, 35, 43, 38

Check out

1 Ruth bought a car in 2015 for $22 000. The estimated value of the car during the following six years is recorded in this table.

Year	Value ($)
2015	22 000
2016	20 500
2017	17 600
2018	15 500
2019	12 000
2020	8000
2021	4300

Draw a line graph for this time series and comment on what it shows.

2 a Draw a stem-and-leaf diagram for the heights, in centimetres, of these 15 people.

154, 147, 181, 149, 163, 168, 171, 175, 170, 174, 157, 154, 164, 167, 168

b What is the mean height?

c What is the range of heights?

d What is the modal height?

Don't forget to include a key and to write the leaves in numerical order.

```
2 | 4  9
3 | 3  4  5  8
4 | 3  3
5 | 5
6 | 2  5
7 | 4
8 | 3  9
9 | 9
```

Key

2 | 4 means 24 cents

mode = 43
median = 43
range = 75
mean = 54

3 How to compare data.

For example:

The lists below show the number of books read in a year in a random sample of 15 Class 7 students and 15 Class 11 students.

List A

29, 18, 16, 32, 41, 26, 50, 54, 7, 19, 24, 28, 32, 23, 9

List B

17, 42, 53, 46, 19, 33, 44, 52, 47, 35, 32, 43, 28, 18, 31

Assuming students in Class 11 do more studying and read more books, which list do you think belongs to Class 11 students and why?

List A

In order:	7, 9, 16, 18, 19, 23, 24, 26, 28, 29, 32, 32, 41, 50, 54
Median:	26
Mean:	7 + 9 + 16 + 18 + 19 + 23 + 24 + 26 + 28 + 29 + 32 + 32 + 41 + 50 + 54 = 408
	408 ÷ 15 = 27.2
Mode:	32
Range:	54 − 7 = 47

List B

In order:	17, 18, 19, 28, 31, 32, 33, 35, 42, 43, 44, 46, 47, 52, 53
Median:	35
Mean:	17 + 18 + 19 + 28 + 31 + 32 + 33 + 35 + 42 + 43 + 44 + 46 + 47 + 52 + 53 = 540
	540 ÷ 15 = 36

3 These frequency diagrams show the data on the left.

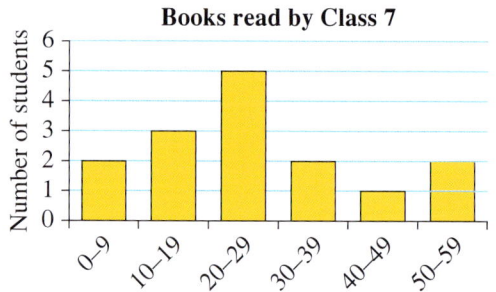

Books read by Class 7

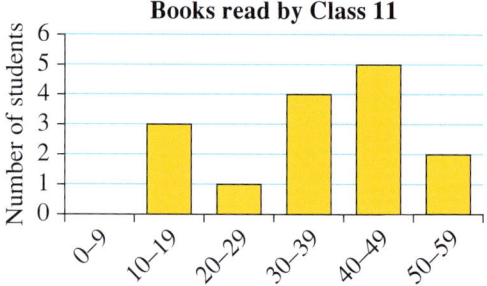

Books read by Class 11

a Why is the mode no good for comparing?

b What does the range tell you?

c Using the frequency diagrams, what is the modal interval for books read for each class?

Mode: none
Range: $53 - 17 = 36$

The best average to compare here is the mean. List B is probably for Class 11, as the mean is higher.

4   How to compare proportions in two pie charts that represent different totals.

For example:
The two pie charts below show the drinks sold in a café on different days.

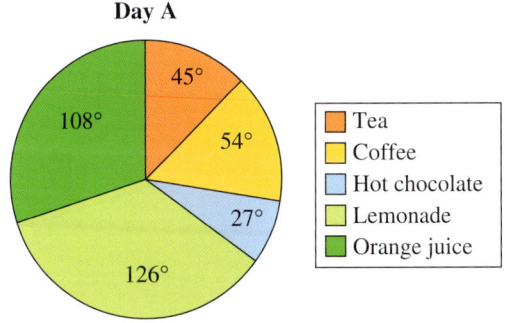

Day A

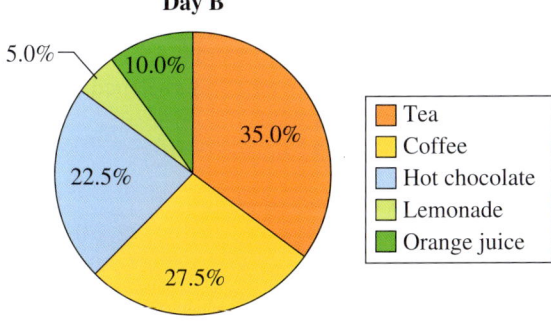

Day B

On Day A, 36 glasses of orange juice were sold.
On Day B, 28 cups of tea were sold.
Find the total number of drinks sold on each day.

Day A
36 glasses of orange juice are represented by 108°.
So one drink is represented by
$108° \div 36 = 3°$
$360° \div 3° = 120$ drinks in total.
Day B
28 cups of tea represents 35%. So one drink is represented by $35\% \div 28 = 1.25\%$
$100\% \div 1.25\% = 80$ drinks in total.

d   Give one advantage of using a stem-and-leaf diagram rather than the diagrams above to represent the data.

4   **a**   Using the pie charts on the left, copy and complete these tables.

Day A

Drink	Number sold
Tea	
Coffee	
Hot chocolate	
Lemonade	
Orange juice	36
Total	120

Day B

Drink	Number sold
Tea	28
Coffee	
Hot chocolate	
Lemonade	
Orange juice	
Total	80

b   One of the days was in the summer and the other was in the winter. Which day do you think was in the summer, A or B? Give a reason for your answer.

1 Find angle c and angle d.

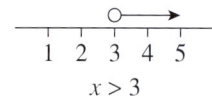

2 Work out:

a $6\frac{1}{3} - 2\frac{1}{6}$ **b** $2\frac{1}{6} - \frac{3}{4}$ **c** $3\frac{1}{2} - 1\frac{3}{5}$

d $4\frac{1}{7} - 1\frac{3}{4}$ **e** $6\frac{1}{9} - 5\frac{3}{4}$ **f** $5\frac{2}{5} - 3\frac{5}{8}$

3 Describe the inequalities represented by these number lines. The first has been done for you.

```
      ○────→
  1  2  3  4  5
      x > 3
```

a
```
  ←●
 ⁻2 ⁻1  0  1
```

b
```
     ○────○
 10 20 30 40 50
```

c
```
   ●────○
 3  4  5  6  7  8  9
```

4 Copy and complete:

a $73 \div \square = 7300$

b $\square \times 0.1 = 0.48$

c $\square \div 10 = 8.9$

d $154 \times \square = 1.54$

e $\square \div 0.01 = 230$

5 Write down a formula to find the total cost T, of x shirts at \$20 per shirt and \$4 delivery charge.

6 A restaurant adds a service charge of 8% onto every bill. What is the total bill if a meal costs

a \$40 **b** \$65 **c** \$9.50?

7 Solve:

a $7x + 5 = 11x + 1$ **b** $5x + 14 = 7x + 4$

c $4x - 3 = 9x - 23$ **d** $2x - 15 = 11x + 12$

e $4x + 3 = 38 - x$ **f** $7x - 5 = 43 - 3x$

g $19 - 3x = 41 - 4x$ **h** $49 - 2x = 7 - 9x$

8 This graph shows the temperature of a cold store after first switching on the cooling unit.

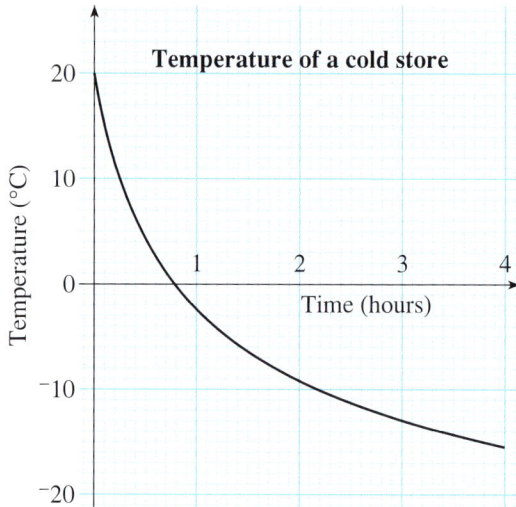

a After how long is the temperature
  i 0°C **ii** ⁻10°C?

b What was the temperature:
  i when the unit was switched on
  ii after 1 hour
  iii after $3\frac{1}{2}$ hours?

9 Without a calculator, work out:

a 350×0.01 **b** 72×0.1

c $68 \div 0.1$ **d** $9 \div 0.01$

e 4100×0.01 **f** 300×0.1

g $0.6 \div 0.1$ **h** $0.25 \div 0.01$

10 Write down all the integers that satisfy the inequality $3 < x \le 7$

11 The perimeter of this isosceles triangle is 39 cm. Construct and solve an equation to find missing side length, t.

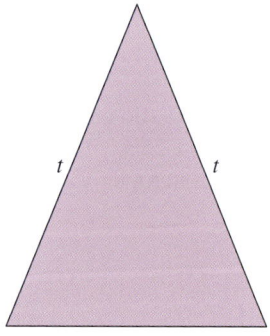

11 cm

12 Work out $\left(3\frac{2}{5}-1\frac{1}{3}\right)\times12$

13 Work out:

   a 2.6×0.5 **b** 4.15×0.4

   c 0.3^2 **d** 3.25×0.09

   e 6.5×0.07 **f** 3.2×0.004

14 Solve these equations.

   a $2x+3=9$ **b** $x+4=2x-2$

   c $3(x+1)=6$ **d** $2x-3=x+6$

15 Calculate the missing angles a and b.

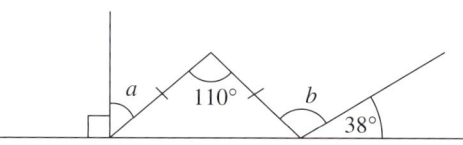

16 What is the selling price of a television made for $1000 and sold at a 25% profit?

17 Using $v=u+at$, find t when:

   a $v=40$, $u=30$ and $a=5$

   b $v=20$, $u=48$ and $a=\,^-7$

18 Copy and complete:

   a $7.35\div0.5=(\square\times10)\div\square$

          $=73.5\div\square=\square$

   b $68.4\div0.4=(\square\times10)\div\square$

          $=\square\div\square=\square$

19 Find the midpoint of the line segment AB where:

   a A $(6,8)$ and B $(12,2)$

   b A $(^-2,3)$ and B $(14,7)$

   c A $(^-5,6)$ and B $(8,^-4)$

20 A water tank holds 100 litres. Mrs Shaw uses $6\frac{3}{4}$ litres. How much water is left in the tank?

21 a Derive the formula for the perimeter, P, of this triangle.

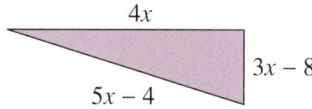

   b Rearrange the formula to make x the subject.

   c Use the rearranged formula to find x when $P=60$.

   d What are the side lengths of the triangle in part **c**?

22 Work out:

   a 0.3×8 **b** $^-6\times2.7$

   c 12×0.04 **d** $0.08\times{}^-9$

   e 0.006×4

23 If $x\geq3$, complete these equivalent inequalities.

   a $x+2\geq\ldots$

   b $3x\geq\ldots$

   c $\ldots-\ldots\geq1$

24 Round:

   a 27.48 to **i** 1 s.f. **ii** 2 s.f.

   b 0.0637 to **i** 1 s.f. **ii** 2 s.f.

   c 46 587 to **i** 1 s.f. **ii** 2 s.f. **iii** 3 s.f.

   d 5999 to **i** 1 s.f. **ii** 2 s.f.

25 Copy and complete this table of equivalent decimals and fractions. Cancel the fractions to their simplest form and, where appropriate, show them as mixed numbers.

Decimal	Fraction
0.8	
	$\frac{1}{4}$
$0.\dot{3}$ (or 0.333...)	
	$\frac{3}{10}$
0.05	
1.06	
	$\frac{11}{100}$
	$2\frac{4}{5}$
	$\frac{7}{9}$
8.2	

26 12 years from now, Ali will be three times older than he was 18 years ago.

   a If Ali is a years old today, write down:

      i   his age 12 years from now

      ii his age 18 years ago

      iii an equation that shows the above information

   b Solve the equation to find Ali's age now.

27 These are the lengths, in centimetres, of a catch of 15 fish.

28, 38, 40, 31, 30, 38, 32, 35
29, 30, 32, 37, 38, 38, 39

a Put the numbers in order and find the mean, median, mode and range of the lengths of the catch.

b A different catch of fish had the data:

$$\begin{aligned}
\text{mean} &= 24.7 \\
\text{median} &= 25 \\
\text{mode} &= 38 \\
\text{range} &= 23
\end{aligned}$$

Compare the two catches.

c Do you think the two catches contain the same species? Why?

28 Complete this calculation.

$17.28 \times 9.9 = 17.28 \times 10 - 17.28 \times \square = \square$

29

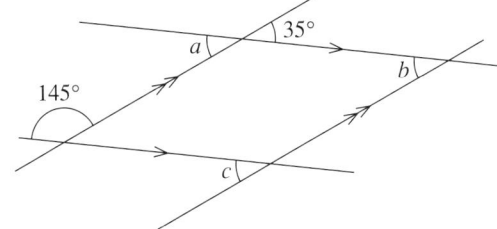

Copy and complete:

a Angle a is . . .° because . . . .
b Angle b is . . .° because . . . .
c Angle c is . . .° because . . . .

30 Achim sells his car at a 30% loss. What is his selling price if he bought the car for $30 000?

31 15 students were asked to try to draw a line 4.5 cm long without measuring.

The lengths of the lines they drew are shown in this stem-and-leaf diagram.

```
2 | 4  9
3 | 1  2  7
4 | 0  1  4  4  5  7
5 | 4  7  9
6 | 1
```

Key

2 | 4 means 2.4 cm

a What was the shortest length drawn?
b What was the longest length drawn?
c What was the
   i mean   **ii** median   **iii** modal length?

32 Work out:

a $4\frac{1}{4} + 1\frac{3}{5} - 2\frac{1}{2}$

b $3\frac{2}{3} + 2\frac{1}{4} - \frac{1}{2}$

c $11\frac{5}{8} - 8\frac{2}{5} + 1\frac{1}{20}$

33 Find angles a and b in these triangles.

a

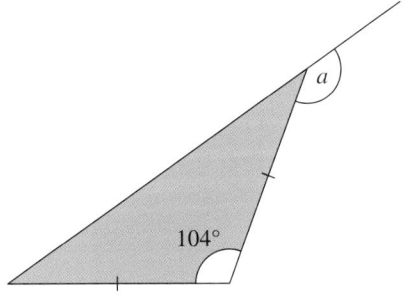

b

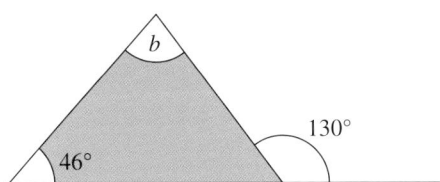

34 Amy has a sweets and Ben has b sweets.
Derive an equation for each of these statements.

a Amy and Ben have a total of 50 sweets.
b Amy has 4 times as many sweets as Ben.
c If Amy gave Ben 10 sweets they would both have the same amount of sweets.

35 Work out:

a $73.4 \div 0.2$
b $1.53 \div 0.3$
c $0.693 \div 0.9$
d $1.32 \div 1.1$

36 Solve these equations.

a $\frac{2}{5}x = 16$

b $\frac{3}{8}x = 15$

c $\frac{6}{11}x = 24$

37 Workers in a glove factory earn $150 a week.
The managers offer them a wage increase of 10%.

a How much would the workers earn if they accepted the managers' offer?
b The workers want an increase of 15%.
How much pay per week would that give them?

38 a Draw the number line to represent the inequality $12 \le x < 17$.
b Write down all the possible integer values of x.

39 The height, in centimetres, of 30 plants was measured after 8 weeks of growth. The results are shown in this table.

Height, h (cm)	Frequency
$5 \le h < 10$	3
$10 \le h < 15$	8
$15 \le h < 20$	11
$20 \le h < 25$	7
$25 \le h < 30$	1

  a Draw a histogram for this data.
  b What is the modal class?

40 Work out $8\frac{2}{15} \times 17 + 21\frac{13}{15} \times 17$

41 Work out $24 \div \left(2\frac{2}{3} - 2\frac{2}{5}\right)$

42 Solve:
  a $6(x + 4) = 5(x + 7)$
  b $3(x - 3) = 7(x - 7)$
  c $4(x - 9) = 2(x + 7)$
  d $10(p - 5) = 5(p + 2)$
  e $7(4x - 10) = 9(2x - 5)$

43 Find the sizes of angles a–c.

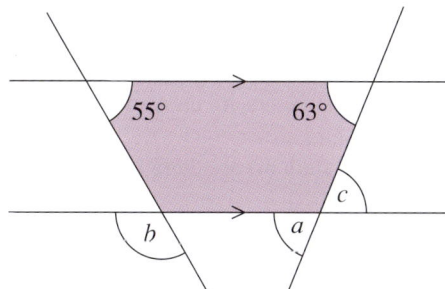

44 The midpoint of a line segment AB is $(4, 7)$. Write down some possible coordinates for A and B

45 Use the associative law to make this calculation easier, then work out the answer.
$3\frac{2}{19} + 2\frac{8}{37} + 2\frac{17}{19}$

46 Work out the new values.
  a Decrease 300 ml by 18%.
  b Increase $340 by 15%.
  c Increase 325 km by 16%.
  d Decrease 4800 cm by 34%.

47 Derive a formula for the perimeter, P, of this rectangle. Use this formula to find y when $P = 96$ and $x = 9$.

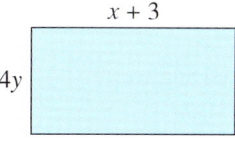

48 In the diagram, find the value of x.

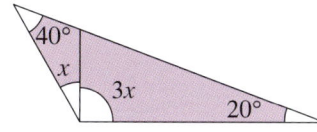

49 In a restaurant, a service charge of 15% is added to the price of the meal. Including the service charge, what will be the bill for a meal costing $140 without the service charge?

50 I started with a number, n. If I subtract 10 from this number and then multiply the result by 7 I get the same number as I do when I add 18 to n and multiply the result by 3. What is n?

51 Work out:
  a $8 \times 1\frac{1}{4}$ **b** $2\frac{2}{3} \times 7$ **c** $6 \times 4\frac{1}{4}$
  d $4 \div \frac{3}{5}$ **e** $5 \div \frac{2}{3}$

52 A cricket ball is thrown up into the air and caught again. The graph shows the height h of the ball at time t.

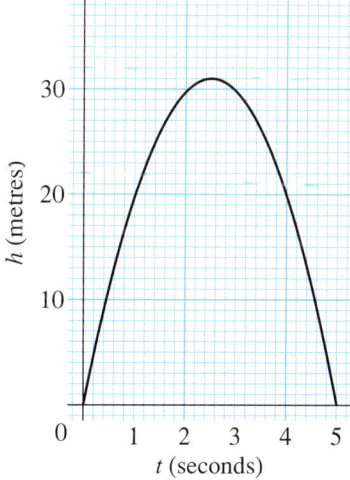

  a How long does the ball take to reach its greatest height?
  b Estimate the height at $t = 3.5\,\text{s}$
  c At what times is $h = 15\,\text{m}$?

53 Solve:
- **a** $5(x - 2) = 3x - 4$
- **b** $2x - 3 = 9 + 4x$
- **c** $15 - 4x = 2(3x + 1)$

54 Find the size of the marked angles.

a

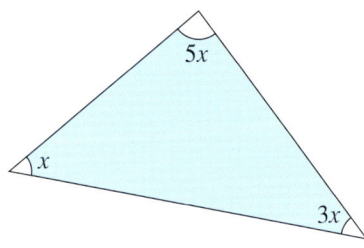

b

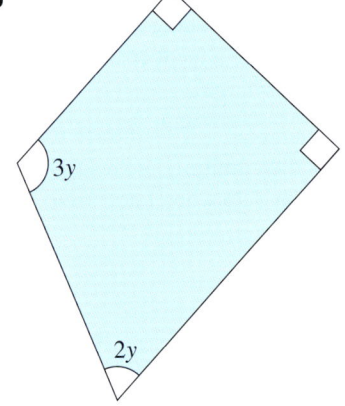

55 Find the area of this rectangle.

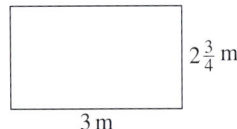

56 Sahil buys n oranges and 3 more mangoes than oranges. Write a formula for the total number, T, pieces of fruit she buys in terms of n in its simplest form.

57 The following pie charts show how two different groups of children travel to school.

How group A travel to school

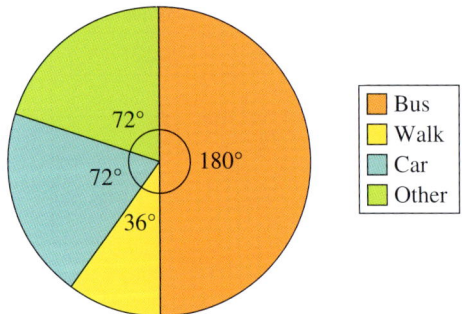

How group B travel to school

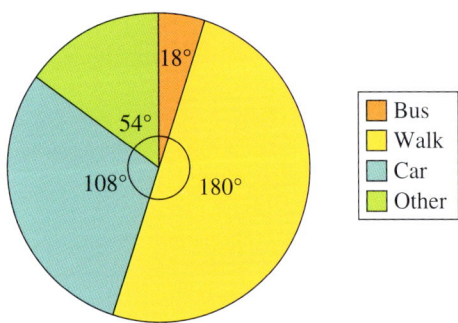

In group A, 15 students get the bus.
In group B, 12 students travel by car.
- **a** How many students are there altogether in group A?
- **b** How many students are there altogether in group B?
- **c** How many students in each group walk to school?
- **d** Write down some comparisons between the two pie charts.
- **e** One group of students are age 15 and the other group are age 9. Which group do you think was made up of 15-year-olds, A or B? Give a reason for your answer.

58 Work out the size of the obtuse angle in this isosceles triangle.

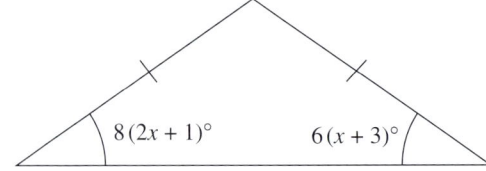

Sequences

Objectives

In this chapter you will learn about:

- term-to-term rules
- *n*th term
- sequences using spatial patterns.

What's the point?

A famous mathematician called Leonardo Fibonacci was well known for writing a book called *Liber Abaci* ('Book of Calculation') early in the thirteenth century. In this book he wrote about a special number sequence now called the Fibonacci sequence. The sequence is 1, 1, 2, 3, 5, 8, . . . , in which the next number is found from adding the previous two. The Fibonacci sequence is commonly seen in nature, for example in the arrangement of leaves on the stem of a plant. In computing, the Fibonacci search technique is faster than a binary search.

Before you start

You should know ...

1 How to work with negative numbers.
For example:

$$2 - 7 = {}^-5$$
$${}^-3 + 5 = 2$$

Check in

1 Work out:
   a $6 - 10$
   b ${}^-3 + 12$
   c ${}^-8 - 5$
   d ${}^-12 + 5$

2 How to find the value of an expression.
For example:

What is the value of $5n - 3$ if $n = 4$?
When $n = 4$,
$$5n - 3 = 5 \times 4 - 3$$
$$= 20 - 3$$
$$= 17$$

3 How to complete simple sequences by looking at the common difference.
For example:

The next two terms of 4, 6, 8, 10, . . . are 12 and 14 (add 2 each time).

2 Find the value of these expressions when $n = 4$.
a $2n + 7$
b $5n - 8$
c $7n - 10$
d $3n + 15$

3 Write down the next two terms in these sequences.
a 15, 20, 25, 30, 35, . . .
b 30, 40, 50, 60, . . .
c 4, 7, 10, 13, 16, . . .

11.1 Rules of sequences

Simple investigations

Mathematical investigations are often open-ended questions or problems, such as these.

- What two numbers sum to 5?
- What are the dimensions of a rectangle with area 15 cm²?
- Which shapes have two lines of symmetry?

There is more than one possible answer to such questions. Further, such questions are often just the starting point of an investigation, as other follow-up questions spring to mind.

Here are some examples.
- What two numbers sum to:
  a 6   **b** 7   **c** 13?
- What are the dimensions of a triangle with area 15 cm²?
- Which shapes have three lines of symmetry? Which have four lines of symmetry?

Exercise 11A

Find at least five solutions to each of these open-ended questions.

1 Which two numbers sum to 8?

2 Which two numbers have a product of 36?

3 You can write the number 7 as the sum of two consecutive whole numbers: $3 + 4 = 7$.

Which other numbers can be expressed as the sum of consecutive whole numbers?

4 Which numbers can be written as the difference between two square numbers?

5 The number 141 reads the same when written backwards as it does when written forwards. It is also a multiple of 3. Which other such numbers are there?

6 Which numbers have exactly four factors?

7 You have an unlimited supply of 3 cm and 5 cm rods. What lengths can you make with them?

8 Which prime numbers can be written as the sum of two squares?

Looking for patterns

In many open-ended questions it is quite easy to see patterns.

Example 1

Which pairs of positive integers sum to 7?

$1 + 6$
$2 + 5$
$3 + 4$
$4 + 3$
$5 + 2$
$6 + 1$

The numbers on the left-hand side increase by one each time while those on the right decrease by one.

Searching and finding patterns is basic to mathematics. When you find a pattern the next step is to see whether you can write down a related rule for the pattern.

Example 2

Write down the next three numbers in this sequence.

6, 10, 14, 18, . . .

....

The difference between 10 and 6 is 4, the difference between 14 and 10 is 4, etc., so the rule is: *add 4*.

The next three numbers will be 22, 26 and 30.

A **term-to-term** rule describes how to get from one term to the next. For the sequence in Example 2 the term-to-term rule is *add 4*. By continuing the pattern we can find terms further along in the sequence. We can find the 8th term by continuing to add 4 until we have 8 numbers:

6, 10, 14, 18, 22, 26, 30, 32

So the 8th term is 32.

A sequence can be found if you know the term-to-term rule and one number in the sequence.

Example 3

The first term of a sequence is 5. The term-to-term rule is: *multiply by 2 then subtract 4*. Work out the first five terms.

....

1st term: 5
2nd term: $5 \times 2 - 4 = 10 - 4 = 6$
3rd term: $6 \times 2 - 4 = 12 - 4 = 8$
4th term: $8 \times 2 - 4 = 16 - 4 = 12$
5th term: $12 \times 2 - 4 = 24 - 4 = 20$

An **arithmetic sequence** or **linear sequence** is one in which the difference between the terms is the same each time. This is called the **common difference**. In Example 2 the difference was 4 each time so 6, 10, 14, 18, . . . is an arithmetic sequence. In Example 3 the sequence is 5, 6, 8, 12, 20, ... Notice that the differences are 1, 2, 4 and 8. While Example 3 is still a sequence it is not an arithmetic sequence, because the common difference is not the same each time.

The common difference doesn't have to be an integer.

In the sequence $3, 5\frac{1}{2}, 8, 10\frac{1}{2}, 13, . . .$ the common difference is $2\frac{1}{2}$.

A **position-to-term** rule describes how to calculate the term from its position in the sequence. This is often more useful than the term-to-term rule, particularly when you want to find terms that are a long way into the sequence (e.g. the 100th term) without having to work out the sequence from the beginning. In the next example you will be using position-to-term rules. You will learn how to work them out later.

Example 4

Find the first five terms of the sequences with the position-to-term rules:

a multiply by 6

b multiply by 3 then add 2

....

a

Position	Sequence
1	$1 \times 6 = 6$
2	$2 \times 6 = 12$
3	$3 \times 6 = 18$
4	$4 \times 6 = 24$
5	$5 \times 6 = 30$

$\times 6$

b

Position	Sequence
1	$1 \times 3 + 2 = 5$
2	$2 \times 3 + 2 = 8$
3	$3 \times 3 + 2 = 11$
4	$4 \times 3 + 2 = 14$
5	$5 \times 3 + 2 = 17$

$\times 3 + 2$

Exercise 11B

Write down the next three numbers in these sequences.

1
  a   3, 8, 13, 18, . . .
  b   2, 3, 5, 8, 12, . . .
  c   1, 2, 4, 8, 16, . . .
  d   600, 300, 150, 75, . . .
  e   12, 26, 40, 54, . . .
  f   89, 86, 83, 80, . . .
  g   243, 81, 27, 9, . . .
  h $3, 4\frac{1}{2}, 6, 7\frac{1}{2}, . . .$
  i $20, 18\frac{3}{4}, 17\frac{1}{2}, 16\frac{1}{4}, . . .$

2 For each of the sequences in question **1**, write down the term-to-term rule.

3 For each of the sequences in question **1**, write down the 10th term.

4 What are the missing numbers in these sequences?
  a 23, 29, 35, □, 47, 53, □, 65, . . .
  b □, 3, 8, 13, 18, □, 28, . . .
  c ⁻2, □, 2, □, □, 8, 10, 12, 14, 16, . . .
  d 5, □, □, ⁻1, ⁻3, ⁻5, ⁻7, □, . . .

5 Write down the first five terms of these sequences.
  a The fourth term is 11, the term-to-term rule is *add 3*
  b The third term is 90, the term-to-term rule is *subtract 10*
  c The second term is 6, the term-to-term rule is *multiply by 2*
  d The fifth term is 1, the term-to-term rule is *divide by 3*
  e The first term is 3, the term-to-term rule is *multiply by 3 then subtract 5*
  f The first term is 4, the term-to-term rule is *multiply by 5 then add 2*
  g The first term is 116, the term-to-term rule is *divide by 2 then add 2*

6 For each sequence in question **5** say whether or not it is an arithmetic sequence.

7 The term-to-term rule of a sequence is *add 3, then double it.*
  Explain why you cannot find the first five terms of this arithmetic sequence.

8 Write down the first five terms of these sequences.
  a The position-to-term rule is *subtract 2*
  b The position-to-term rule is *add 7*
  c The position-to-term rule is *multiply by 3 then add 4*
  d The position-to-term rule is *multiply by 10 then subtract 1*
  e The position-to-term rule is *multiply by 5 then subtract 3*
  f The position-to-term rule is *multiply by 4 then add 2*

9 You should have found that the sequences in question **8** are all arithmetic sequences. For each sequence write down the term-to-term rule. Can you see any connection between the term-to-term rule and the position-to-term rule?

10 Here is a page from a calendar.

August						
Mon	**Tue**	**Wed**	**Thu**	**Fri**	**Sat**	**Sun**
			1	2	3	4
5	6	7	8	9	10	11
12	13	14	15	16	17	18
19	20	21	22	23	24	25
26	27	28	29	30	31	

Write down as many linear sequences as you can find. Do they have anything in common?

11 a Write down three different arithmetic sequences with a third term of 10.
  b Write down three different arithmetic sequences with a fifth term of 100.
  c Try to find position-to-term rules for the sequences you have made in parts **a** and **b**.

12 The 10th, 12th and 15th terms of an arithmetic sequence are 35, 43 and 55. Write down the first five terms of this sequence and the position-to-term rule.

11.2 The *n*th term

When we use algebra to describe terms using a position-to-term rule, this is known as finding the **nth term**.

If you wish to predict the 10th or the 52nd number in a sequence you will need to write the rule algebraically. This can be a more challenging task.

You need to identify the position of each term of the sequence.

Example 5

What is the 75th term in the sequence
    $^-2, ^-4, ^-6, ^-8, \ldots$?
What is the nth term?

Position	Term
1	$^-2$
2	$^-4$
3	$^-6$
4	$^-8$

This table clearly shows that each term is double the position and marked negative, so the 75th term will be

$$75 \times {}^-2 = {}^-150$$

The position-to-term rule is $\times\ ^-2$. Using algebra, the nth term will be ^-2n.

Finding the nth term or the algebraic formula for the sequence is not always as easy.

Example 6

What is the nth term in the sequence
    $2, 5, 8, 11, \ldots$?

Look at the differences:
    $+3 \quad +3 \quad +3$
    $2 \quad 5 \quad 8 \quad 11$

The common difference is 3 so the sequence will be related to the 3-times table.

Put the terms in a table.

Position	Term
1	2
2	5
3	8
4	11

Compare the sequence with the 3-times table:

$$2 \quad 5 \quad 8 \quad 11$$
$$3 \quad 6 \quad 9 \quad 12$$

You can see that each term in the sequence is just 1 less than the 3-times table so the position-to-term rule is $\times 3$ then -1

Using algebra, the nth term is $3n - 1$.

There is another way to do Example 6 that doesn't involve comparing with the 3-times table. This is the 'position zero' method. Put an extra row at the start of your table with position 0 in it. Continue the sequence backwards by subtracting 3: you get $^-1$ for position zero. The term in position zero is what you need to add or subtract to the $3n$ part.

Position	Term
0	$^-1$
1	2
2	5
3	8
4	11

-3
$+3$

nth term is $3n - 1$

Example 7

What is the nth term in this sequence?
$28, 24, 20, 16, 12, \ldots$

Look at the differences:

$^-4 \quad\quad ^-4 \quad\quad ^-4 \quad\quad ^-4$
$28, \quad 24, \quad 20, \quad 16, \quad 12$

> This is what is written in front of the n.

Notice you subtract 4 each time.

> Work backwards by adding 4 to the first term to get to the term in position zero. This is written after the ^-4n.

Position	Term
0	32
1	28
2	24
3	20
4	16
5	12

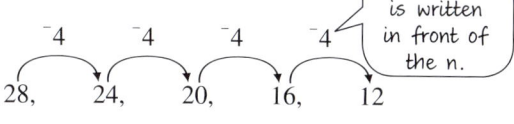

$+4$
-4

nth term is $^-4n + 32$

This can also be written with the positive number first. So the nth term is $32 - 4n$.

If you know the nth term, you can find any term in the sequence using substitution.

Example 8

Find the 20th term of the sequence with the nth term $42 - \frac{1}{2}n$

Substitute $n = 20$ into the nth term $42 - \frac{1}{2}n$

$42 - \frac{1}{2} \times 20 = 42 - 10 = 32$

You can use the nth term rule to decide whether a number is part of a sequence.

Example 9

The nth term of a sequence is $4n - 3$. Are the numbers 85 and 195 in this sequence?

You need to write the equation $4n - 3 = 85$ and solve this equation to see what term number this is.

$4n - 3 = 85$ $[+3]$

$4n = 88$ $[\div 4]$

$n = 22$

So 85 is in the sequence, and it is the 22nd term.

Now do the same with 195.

$4n - 3 = 195$ $[+3]$

$n = 198$ $[\div 4]$

$n = 49.5$

This time n is not a whole number, so 195 is not in the sequence.

Exercise 11C

1 Find the nth term and hence the 43rd term in each of these arithmetic sequences.

    a $3, 5, 7, 9, \ldots$

    b $207, 205, 203, 201, \ldots$

    c $7, 10, 13, 16, \ldots$

    d $3, 8, 13, 18, \ldots$

    e $34, 31, 28, 25, \ldots$

    f $5, 9, 13, 17, \ldots$

    g $511, 498, 485, 472, \ldots$

    h $8, 9\frac{1}{2}, 11, 12\frac{1}{2}, 14, \ldots$

    i $7.5, 6.25, 5, 3.75, 2.5, 1.25, \ldots$

2 Find the nth term for the multiples of 19.

3 Write down some arithmetic sequences of your own and find the nth term for each one.

4 Write down the first five terms of the sequence with nth term $7n + 3$.

5 Match the cards. The first one is done for you.

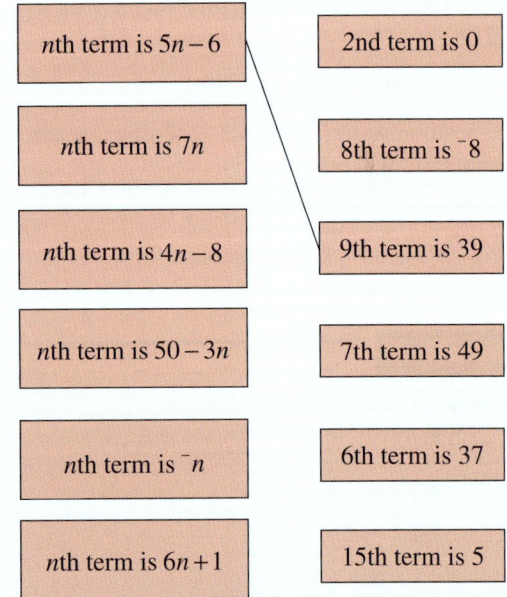

6 Which of these sequences are linear?

    a $7, 8.5, 10, 11.5, 13, \ldots$

    b $^-2, ^-10, ^-18, ^-26, ^-34, \ldots$

    c $1, 2, 4, 8, 16, 32, \ldots$

    d $\frac{7}{2}, 5, \frac{13}{2}, 8, \frac{19}{2}, 11, \ldots$

7 Here is a linear sequence.
    $1, 8, 15, 22, 29, \ldots$
    Bilal says, 'The 5th term of this sequence is 29, so the 50th term will be 10 times 29, which is 290.'
    Explain why Bilal is wrong.

8 A sequence has nth term $3.1 + 2.5n$.
    a   Find the first six terms in the sequence.
    b   What do you notice about the numbers in the sequence?
    c   Is 483.2 in the sequence? Explain how you know.

9 How many linear sequences have the first two terms 40, 45?

10 Sadie says, 'None of the terms in the sequence with nth term $5n-1$ will end in 5.' Is she correct? Explain how you know.

11 Here are the first five terms of two sequences. Only one sequence is linear.

   a   38, 48, 58, 68, 78, . . .
   b   0.5, 1, 2, 4, 8, . . .

Which sequence will reach 128 in the fewest terms? Show your working.

12 Find the nth term and hence the 43rd term in each of these sequences.

   a 1, 4, 9, 16, . . .
   b 3, 6, 11, 18, . . .

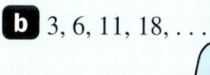

> Note that the nth terms in these sequences are not linear. This is extension work.

Many sequences are formed naturally or grow from simple patterns. In each case the mathematician's task is to find the underlying structure. This structure is usually expressed as an algebraic relationship or a formula. The relationship or formula can then be used to predict terms.

Example 10

Look at the sequence generated by these shapes.

a How many sticks are needed for the 4th and 5th shapes?
b What is the rule for generating these shapes? Use it to find the number of sticks required for the 76th shape.

....

a The 4th shape is:
It has 9 sticks.

a The 5th shape is:
It has 11 sticks.

---

b Put the terms in a table.

Shape	Number of sticks
1	3
2	5
3	7
4	9
5	11

The sequence is:

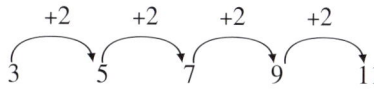

The common difference is 2 so the terms are related to the 2-times table:

2	4	6	8	10
3	5	7	9	11

In fact the terms in the sequence are 1 more than the 2-times table.

The nth shape will have $2n+1$ sticks.

Hence, the 76th shape will have $2\times76+1=153$ sticks.

It is helpful to be able to understand and justify where the nth term comes from using the picture.

You will see from the diagram that there is one blue stick to start the pattern. This is the +1 part of the formula. Then when each pattern is formed two extra sticks are added – shown in red, green, and black – which is where the $2n$ comes from.

Exercise 11D

1 Look at the sequences generated by these shapes.

a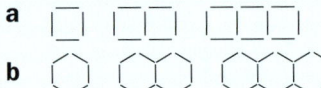

b

In each case find the number of sticks required for:

i the 4th and 5th shapes

ii the nth shape

iii the 58th shape

iv Explain where the nth term comes from using the picture.

2 This table shows a sequence of shapes made from equilateral triangles with sides of one unit.

Shape	Area of shape (in triangles)	Perimeter of shape
△	1	3
◹	2	4
◹◹	3	5

a Draw the next three shapes that continue the sequence.

b Copy and continue the table for the next three shapes, indicating their area and perimeter.

c A shape has area 9 triangles. What is its perimeter?

d If the perimeter of a shape is 61 triangles what is its area?

e What would be the perimeter if the area was n triangles?

f Explain your answer to part **c** using the picture to justify where the nth term comes from.

g Write a formula for the perimeter, p, of the shape when the area is n.

3 This table shows a sequence of octagons made from line segments.

Shape	Number of line segments
⬡	8
⬡⬡	15
⬡⬡⬡	22

a Draw the next two shapes that continue the sequence.

b Continue the table for the next two shapes.

c How many line segments are needed to make the 8th shape in this sequence?

d Write down a formula to show the number of line segments, ℓ, needed to make the nth shape in the sequence.

e Use your formula to find the position in the sequence of the shape with 351 line segments.

4 Look at the tile pattern shown in this table.

Tile pattern	Number of tiles
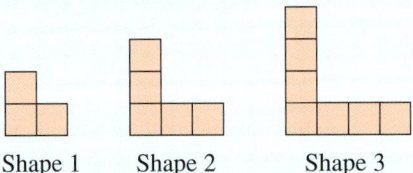	4
	7
	10

a Copy and continue the table for the next three rows.

b How many tiles will be in the 10th tile pattern?

c How many tiles are needed for the nth tile pattern?

d In which position in the sequence is the tile pattern that has 334 tiles?

e Explain where the nth term comes from using the picture.

5 Look at this tile pattern.

Shape 1   Shape 2   Shape 3

a Draw the next two shapes.

b How many tiles will be needed for the 10th shape?

c How many tiles will be needed for the nth shape?

d Explain where the nth term comes from using the picture.

e Which shape contains 121 tiles?

6 Look at the pattern of blue circles in these diagrams.

Diagram 1   Diagram 2   Diagram 3

a Write down the *n*th term for the number of blue circles in each diagram.
b How many blue circles will there be in the 100th pattern?
c If a diagram in the sequence contains 82 blue circles, which diagram is it?

7 Look at the pattern of blue circles in these diagrams.

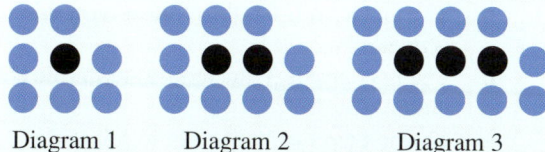

Diagram 1   Diagram 2   Diagram 3

a Write down the *n*th term for the number of blue circles in each diagram.
b How many blue circles will there be in the 50th pattern?
c If a diagram in the sequence contains 65 blue circles, which diagram is it?

8 Draw some patterns of your own with blue and black counters. (Make sure there is one black counter in the first diagram, two in the second diagram and so on.) Find the *n*th term for the number of blue counters in your diagrams.

9 This table shows a sequence of shapes made from dots.

Shape	Number of dots
•	1
••	3
•••	6

a Draw the next two shapes that continue the sequence.
b Continue the table for the next two shapes.
c How many dots has the 10th shape in the sequence?
d A shape is made up of 78 dots. Where does this shape come in the sequence?
e Copy and complete:

Number of dots in *n*th shape is
$$= \frac{1}{2}n \times (\ldots)$$

Consolidation

Example 1

Find the first five terms of these sequences.
a The first term is 5, the term-to-term rule is *add 3*.
b The position-to-term rule is *multiply by 2 then add 5*.

a

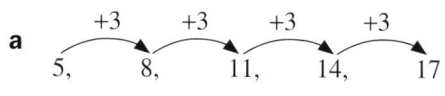

b

Position	1	2	3	4	5
Sequence	1×2 $+5 =$ 7	2×2 $+5 =$ 9	3×2 $+5 =$ 11	4×2 $+5 =$ 13	5×2 $+5 =$ 15

$\times 2 + 5$

So the first five terms of the sequence are
7, 9, 11, 13, 15

Example 2

a What is the sequence of the areas made by the shapes below?
b What is the *n*th term of the areas made by the shapes below?
c Justify the *n*th term by relating it back to the diagram.

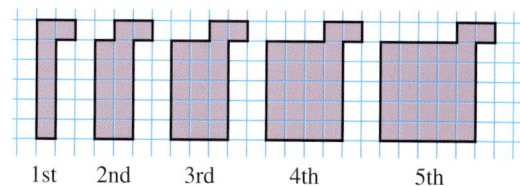

1st   2nd   3rd   4th   5th

a The sequence in the areas of the shapes is
7, 12, 17, 22, 27, . . .

b To find the *n*th term look at the differences:

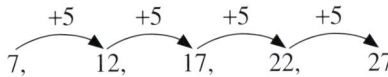

+5   +5   +5   +5
7,   12,   17,   22,   27

The difference is *add 5* each time so it is a linear relationship to do with multiples of 5.
Compare the terms of the sequence with the sequence 5*n*:

7,   12,   17,   22,   27   Sequence
5,   10,   15,   20,   25   5*n* (or multiples of 5)

The terms in the sequence are each 2 more than the corresponding multiple of 5. The *n*th term is therefore $5n + 2$.

c Each time you change diagrams you have 5 extra squares added. This is where the 5*n* comes from. The $+2$ comes from the extra two squares at the top of the final column of squares.

Exercise 11

1 Write down the next three terms in each arithmetic sequence.
  a 7, 15, 23, 31, …
  b 104, 93, 82, 71, …
  c 10, 20, 30, 40, …
  d 5, 11, 17, 23, …
  e 17, 13, 9, 5, …

2 Find the *n*th term of each of the sequences in question **1**.

3 Look at this tile pattern.

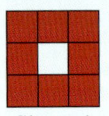  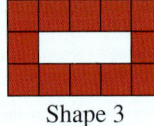

  Shape 1   Shape 2   Shape 3

a How many red tiles are needed for the 5th shape?
b How many red tiles are needed for the 15th shape?
c How many red tiles are required for the *n*th shape?
d If the *m*th shape has 100 red tiles, what is the value of *m*?
e Justify your *n*th term by using the diagrams to explain where it comes from.

4 Write down the first five terms of these sequences.
  a The first term is 6, the term-to-term rule is *add 4*.
  b The first term is 72, the term-to-term rule is *subtract 10*.
  c The first term is 64, the term-to-term rule is *divide by 2*.
  d The first term is 3, the term-to-term rule is *multiply by 3*.
  e The first term is 3, the term-to-term rule is *multiply by 4 then add 2*.

5 For each of the sequences in question **4**, say whether it is an arithmetic sequence or not.

6 Write down the first five terms of these sequences.
  a The position-to-term rule is *subtract 5*.
  b The position-to-term rule is *add 6*.
  c The position-to-term rule is *multiply by 4 then add 7*.
  d The position-to-term rule is *multiply by 8 then subtract 3*.

Summary

You should know ...

1 How to find terms in a sequence.
For example:
a Find the next four terms of the sequence with first term 81 and term-to-term rule *subtract 4*.
b Find the first five terms of the sequence with position-to-term rule *multiply by 10 then add 5*.

a

So the next four terms are 77, 73, 69, 65

b

Position	1	2	3	4	5
Sequence	1×10 $+5$ $=15$	2×10 $+5$ $=25$	3×10 $+5$ $=35$	4×10 $+5$ $=45$	5×10 $+5$ $=55$

$\times 10 + 5$

So the first five terms are 15, 25, 35, 45, 55

2 How to find the rule for a sequence.
For example:
6, 13, 20, 27, . . .

$$6 \xrightarrow{+7} 13 \xrightarrow{+7} 20 \xrightarrow{+7} 27$$

7 is added to the previous term.
The sequence is related to the 7-times table:
7   14   21   27
Each term is one less than the 7-times table,
so the *n*th term is $7n - 1$

Check out

1 a Find the next three terms in these sequences.
  i 17, 22, 27, 32, . . .
  ii 3, 6, 11, 18, . . .
b Find the first five terms of the sequence with first term 7 and term-to-term rule *add 11*.
c Find the first five terms of the sequence with position-to-term rule *multiply by 6 then subtract 10*.

2 For the following sequences, find the:
  i term-to-term rule
  ii *n*th term
a 3, 5, 7, 9, . . .
b 19, 22, 25, 28, . . .
c 88, 81, 74, 67, . . .

12 Ratio and proportion

Objectives

In this chapter you will learn about:

- using different units of measurements in ratios
- simplifying ratios
- understanding the relationship between ratio and proportion
- using equivalent ratios
- sharing in a given ratio.

What's the point?

People use ratio and proportion every day when they are cooking. For example, to bake a cake you might need 2 eggs and 125 g each of sugar, flour and butter. To make a cake twice as big all the ingredients need to be doubled so that they stay in the same proportion and the recipe will still work.

Before you start

You should know ...

The metric abbreviations:

mm – millimetres
cm – centimetres
m – metres
km – kilometres
g – grams
kg – kilograms
t – tonnes
ml – millilitres
ℓ – litres

Check in

From the list on the left, which are measurements of:

a length
b mass
c capacity?

12.1 Units of measurement

You need to be able to recall relationships between units of measurement.

- Measurements of length:
  10 mm = 1 cm
  100 cm = 1 m
  1000 m = 1 km

- Measurements of mass:
  1000 g = 1 kg
  1000 kg = 1 t

- Measurements of capacity:
  1000 ml = 1 ℓ

- Measurements of time:
  60 seconds in a minute
  60 minutes in an hour
  24 hours in a day
  7 days in a week
  365 days in a year (366 in a leap year)

- Measurements of area:
  $100 \text{ mm}^2 = 1 \text{ cm}^2$
  $10\,000 \text{ cm}^2 = 1 \text{ m}^2$
  $1\,000\,000 \text{ m}^2 = 1 \text{ km}^2$

- Measurements of volume:
  $1000 \text{ mm}^3 = 1 \text{ cm}^3$
  $1\,000\,000 \text{ cm}^3 = 1 \text{ m}^3$

Exercise 12A

1 Copy and complete:
- **a** 0.31 km = ☐ m
- **b** 48 hours = ☐ days
- **c** 68 000 ml = ☐ ℓ
- **d** 4300 m = ☐ km
- **e** 0.7 t = ☐ kg
- **f** 300 minutes = ☐ hours
- **g** 500 mm = ☐ cm
- **h** 16 ℓ = ☐ ml
- **i** 0.8 kg = ☐ g

2 Copy and complete this crossnumber puzzle using the clues given.

Across

1 Akanni has a mass of 56 000 g. What is his mass in kg?

2 A piece of metal 5.1 m long is to be divided into three equal pieces. How many centimetres is each equal part?

3 4 tins of tomatoes have a total mass of 1.5 kg. What is the mass, in grams, of 1 tin?

Down

1 A piece of material is 6 m long. A 33 cm length is cut from this material. How many centimetres of material are left?

2 Anna is 1.37 m tall. What is her height in centimetres?

3 8 bottles contain a total of 2.6 ℓ of water. How many millilitres of water are there each bottle?

3 Copy and complete:
- **a** $3 \text{ m}^2 = ☐ \text{ cm}^2$
- **b** $1.7 \text{ km}^2 = ☐ \text{ m}^2$
- **c** $2.6 \text{ m}^3 = ☐ \text{ cm}^3$
- **d** $990\,000 \text{ cm}^2 = ☐ \text{ m}^2$
- **e** $6\,000\,000 \text{ mm}^2 = ☐ \text{ m}^2$

12.2 Simplifying ratio

A ratio compares the size of two quantities.

In this diagram the ratio of triangles to squares is 3 : 5.

The ratio of squares to triangles is 5 : 3.

We can also write these fractions.

$\frac{3}{8}$ of the shapes are triangles.

$\frac{5}{8}$ of the shapes are squares.

> Where 8 is the total ratio.

Ratios are **equivalent ratios** if they are in the same proportion.

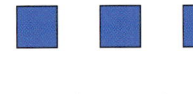

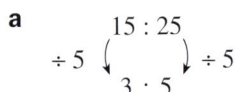

The ratio of triangles to squares in this diagram is 4 : 6. This means that for every 4 triangles there are 6 squares. However, you can see from the red and blue shapes that for every 2 triangles there are 3 squares, so the ratio 4 : 6 can be simplified to 2 : 3.

To write a ratio in its simplest form you divide by the HCF (highest common factor) of the numbers in the ratio.

Example 1

Simplify: **a** 15 : 25   **b** 4 : 8 : 12   **c** 1.2 : 2.4 : 6

a
$$15 : 25$$
$\div 5 \Big(\quad \Big) \div 5$
$$3 : 5$$

b
$$4 : 8 : 12$$
$\div 4 \Big(\quad \Big) \div 4 \Big) \div 4$
$$1 : 2 : 3$$

c
$$1.2 : 2.4 : 6$$
$\Big(\times 10 \Big) \times 10 \Big) \times 10$
$$12 : 24 : 60$$
$\Big(\div 12 \Big) \div 12 \Big) \div 12$
$$1 : 2 : 5$$

> A ratio in its simplest form contains no decimals or fractions.

Exercise 12B

1 Write each ratio in its simplest form.
  a 30 : 40   **b** 35 : 25
  c 10 : 40 : 25 **d** 18 : 36
  e 52 : 13   **f** 24 : 84 : 108
  g 17 : 51   **h** 140 : 200 : 80

2 In what ratio are the side lengths of these triangles? Write the smallest number first and largest last. Don't forget to simplify the ratio.

a
42   49
21

b
10
12

c
7

3 Which pairs of ratios are equivalent?
18 : 27, 2 : 3
22 : 77, 200 : 700
8 : 16, 2 : 3
34 : 51, 4 : 6
95 : 38, 15 : 6
32 : 56, 7 : 4
24 : 12 : 18, 480 : 240 : 150
2.1 : 4.2 : 7, 12 : 24 : 40

4 Construct two different triangles with sides in the ratio 3 : 4 : 5. What sort of triangles are they?

5 Prashant knows that you cannot have decimals in a ratio in its simplest form. Here is his method to write 0.4 : 0.8 : 6 in its simplest form.

$$0.4 : 0.8 : 6$$
$\times 10 \Big(\times 10 \Big(\quad \Big) \times 10$
$$4 : 8 : 60$$

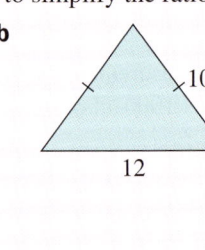

> Multiply by 10 to clear the decimal.

Explain the mistake Prashant has made.

6 Write each of these as a ratio in its simplest whole number form.
   a 0.5 : 7
   b 4.9 : 1.4
   c 5 : 6.5 : 3.5
   d 14 : 8.2
   e 7.5 : 10 : 2.5

7 Write each of these as a ratio in its simplest whole number form.
   a $\frac{1}{4} : 5$ **b** $2.2 : \frac{1}{5}$
   c $36\% : 0.6 : \frac{3}{25}$ **d** $\frac{2}{5} : \frac{3}{4} : \frac{2}{3}$
   e $1\frac{1}{2} : 2\frac{1}{4} : \frac{3}{4}$

Using the same units

• When writing a ratio, both quantities must be in the **same units**.

Example 2

The length of line P is 2 cm: _____ P
The length of line Q is 13 mm: _____ Q
Write down the ratio that compares the length of line P to the length of line Q.

....

The ratio comparing the length of P to the length of Q is:

$$2\,\text{cm} : 13\,\text{mm}$$
$$or \ \ 20\,\text{mm} : 13\,\text{mn}$$
$$or \ \textbf{20} : \textbf{13}$$

The ratio itself has **no units**.

Example 3

It takes James 2 hours 14 minutes to walk from Lake Pupuke to Auckland Zoo.
It takes him 50 minutes to cycle there.
What is the ratio between his walking time and his cycling time?

....

The ratio of walking time to cycling time is:

$$2 \text{ hours } 14 \text{ minutes} : 50 \text{ minutes}$$
$$or \ 134 \text{ minutes} : 50 \text{ minutes}$$
$$or \ \textbf{134} : \textbf{50}$$
$$or \ \textbf{67} : \textbf{25}$$

Exercise 12C

1 a Which bag contains the most sugar?
   b Is it correct to use the ratio 100 : 5 to compare the two quantities of sugar? Why?
   c Express 5 kilograms in grams.
   d Now write a ratio to compare the quantities of sugar in (X) and (Y).

2 Write a ratio to compare the mass of the first object with the mass of the second.
   a

   b

3 Use a ratio to compare these quantities.
   a 1 m 10 cm; 57 cm   **b** 100 mm; 1 cm
   c 1.3 cm; 18 mm   **d** 100 mm; 1 m
   e 1.2 kg; 311 g   **f** 5 min; 40 s
   g 350 ml; 1.1 litres   **h** 5; three dozen
   (**Hint:** a 'dozen' means 12.)

4 Use a ratio to compare these quantities.
   a 1 hour; 13 min   **b** 1 week; 4 days
   c 0.8 cm; 15 mm   **d** 980 kg; 1 t
   e 1.4 t; 700 kg

5 This was Farhan's homework on simplifying ratios.

	Question	Working	Answer
a	3 g : 3 kg	divide by 3	1 : 1
b	17 cm : 85 cm	divide by 17	5 : 1
c	10 m : 70 cm	divide by 10	1 : 7
d	150 ml : 50 ml	divide by 10	15 : 5

Farhan has made some mistakes. Mark and correct his homework.

6 Use a ratio to compare these quantities.
   a $5\,\text{cm}^2, 250\,\text{mm}^2$
   b $0.4\,\text{m}^2, 16\,000\,\text{cm}^2$
   c $1.6\,\text{cm}^3, 20\,000\,\text{mm}^3$
   d $0.25\,\text{m}^3, 75\,000\,\text{cm}^3$

12.3 Sharing in a given ratio

You can divide a quantity into amounts using ratio to decide how much is in each amount.

Example 4

Share 50 pens among Jo, Minny and Sam in the ratio 1 : 2 : 7.

....

$1 + 2 + 7 = 10$ total parts

There are 50 pens. So there are 5 pens in each part.

Jo gets **5 pens**
Minny gets $5 \times 2 = $ **10 pens**
Sam gets $5 \times 7 = $ **35 pens**

Exercise 12D

1 A cake is shared between Annabel, Ria and Faith in the ratio 1 : 2 : 3.

a If Annabel gets 1 part, how many parts should Ria get? How many parts should Faith get?

b Into how many equal slices should you cut the cake?

2 Share the following among A, B and C.
  a 60 nuts, in the ratio 4 : 1 : 7
  b 100 ml, in the ratio 2 : 5 : 3
  c 75 m, in the ratio 8 : 2 : 5
  d 1.8 m, in the ratio 1 : 7 : 1
  e $25, in the ratio 1 : 14 : 5
  f $54, in the ratio 8 : 5 : 14
  g $100, in the ratio 2 : 4 : 19
  h 1000 kg, in the ratio 117 : 62 : 21

3 **a** Draw a line 12 cm long. Divide it in the ratio 2 : 1 : 3.
  b Draw a line 72 mm long. Divide it in the ratio 2 : 4 : 3.

4 Share:
  a 0.9 kg, in the ratio 3 : 17 : 10
  b $24.50, in the ratio 1 : 1 : 12
  c 1.5 kg, in the ratio 13 : 2 : 5

5 The angles A, B and C in this triangle are in the ratio 3 : 5 : 7. Work out the size of each angle.

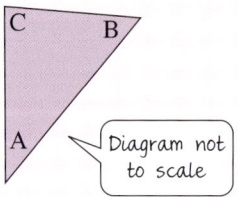

Diagram not to scale

6 The angles of a quadrilateral are in the ratio 1 : 6 : 3 : 5. What are the angles?

7 Amy has two numbers.
What are the two numbers if the ratio of their difference : their sum : their product is:
  a 1 : 5 : 18
  b 1 : 3 : 10

12.4 Proportion

You need to be able to express a ratio as a proportion.

When Wasim and Virat share $400 in the ratio 7 : 3, the proportion of the money that Wasim gets is $\frac{7}{7+3} = \frac{7}{10}$ and the proportion of the money that Virat gets is $\frac{3}{7+3} = \frac{3}{10}$.

$\frac{7}{10} \times 400 = \$280$ and $\frac{3}{10} \times 400 = \$120$

Example 5

The ratio of blue beads to red beads to green beads is 3 : 15 : 2.
What proportion of the beads are:
a blue   **b** red   **c** green?
Give your answers as percentages.

....

a $\frac{3}{3 + 15 + 2} = \frac{3}{20} = \frac{15}{100} = 15\%$

b $\frac{15}{3 + 15 + 2} = \frac{15}{20} = \frac{3}{4} = 75\%$

c $100\% - 15\% - 75\% = 10\%$

Exercise 12E

1 Josie and Max share 60 sweets in the ratio 2 : 3.
Josie says, 'I get $\frac{2}{3}$ of the sweets.'

Is Josie correct? Explain your answer.

2 The ratio of $10 bills to $5 bills in a purse is 13 : 27.

a What fraction of the bills are $10?

b What percentage of the bills are $5?

c Sam thinks there must be a total of 40 bills in the purse. Is he correct? Explain your answer.

3 There are some red and blue pens. $\frac{3}{10}$ of the pens are red. Find the ratio of red pens to blue pens.

4 A recipe requires flour, margarine and butter to be mixed in the ratio 4 : 2 : 1 by mass. What proportion of these ingredients is flour?

5 Ida has yellow counters and green counters in the ratio 11 : 31. Elin has yellow counters and green counters in the ratio 24 : 73.

a Who has the greatest proportion of yellow counters? Justify your answer.

b Is this easier to see when you express the proportion as a fraction or as a percentage?

6 Pink paint is made from mixing red paint and white paint. The higher the proportion of red paint, the darker the paint.
Paint A has a ratio of red paint to white paint of 11 : 7.
Paint B has ratio of red paint to white paint of 16 : 9.
Show that paint B is darker.

7 There are two different types of boxes, box A and box B. All boxes of type A have the same mass as each other and boxes of type B have the same mass as each other.
16 boxes of type A have a mass of 2 kg.
5 boxes of type B have a mass of 620 g.
Sho thinks box B has a greater mass than box A.
Is he correct? Explain your answer.

8 There are two different types of drinking glasses – clear glasses and coloured glasses.
3 clear glasses hold 1305 ml of water.
8 coloured glasses hold 3.4 litres of water.

a Which type of glass holds the most water? Justify your answer.

b What assumption have you made answering this question?

9 Triangle A has angles in the ratio 11 : 4 : 9.
Triangle B has angles in the ratio 4 : 3 : 1.

Which triangle contains the largest angle? Justify your answer.

10 The ratio of black balls to red balls to green balls in bag A is 2 : 17 : 1.

The ratio of black balls to red balls to green balls in bag B is 3 : 15 : 2.

Here is Omar's working to find which bag has the greater proportion of black balls.

Bag A: $\frac{2}{2 + 17 + 1} = \frac{2}{20} = 10\%$

Bag B: $\frac{3}{3 + 15 + 2} = \frac{3}{20} = 15\%$

So bag B has more black balls.

What mistake has Omar made?

(**Hint:** Look closely at Omar's conclusion.)

11 Pablo guesses the lengths of some lines. The results are shown in the table.

Line	Real length	Guess
A	5.4 cm	5.2 cm
B	6 cm	6.3 cm
C	3.5 cm	3.1 cm
D	8.2 cm	8.6 cm

For which line was the guess:

a the most accurate

b the least accurate?

Explain your answer.

12 The cakes in a full box have a total mass of
2.1 kg. Each cake has the same mass.
25 of the cakes are sold.
The cakes left in the box have a total
mass of 350 g.
How many cakes were in the original box?

Investigation

Take a two-digit number. Reverse the digits to make
a new number. Which numbers, when its digits are
reversed, will be one and three quarters times as
big as the original number?

Is it possible for the reversed number to be one
and a half times as big as the original number?

Consolidation

Example 1

Write as a ratio:

a 2 m 15 cm : 85 cm   **b** 2 hours : 50 minutes

....

a 2 m 15 cm : 85 cm
   $= 215$ cm : 85 cm
   $= 215 : 85$

b 2 hours : 50 minutes
   $= 120$ minutes : 50 minutes
   $= 120 : 50$

Example 2

Write 5 : 25 : 45 as a ratio in its simplest form.

....

   Divide by 5:

   $5 : 25 \qquad : 45$

   $\div 5 \Big\downarrow \qquad \Big\downarrow \div 5 \qquad \Big\downarrow \div 5$

   $1 : 5 \qquad : 9$

Example 3

Share 80 marbles among Alan, Kamil and Yasmin in
the ratio 2 : 3 : 5.

....

There are $2 + 3 + 5 = 10$ parts
Each part = 80 marbles $\div 10 = 8$ marbles
Alan gets $2 \times 8 = 16$ marbles
Kamil gets $3 \times 8 = 24$ marbles
Yasmin gets $5 \times 8 = 40$ marbles

Example 4

The ratio of red sweets to yellow sweets to green
sweets is 2 : 13 : 10.

a What fraction of the sweets are yellow?

b What percentage of the sweets are green?

....

a $\dfrac{13}{2+13+10} = \dfrac{13}{25}$

b $\dfrac{10}{2+13+10} = \dfrac{10}{25} = \dfrac{40}{100} = 40\%$

Exercise 12

1 Compare these quantities as ratios.

   a $\frac{1}{2}$ hour; 15 minutes

   b 2 kg; 300 g

   c 250 m; 3 km

   d $6\frac{1}{2}$ hours; 100 minutes

   e 25 mm; 2 m

   f 18 cm; 4 km

2 Write these ratios in their simplest form.

   a 2 : 20 : 200   **b** 36 : 40

   c 85 : 17 : 51   **d** 25 : 625 : 75

   e 184 : 56   **f** 91 : 65 : 117

3 Divide:

   a 0.3 m in the ratio 3 : 2

   b 0.42 kg in the ratio
        6 : 5 : 3

   c 2.7 l in the ratio 2 : 4 : 3

   d 2 hours in the ratio 3 : 7 : 5

> Hint: you
> may want to
> change units
> first.

4 Grey paint is made from mixing white paint
   and black paint in the ratio 15 : 2.
   If 5.1 litres of grey paint is made, how many
   millilitres of black paint is used?

5 The ratio of black counters to white counters
   to grey counters is 3 : 5 : 8.

   a What fraction of the counters are grey?

   b What percentage of the counters are white?

6 To make a fruit drink, pineapple juice, orange
   juice and mango juice are mixed in the ratio
   5 : 3 : 4. How much of each juice do you need
   to make 0.6 litres of the fruit drink?

7 Two different types of grey paint are mixed from black and white paint in the ratios shown.

	Black : White
Paint A	5 : 8
Paint B	2 : 3

Which paint is darker?

8 To make 3 glasses of orange squash you need 825 ml of water and 75 ml of orange cordial.
  a How much water do you need to make 10 glasses of orange squash?
  b How many glasses of squash can you make with 0.2 ℓ of orange cordial?

9 Maduka plays tennis. During one practice, the ratio of the serves he hit in to the serves he hit out was 23 : 2.
  a Which of these statements are definitely true or could be true?

    i He hit 23 times as many serves in as he hit out.
    ii 92% of his serves were in.
    iii For every 2 of his serves that were in, 23 serves were out.
    iv $\frac{2}{23}$ of his serves were out.
    v He hit 69 serves in and 6 serves out.

  b For any incorrect statement in part **a**, write the correct statement.

Summary

You should know …

1 To compare two quantities using a ratio, the quantities must be in the same units.

For example:
2 hours : 15 minutes
$= 120 \text{ min} : 15 \text{ min}$
$= 120 : 15$
$= 8 : 1$
A ratio itself has no units.

2 How to simplify a ratio by dividing the numbers in the ratio by the HCF of the numbers.

For example:
The ratio 8 : 12 : 20 = 2 : 3 : 5 (divide all numbers by 4)

3 How to share something in a given ratio.

For example:
Share $30 in the ratio 2 : 3 : 1
There are $2 + 3 + 1 = 6$ parts
Each part is worth $30 ÷ 6 = $5
so 2 parts are worth $2 \times $5 = $10
and 3 parts are worth $3 \times $5 = $15
so $10 to $15 to $5

Check out

1 Use a ratio to compare these quantities.
  a 1 hour; 45 minutes
  b 25 mm; 2 cm
  c 3 kg; 200 g

2 Write each ratio in its simplest form.
  a 6 : 9
  b 28 : 70
  c 24 : 9 : 33
  d 40 : 15 : 35
  e $2\frac{1}{4}$ hours : 30 minutes

3 Share $80 in the ratio
  a 4 : 1 : 5
  b 5 : 3
  c 1 : 9
  d 8 : 7 : 5
  e 1 : 1 : 3
  f 3 : 5 : 12

4   How to use proportion to compare ratios.

For example:
Hanni has yellow sweets and red sweets in the ratio 7 : 5.
Lida has yellow sweets and red sweets in the ratio 4 : 3.
Who has the greatest proportion of yellow sweets?

Hanni has $\dfrac{7}{7+5} = 7 \div 12 \times 100\% = 58\%$ yellow

Lida has $\dfrac{4}{4+3} = 4 \div 7 \times 100\% = 57\%$ yellow

 It is easier to compare proportions using percentages in this example.

Hanni has the greater proportion of yellow sweets.

4   Dmitry has boxes and bags in the ratio 8 : 15.
Tamar has boxes and bags in the ratio 7 : 13.
Who has the greatest proportion of bags?

13 Transformations

Objectives

In this chapter you will learn about:

- translations, including using vectors
- reflections in horizontal, vertical or diagonal lines
- rotations
- rotational and reflectional symmetry
- enlargements.

What's the point?

Transformations are an important part of the real world. Simple examples are the hands of a clock rotating to show us the time, and company logos being enlarged to lots of different sizes depending where they are displayed. More complex examples include transformations as a vital part of graphics in computer games.

Before you start

You should know ...

1   About rotational symmetry.
    For example:

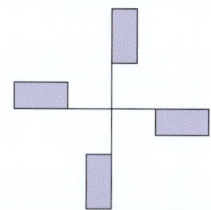

This shape has rotational symmetry of order 4 because it fits back on itself four times when turned through one complete turn.

Check in

1   What is the order of rotational symmetry of these shapes?

a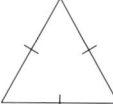

b

c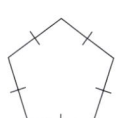

d

2 About lines of symmetry.
For example:

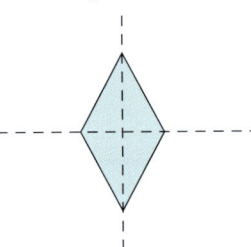

This rhombus has two lines of symmetry.

3 How to draw simple lines given their equations.
For example:

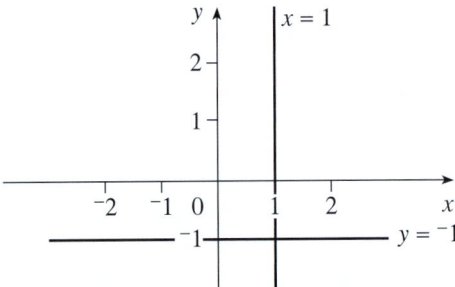

Points on the line $y = {}^-1$ always have a
y-coordinate of $^-1$.

4 How to enlarge a shape using a given scale
factor.
For example:

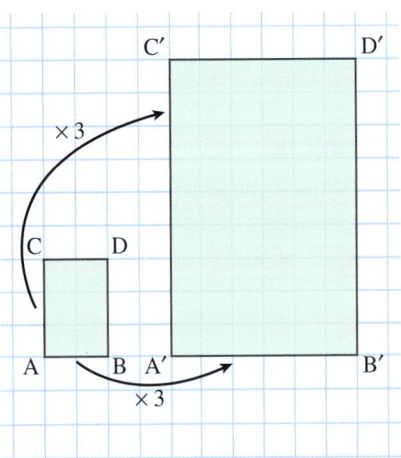

Enlarging rectangle ABCD by scale factor
3 produces a rectangle with sides 3 times
as long.

2 How many lines of symmetry do these
shapes have? Copy each shape and draw its
lines of symmetry.

a 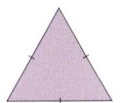 **b**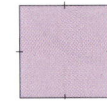

c  **d**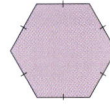

3 Draw the graphs of these lines.
  a $x = 2$ **b** $x = {}^-3$
  c $y = 3$ **d** $y = {}^-2$

4 Draw an enlargement of each of these
shapes with scale factor 2.

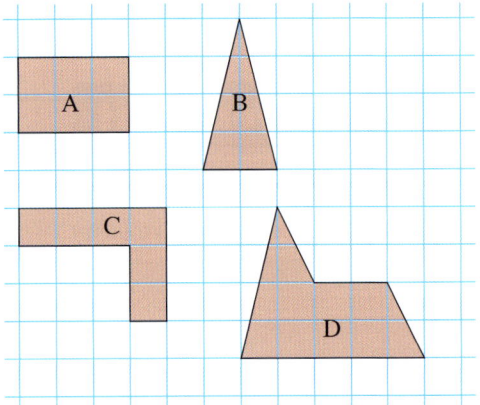

13.1 Translations

You will need graph paper.

A **translation** maps a shape to another position by a movement that is equivalent to sliding without turning.

In the diagram the quadrilateral ABCD has been translated to its **image**, A′B′C′D′.

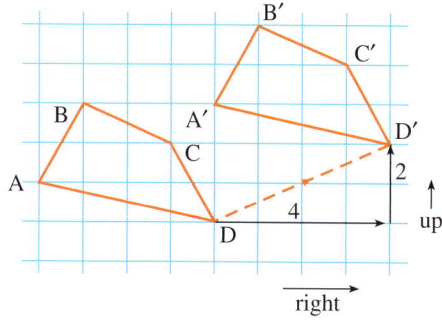

Notice that each point has been moved 4 squares right and 2 squares up.

This translation can be represented by the **column vector** $\mathbf{T} = \begin{pmatrix} 4 \\ 2 \end{pmatrix}$.

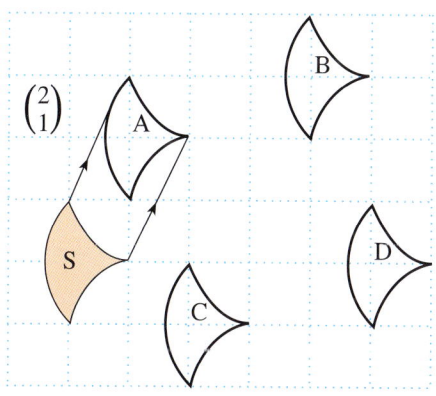

Look at the diagram above. The shaded shape S is mapped on to shape A by a translation described by the column vector $\begin{pmatrix} 1 \\ 2 \end{pmatrix}$.

This vector means move 1 square right and 2 squares up.

Use this diagram for Exercise 13A, which follows.

Exercise 13A

1 Use a column vector to describe the translation that maps shape A in the diagram onto:
   a shape B **b** shape C **c** shape D

2 Which shape does shape C map onto if the translation is described as follows?

   a $\begin{pmatrix} 3 \\ 1 \end{pmatrix}$ **b** $\begin{pmatrix} 2 \\ 4 \end{pmatrix}$

   c $\begin{pmatrix} -2 \\ 1 \end{pmatrix}$ **d** $\begin{pmatrix} -1 \\ 3 \end{pmatrix}$

3 What can you say about the translation that maps shape A onto shape B and the translation that maps shape C onto shape D?

4 Copy and complete this table.

	Translation	Vector	Translation	Vector
a	A → B	( )	B → A	( )
b	B → D	( )	D → B	( )
c		$\begin{pmatrix} 2 \\ 4 \end{pmatrix}$		$\begin{pmatrix} -2 \\ -4 \end{pmatrix}$

5 What is the connection between the vectors in each part of question **4**?

You should have found in question **5** Exercise 13A that the signs were opposite for the reverse transformation, which is also known as the inverse translation.

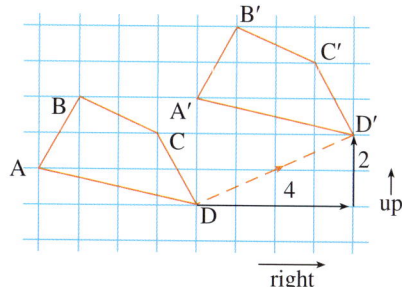

vector $\mathbf{T} = \begin{pmatrix} 4 \\ 2 \end{pmatrix}$.

The **inverse translation** represented by
$\mathbf{T}^{-1} = \begin{pmatrix} -4 \\ -2 \end{pmatrix}$

maps A′B′C′D′ back to ABCD.

The negative values tell us to move left and down.
$\begin{pmatrix} -4 \\ -2 \end{pmatrix}$ means move 4 squares left and 2 squares down.

Given the coordinates of a shape and a translation, you can easily find its image.

Notice that ABCD and A′B′C′D′ are the same shape and size. All side lengths and angles in the image are the same as in the original shape. The two shapes are said to be **congruent**.

Example 1

Find the image A′B′C′ of triangle A (2,1), B (5,1),

C (1,4) under the translation $\begin{pmatrix} ^-5 \\ 3 \end{pmatrix}$.

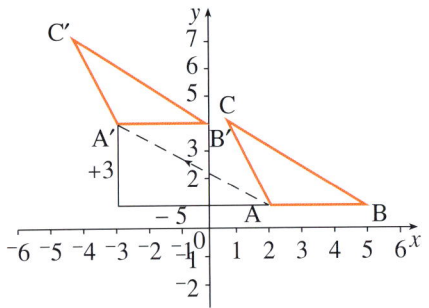

You can write:

$$A\,(2,1) \xrightarrow{\begin{pmatrix} ^-5 \\ 3 \end{pmatrix}} A'\,(^-3, 4)$$

$$B\,(5,1) \xrightarrow{\begin{pmatrix} ^-5 \\ 3 \end{pmatrix}} B'\,(0, 4)$$

$$C\,(1,4) \xrightarrow{\begin{pmatrix} ^-5 \\ 3 \end{pmatrix}} C'\,(^-4, 7)$$

So triangle ABC has been translated into triangle A′ (⁻3, 4), B′ (0, 4), C′ (⁻4, 7).

Note: triangle ABC and triangle A′B′C′ are the same shape and the same size.

Given the coordinates of the image and the translation vector, you can find the original shape (the **object**).

Example 2

The triangle ABC is mapped onto triangle A′B′C′ with vertices A′(3, 1), B′(1, ⁻2), C′(1, 2) under the translation $\mathbf{T} = \begin{pmatrix} 2 \\ 3 \end{pmatrix}$.

Find the vertices of triangle ABC.

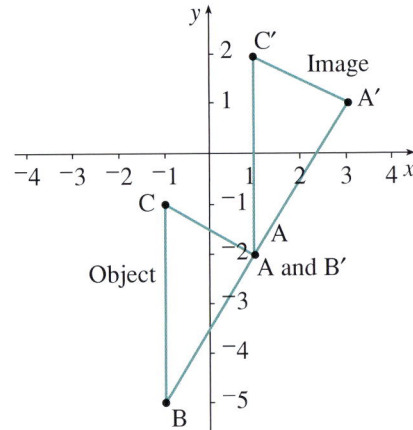

The translation $\mathbf{T} = \begin{pmatrix} 2 \\ 3 \end{pmatrix}$ maps ABC → A′B′C′.

So the inverse translation, $\mathbf{T}^{-1} = \begin{pmatrix} ^-2 \\ ^-3 \end{pmatrix}$ will map A′B′C′ → ABC

Thus

$$A'\,(3, 1) \xrightarrow{\begin{pmatrix} ^-2 \\ ^-3 \end{pmatrix}} A\,(1, ^-2)$$

$$B'\,(1, ^-2) \xrightarrow{\begin{pmatrix} ^-2 \\ ^-3 \end{pmatrix}} B\,(^-1, ^-5)$$

$$C'\,(1, 2) \xrightarrow{\begin{pmatrix} ^-2 \\ ^-3 \end{pmatrix}} C\,(^-1, ^-1)$$

So triangle ABC has vertices A (1, ⁻2), B $\left(^-1, ^-5 \right)$, C $\left(^-1, ^-1 \right)$.

Exercise 13B

1   The vertices of triangle PQR are (⁻4, 2), (⁻2, 3) and (⁻3, 5) respectively. Determine the vertices of the image triangle P′Q′R′ under the translation $\mathbf{T} = \begin{pmatrix} 4 \\ ^-3 \end{pmatrix}$.

2   The vertices of a parallelogram ABCD are (1, 1), (4, 1), (5, 3) and (2, 3) respectively. Find the vertices of the image of the parallelogram under the translation $\mathbf{T} = \begin{pmatrix} ^-3 \\ ^-4 \end{pmatrix}$.

3 The diagram shows a triangle ABC.

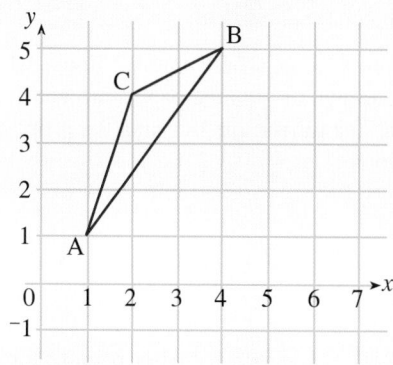

Draw a copy of the diagram.

On your copy draw also the triangles with vertices as listed in parts **a**–**e**.

In each case, if the triangle is a translation of ABC, write down the column vector of the translation.

a $(4, {}^-1)$, $(7, 3)$, $(5, 2)$

b $(2, 1)$, $(4, 2)$, $(5, 5)$

c $(0, 0)$, $(0, 3)$, $(2, 4)$

d $(0, 1)$, $(1, 4)$, $(3, 5)$

e $(2, 0)$, $(3, 3)$, $(5, 4)$

4 a Are all the triangles in question **3** the same shape and size as triangle ABC?

b If two triangles are the same shape and size as each other, must one be a translation of the other?

5 The vector $\begin{pmatrix} 4 \\ 3 \end{pmatrix}$ maps the triangle ABC to its image PQR. Point A maps to P, as shown in the diagram below.

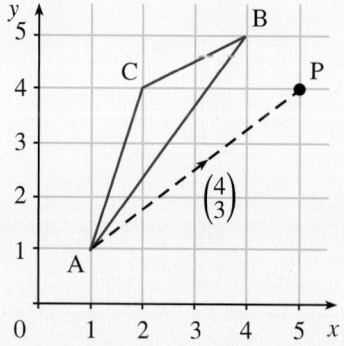

a What are the coordinates of A?

b What are the coordinates of P?

c Given the translation vector and the coordinates of A, explain how to find the coordinates of P without a diagram.

6 a For question **5**, copy and complete:

$$A \xrightarrow{\begin{pmatrix} 4 \\ 3 \end{pmatrix}} P$$
$$(1, 1) \longrightarrow (\quad)$$

b Show the mappings B → Q and C → R in a similar way.

c On squared paper, draw a pair of axes marked from $^-1$ to 10. Mark on triangles ABC and PQR.

d Does a translation map all points the same distance and in the same direction?

e Is triangle ABC the same shape and size as its image, PQR?

7 Another translation maps the triangle PQR from question **5** to STU.

The vector for the translation is $\begin{pmatrix} 1 \\ -1 \end{pmatrix}$

a Show STU on your diagram for question **6**.

b Write down the coordinates of S, T and U.

8 What is the translation vector that maps triangle ABC to triangle STU (question **7**)?

Complete: $\begin{pmatrix} 4 \\ 3 \end{pmatrix} + \begin{pmatrix} 1 \\ -1 \end{pmatrix} = \begin{pmatrix} \quad \end{pmatrix}$.

9 X, Y and Z are three points. The vector $\begin{pmatrix} 2 \\ 7 \end{pmatrix}$ maps X to Y and $\begin{pmatrix} 2 \\ -4 \end{pmatrix}$ maps Y to Z.

a What single vector maps X to Z?

b What vector would map Z to X?

10 A kite ABCD is mapped under a translation $\mathbf{T} = \begin{pmatrix} 4 \\ 5 \end{pmatrix}$ to A′ (2, 6), B′ (4, 8), C′ (6, 6) and D′ (4, 2). Determine the vertices of the kite ABCD.

11 On graph paper, draw triangle MNO with vertices M (1, 1), N (3, 3) and O (3, $^-1$).

Translate triangle MNO using the vector $\begin{pmatrix} 2 \\ -4 \end{pmatrix}$. Label the vertices of the image M′N′O′.

Translate triangle M′N′O′ using the vector $\begin{pmatrix} -1 \\ -3 \end{pmatrix}$. Denote this new image M″N″O″. Label the vertices of M″N″O″.

Give the vectors describing the translations that map:

a △MNO to △M″N″O″

b △M′N′O′ to △MNO

c △M″N″O″ to △MNO.

12 On graph paper, draw the rectangle ABCD with vertices A $(2, 1)$, B $(2, ^-1)$, C $(^-2, ^-1)$ and D $(^-2, 1)$. Translate ABCD with the vector $\mathbf{T} = \begin{pmatrix} ^-3 \\ 1 \end{pmatrix}$. Denote the image of ABCD as $A_1B_1C_1D_1$. Then translate $A_1B_1C_1D_1$ using the vector $\mathbf{T} = \begin{pmatrix} ^-2 \\ ^-2 \end{pmatrix}$. Denote this new image $A_2B_2C_2D_2$. State the vertices of $A_1B_1C_1D_1$ and $A_2B_2C_2D_2$. Give the vectors that translate:

a ABCD to $A_2B_2C_2D_2$

b $A_2B_2C_2D_2$ to ABCD

c $A_2B_2C_2D_2$ to $A_1B_1C_1D_1$

13 $\mathbf{T}$ is the translation $\begin{pmatrix} 4 \\ ^-3 \end{pmatrix}$. ABCD is a square with vertices at $(2, 0)$, $(2, 2)$, $(4, 2)$ and $(4, 0)$ respectively.

a Draw ABCD on graph paper.

b Draw the image of ABCD under $\mathbf{T}$, that is, $\mathbf{T}$(ABCD).

c Show that $\mathbf{T}^{-1} = \begin{pmatrix} ^-4 \\ 3 \end{pmatrix}$.

d Translate the image of ABCD under $\mathbf{T}$ once more, to get $\mathbf{T}^2$(ABCD).

13.2 Reflections

You will need squared paper and compasses or a set square.

The image of the boat reflected in the water looks identical but is upside down.

The idea of reflection is used in mathematics.

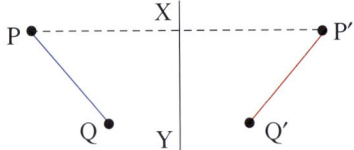

$P'Q'$ is the **reflection** of PQ in the line XY.

XY is called the **mirror line**.

P and P' are the same distance from XY so $PX = XP'$ and PP' is perpendicular to XY.

$P'Q'$ is the same length as PQ.

As with translation and rotation, the shape and size of the object are unchanged under a reflection: the image and the object are congruent.

It is often easy to see where a mirror line is when shapes are drawn on squared paper. But you can always find the mirror line by using the properties of reflections.

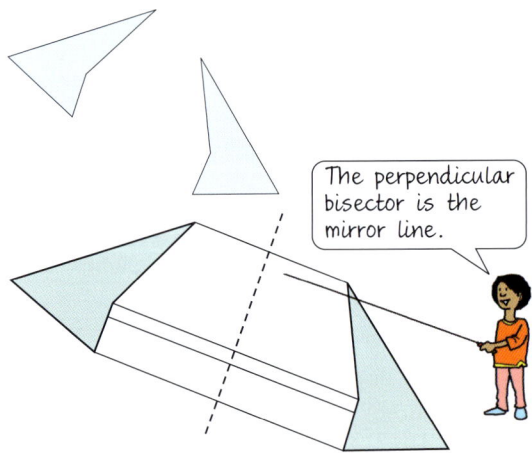

The perpendicular bisector is the mirror line.

To find the mirror line, join corresponding points and find the midpoints of the lines.

Example 3

R is a reflection in the line $y = 3$.
K is the triangle with vertices $(^-2, 5)$, $(0, 6)$, $(3, 4)$.
Draw R(K), the reflection of K.

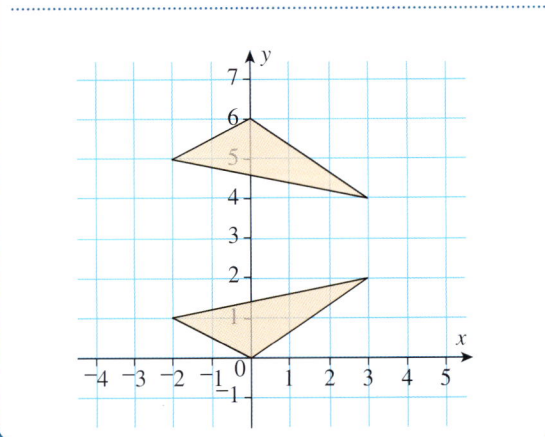

Exercise 13C

1 What is the equation of the mirror line for this reflection of the image of a horseman?

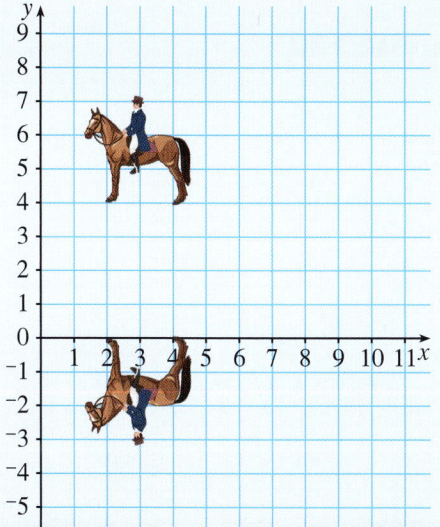

2 Draw a line (to use as a mirror line) and mark two points P and Q anywhere on one side of the line. Use compasses or a set square to draw lines from P and Q perpendicular to the mirror line and so find the reflections P′ and Q′ by measurement.

Measure the lengths PQ and P′Q′. Are they the same?

3 **M** is a reflection in the x-axis. L is the line joining (4, 2) and (⁻3, ⁻1).
a Draw the line L on squared paper.
b Draw **M**(L) by reflecting your line.

4 **R** is a reflection in the line $x = 2$. J is the triangle with vertices (⁻1, ⁻1), (⁻2, 2), (⁻3, 1).
a Draw the triangle J on squared paper.
b Draw **R**(J), the reflection of J.

5 T is the triangle with vertices (3, 1), (2, 1) and (2, 3). **R** is the reflection in the line $x = ⁻1$ and **N** is the reflection in the line $x = 4$.
a Draw the triangle T on squared paper.
b Draw **R**(T).
c Draw **N**(T).

6 Q is the quadrilateral with vertices at (⁻1, 2), (1, 3), (0, 4) and (⁻3, 5).
a Reflect Q in the line $y = ⁻x$ to get the image Q′.

b Copy and complete this table.

Coordinates of Q	Coordinates of Q′
(⁻1, 2)	
(1, 3)	
(0, 4)	
(⁻3, 5)	

c What do you notice about the coordinates of Q compared to Q′?

7 A reflection in the y-axis maps triangle ABC to its image triangle PQR.

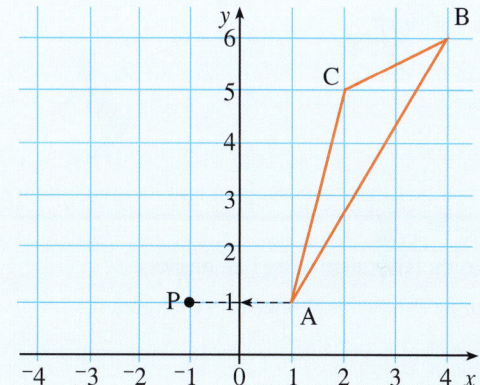

a The position of A is (1, 1). What is the position of P?
b Copy and complete: A (1, 1) → P(   ).
c Write down similar statements for B → Q and C → R.
d If (x, y) is any point, copy and complete this statement for a reflection in the y-axis: $(x, y) → ($ $)$.

8 Copy triangle ABC in question **7** and show its image after a reflection in the line:
a $x = 3$ b $y = 3$ c $y = ⁻x$

Each time, write down the vertices of the image triangle.

9 Under a reflection, are a shape and its image the same shape and size?

10 A reflection maps the following points.
(5, 2) → (5, 2)
(2, 5) → (2, ⁻1)
(0, 0) → (0, 4)

Without drawing a diagram, what is the equation of the mirror line of the reflection?

11 On squared paper plot two triangles with these vertices.
(1.5, ⁻2), (7, ⁻1), (0.5, ⁻9)
(⁻1.5, 2), (1, 7), (⁻8.5, 3)

Show that one is a reflection of the other by finding the equation of the mirror line.

12 Copy triangle ABC from question **7**. Show its image XYZ with vertices at ($^-$4, 6), (1, 9) and (0, 7). Find the equation of the mirror line.

13.3 Rotations

You will need squared paper and tracing paper.

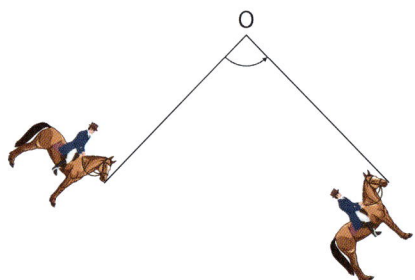

Here is a picture of a horseman and its image under a **rotation**.

A rotation has a **centre** and an **angle**.

Rotations can be clockwise or anticlockwise.

This rotation is 90° anticlockwise about centre O.

To completely define a rotation you need to know:

1 the angle of rotation
2 the direction of rotation (clockwise or anticlockwise)
3 the centre of rotation.

If the angle of rotation is 180° you do not need to state the direction.

Example 4

The triangle with vertices at A (1, 1), B (4, 1) and C (1, 2) is rotated through 90° anticlockwise about the centre (0, 0). State the coordinates of the image vertices, A′, B′ and C′.

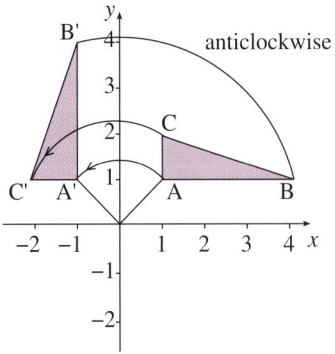

The vertices are A′ ($^-$1, 1) and B′ ($^-$1, 4) and C′ ($^-$2, 1).

Notice that triangles ABC and A′B′C′ in Example 4 are the same shape and size; that is, shape and size stay the same under rotation, and the triangles are congruent.

Exercise 13D

You may use tracing paper throughout this exercise.

1 The following diagrams show a triangle ABC and its image after a rotation. In each case, write down the angle, direction and centre of rotation.

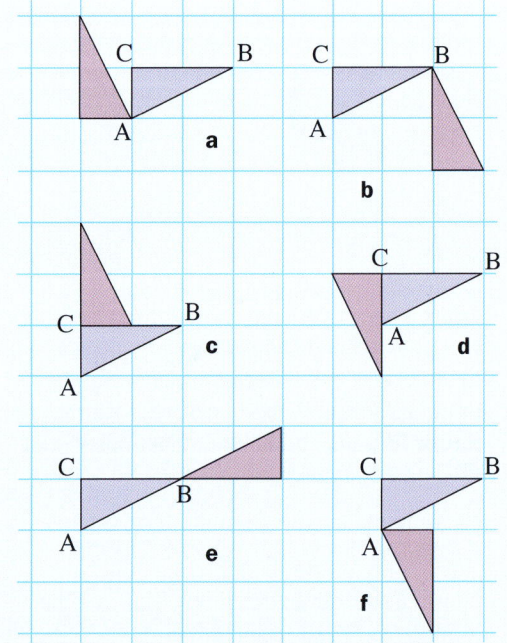

2 In this diagram, the mapping of triangle ABC to triangle RBQ is a rotation of 90° anticlockwise, centre B.

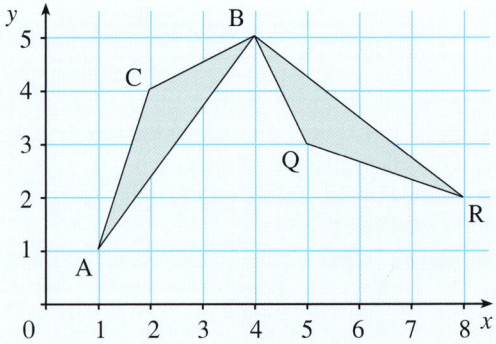

Draw a copy of triangle ABC using squared paper.

Now draw its image after a rotation of:
a 180°, centre B
b 90° anticlockwise, centre C
c 90° anticlockwise, centre A

Each time, write down the coordinates of the vertices of the image of triangle ABC.

3 Repeat question **2** for a rotation of:
a 90° anticlockwise, centre (5, 4)
b 180°, centre (6, 5)

4 Draw the trapezium with vertices (0, 0), (3, 0), (2, 1), (1, 1).

Now draw the image of the trapezium after anticlockwise rotations, centre (0, 0), of 90°, 180° and 270°.

What can you say about the symmetry of the finished drawing?

5 A quadrilateral ABCD has vertices A (1, ⁻3), B (3, ⁻1), C (6, ⁻2) and D (6, ⁻5).
a Draw ABCD on squared paper.
b Find the coordinates of the vertices of the image of the quadrilateral ABCD under a rotation of 90° clockwise about the origin.

6 A kite, PQRS, has vertices P (3, ⁻4), Q (1, 1), R (3, 2) and S (5, 1).
a Draw, on squared paper, the kite PQRS.
b Find the coordinates of the vertices of the image of the kite under a rotation of 180° about the origin.

⑦ One triangle is rotated to make the other triangle. What are the coordinates of A′?

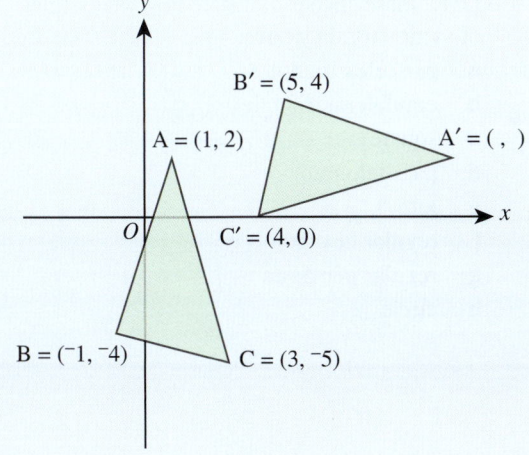

13.4 Symmetry

Reflectional symmetry

You will need plain white paper to cut up, tracing paper and scissors.

● A shape has **reflectional symmetry** if you can fold it along the **line of symmetry** and make two halves that match exactly.

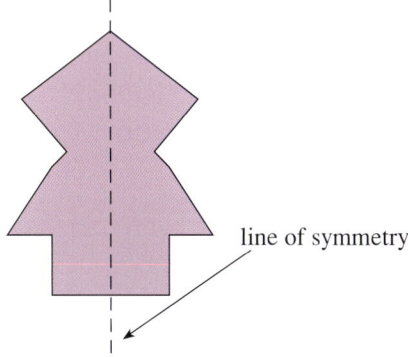

line of symmetry

You can say that one side of the shape is the **mirror image** of the other.

Lines of symmetry are easily found by folding.

Example 5

How many lines of symmetry does a rectangle have?

Cut out a rectangle and fold it vertically across its middle. You will see the two halves match exactly.

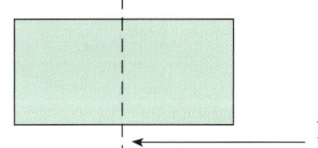

line of symmetry

Repeat, folding horizontally.

line of symmetry

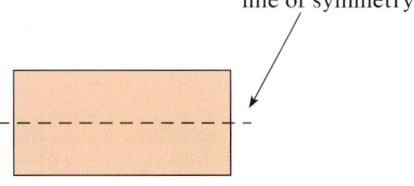

You should see the rectangle has two lines of symmetry.

Some shapes have three lines of symmetry.

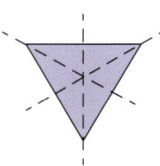

Some have four lines of symmetry.

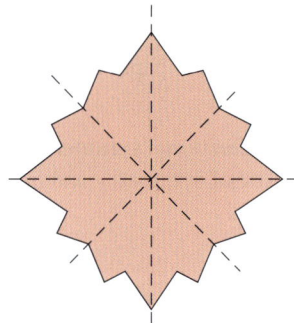

Others have lots of lines of symmetry.

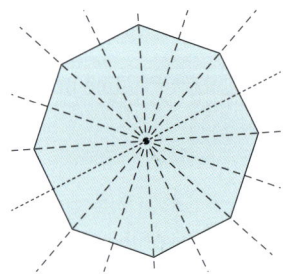

Exercise 13E

1 Make paper cut-outs of these shapes.

a    **b**

c    **d**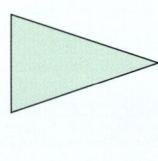

By folding, find out how many lines of symmetry each has.

2 Half of each shape is drawn, together with its line of symmetry. Trace and complete each shape.

a 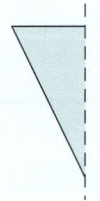   **b**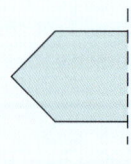

c    **d**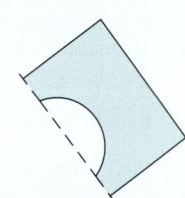

e    **f**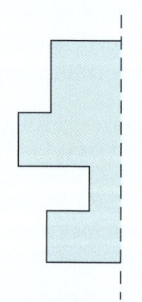

3 a Which of these road signs have reflectional symmetry?

i 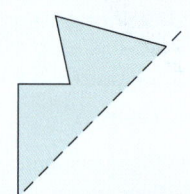   **ii**

iii    **iv**

b How many lines of symmetry does each sign in part **a** have?

4 Draw these shapes and state how many lines of symmetry they have.
  a isosceles triangle
  b equilateral triangle
  c square
  d parallelogram
  e kite
  f regular hexagon
  g regular pentagon
  h circle

Rotational symmetry

- A shape that fits into the same position more than once when rotated through 360° has **rotational symmetry**.

The letter H has an **order of rotational symmetry** of 2, because it fits into the same position twice when turned through 360°.

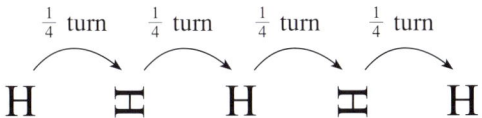

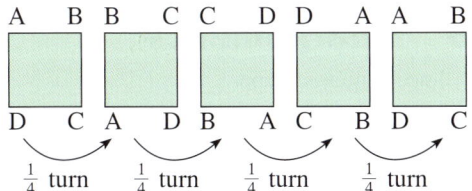

It fits into the same position after a $\frac{1}{2}$ turn and after a full turn.

A square has order of rotational symmetry 4.

A shape that fits into the same position only once it has turned 360° (that is, back to the start again) is *not* described has having an order of rotational symmetry of 1. We say it has **no rotational symmetry**.

Example 6

What is the order of rotational symmetry of these shapes?

a 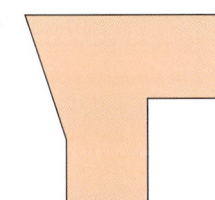 **b**

....

a This shape fits into the same position 6 times, so it has order of rotational symmetry 6.

b This shape looks the same only after a full rotation, so it has no rotational symmetry.

Exercise 13F

1 The triangle ABC is equilateral. Its centre is G. Trace the triangle and place your tracing over it. Place the point of your pencil at G, and rotate the tracing in an anticlockwise direction.

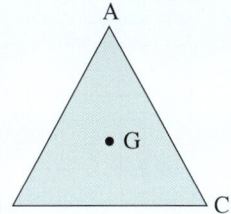

a After what angle of rotation does the triangle fit into the same position?
b How many times does the triangle fit into the same position in one complete turn?
c What is the order of rotational symmetry of the triangle?

2 Write down the order of rotational symmetry of each shape.

a  **b**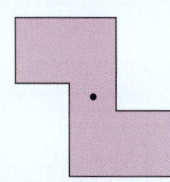

c 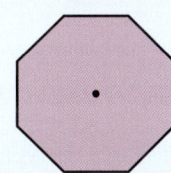 **d**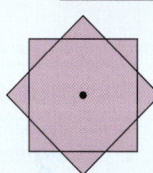

3 What is the order of rotational symmetry of:
a a square
b a parallelogram
c an isosceles triangle
d a rhombus
e a kite
f a rectangle
g an isosceles trapezium
h a regular pentagon
i a circle?

4 a Find some designs with rotational symmetry.
b Copy them and identify the order of rotational symmetry of each design.

5 Copy and complete these sentences.

The number of lines of symmetry of a regular polygon is equal to . . . .
The order of rotational symmetry of a regular polygon is equal to . . . .

6 A shape has been started below. Copy and complete it so that it has 2 lines of reflectional symmetry and rotational symmetry of order 2.

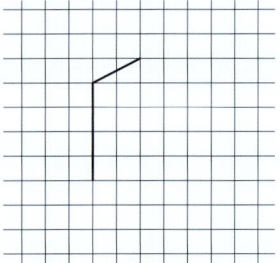

7 Shade in two more squares in this diagram so that it has no rotational symmetry and one line of reflectional symmetry.

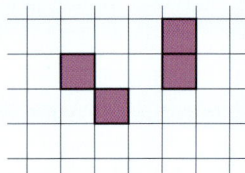

8 Draw several copies of this 2-by-2 grid.

You need to shade 0, 1, 2, 3 or 4 of the squares inside the grid.

a How many different shapes can you make with:
  i 1 line of symmetry
  ii 2 lines of symmetry
  iii 0 lines of symmetry
  iv 4 lines of symmetry
  v 3 lines of symmetry?

b How many different shapes can you make with:
  i order of rotational symmetry 2
  ii no rotational symmetry
  iii order of rotational symmetry 4
  iv order of rotational symmetry 3?

c Can you ever get order of rotational symmetry 3 or 3 lines of symmetry in patterns based
  i inside squares
  ii using squares?

Investigation

Artists have long used ideas of symmetry to make beautiful paintings and patterns.

The Dutch artist M.C. Escher often used symmetry in his works of art.

Search online for 'Escher patterns' to view some of his work.

You should have found the following results in Exercise 13F question **5**.

● The number of lines of symmetry of a regular polygon is equal to the number of sides.

● The order of rotational symmetry of a regular polygon is equal to the number of sides.

13.5 Enlargements

You will need squared paper.

In an **enlargement** the size of an object is either increased or reduced.

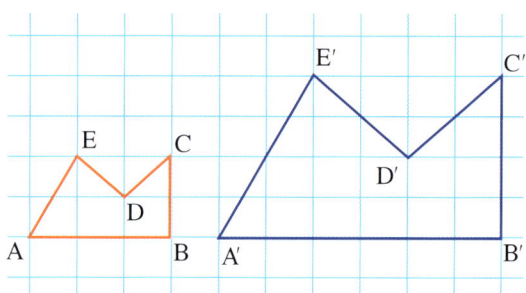

In the diagram the shape ABCDE is enlarged by **scale factor** 2 into A'B'C'D'E'.

Note: (1) the shape remains the same
(2) the angles remain the same
but (3) the length of every side of the image is original × scale factor

The shapes ABCDE and A'B'C'D'E' are **similar** but not congruent. The word 'similar' is a mathematical word meaning one shape is an enlargement of another, preserving shape and angles but increasing or decreasing side lengths.

To completely define an enlargement you need to know the:
● **scale factor** of the enlargement
● **centre of enlargement**.

All parts of the image are enlarged from the centre of enlargement.

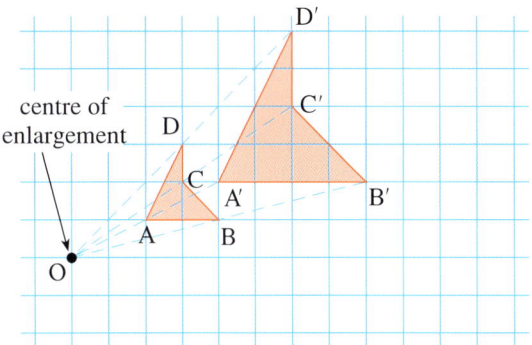

The diagram shows ABCD enlarged with scale factor 2 and centre of enlargement O.

To find the scale factor pick two corresponding sides on the object and the image, divide the image side length by the object side length. Notice that it does not matter which pair of sides you pick because you will always get the same scale factor of 2, e.g.

$$\frac{C'D'}{CD} = \frac{2}{1} = 2$$

Example 7

Find the image A′B′C′ of triangle A $(2,1)$, B $(2,3)$, C $(1,1)$ under an enlargement of scale factor 2 with centre of enlargement O $(0,0)$.

....

Join the vertices A, B, C of the triangle to the centre of enlargement, O. Extend OA to A′ so that

$$OA' = 2 \times OA$$

Repeat for B′ and C′.

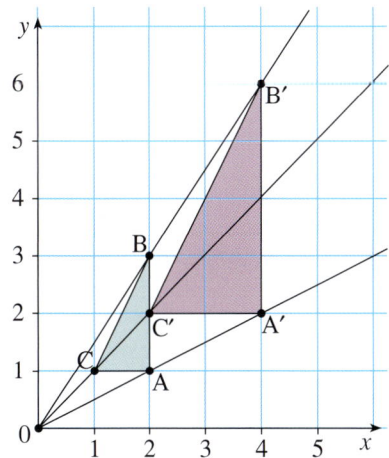

The coordinates of the enlarged triangle are A′ $(4,2)$, B′ $(4,6)$ and C′ $(2,2)$.

Notice that each side of the original triangle is multiplied by 2, so triangle ABC is **similar** to triangle A′B′C′.

Exercise 13G

1 In this diagram, a rectangle has been enlarged by a scale factor 2, using the origin, O, as centre.

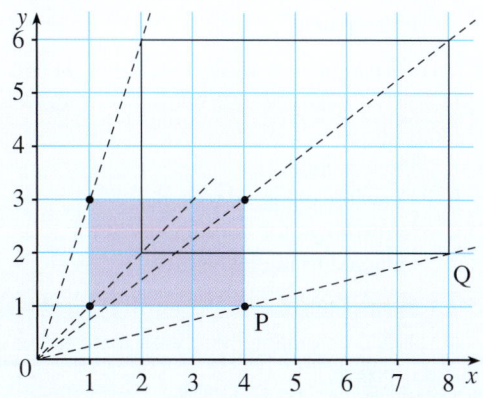

  a In the enlargement, the image of P is Q. What can you say about the lengths of OP and OQ?

  b What are the lengths of the sides of the small rectangle? What are the lengths of the sides of the large rectangle?

  c How does an enlargement scale factor 2 change the lengths of the sides?

2 In this diagram, triangle ABC is mapped to triangle DEF by an enlargement, centre S.

  SD = 3SA and SE = 3SB

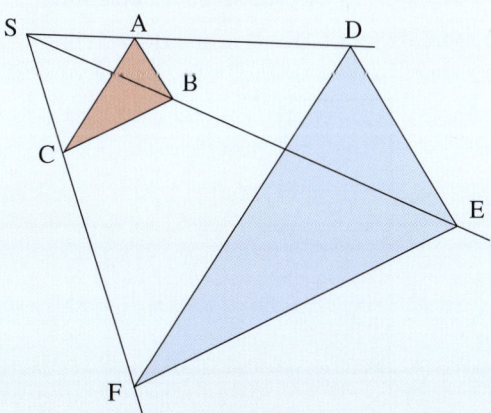

  a What is the scale factor of the enlargement?

  b Copy and complete: SF = …

  c If AB = 3 cm, what is the length of DE?

  d If AC = 5 cm, what is the length of DF?

3 On squared paper, draw the triangles ABC and PQR with:
A (1, 1), B (4, 5), C (2, 4)
P (2, 0), Q (8, 8), R (4, 6).

  a Is triangle PQR an enlargement of triangle ABC?

  b Draw lines through AP, BQ and CR to find the centre of enlargement.

  c What is the scale factor of the enlargement?

4 T is a triangle with vertices (2, 1), (4, 1) and (4, 2). T is enlarged with centre (0, 0) and scale factor 2.

  a Draw triangle T.

  b Draw the enlarged triangle.

  c Write down the coordinates of the enlarged triangle's vertices.

5 A is the triangle with vertices (3, 1), (1, 2) and (1, 1). **E** is an enlargement with scale factor 2 and centre (3, 0).

  a Draw triangle A.

  b Draw the enlargement, **E**(A).

  c Write down the coordinates of the enlarged triangle's vertices.

6 On the same drawing as question **3** draw the triangle KLM with K (5, 1), L (8, 5), M (6, 4).

  a What is the centre of the enlargement that maps KLM to PQR?

  b What transformation maps KLM to ABC?

  c What transformation maps ABC to KLM?

7 In this diagram, rectangle KLMN is an enlargement of rectangle ABCD with S as centre. (S isn't shown on the diagram.)

MN = 12 cm, CD = 4 cm and BC = 3 cm.

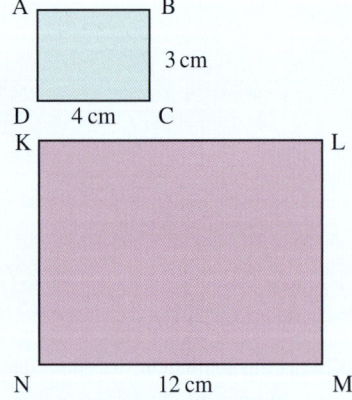

a What is the scale factor of the enlargement?

b What is the length of LM?

c If SA = 3 cm what is the length of SK?

d AC = 5 cm; find the length of KM.

8 Here are several different diagrams showing an enlargement scale factor 2 of quadrilateral ABCD. In each case the centre of enlargement is E, F, G and H respectively.

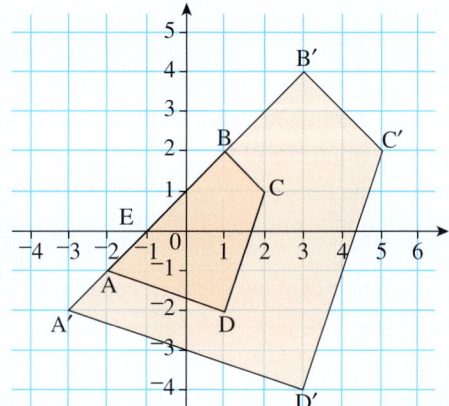

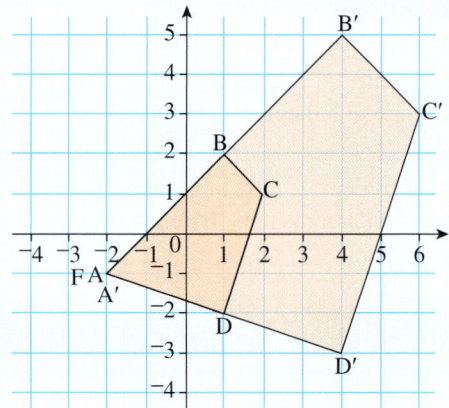

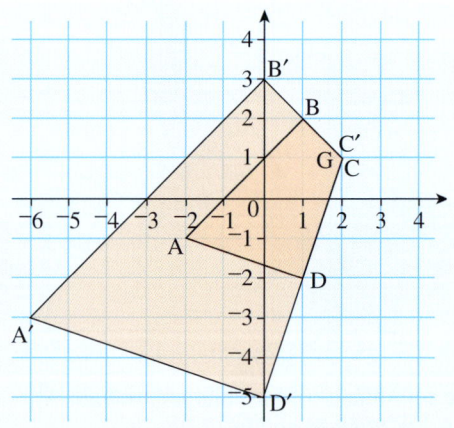

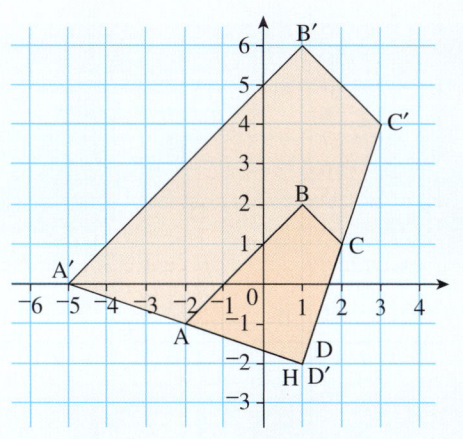

a Write down the coordinates of the invariant point for each diagram.

A point that is the same on the object as on the image is invariant.

b What do you notice?

c When will there be no invariant point?

9 **a** Construct a rectangle ABCD in which AB is 2 cm and BC is 4 cm. On BC construct a triangle EBC outside the rectangle so that EB = 2 cm and EC = 4 cm.

b Find the image of triangle EBC under an enlargement with centre A, given that the scale factor is 2.

c Find the image of triangle EBC under an enlargement with centre D, given that the scale factor is 2.

d What can you say about the two images in parts **b** and **c**?

Consolidation

Example 1

What is the image of the quadrilateral
A $(^-2, 3)$, B $(^-1, ^-3)$, C $(2, ^-4)$, D $(4, 2)$ under the translation $\begin{pmatrix} 2 \\ 1 \end{pmatrix}$?

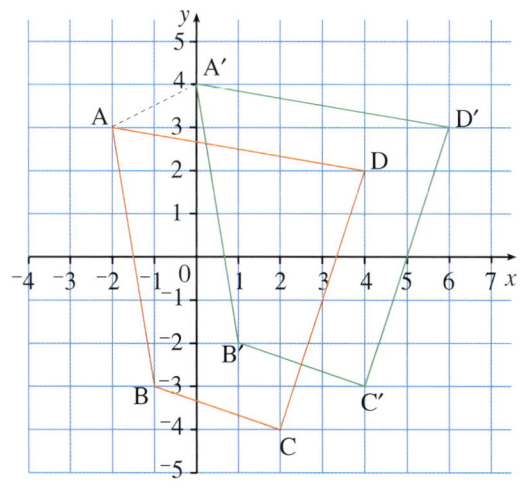

Under $\begin{pmatrix} 2 \\ 1 \end{pmatrix}$:

A $(^-2, 3) \rightarrow$ A$'(0, 4)$, B$(^-1, ^-3) \rightarrow$ B$'(1, ^-2)$
C $(2, ^-4) \rightarrow$ C$'(4, ^-3)$, D $(4, 2) \rightarrow$ D$'(6, 3)$

Example 2

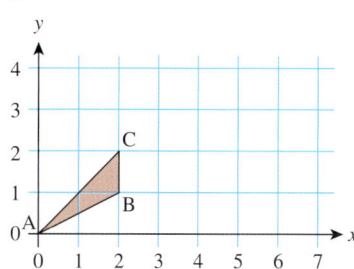

Find the image of triangle ABC after reflection in the line $x = 3$.

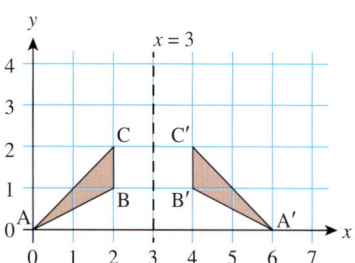

Example 3

a Draw the rotation of the shaded triangle 270° anticlockwise about centre (3, ‾2).

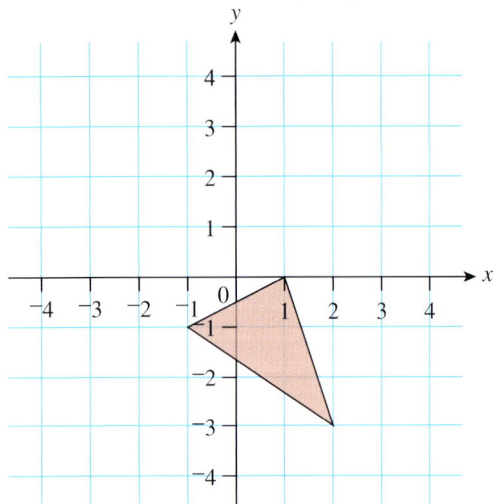

b Describe a different rotation that this is equivalent to.

....

a

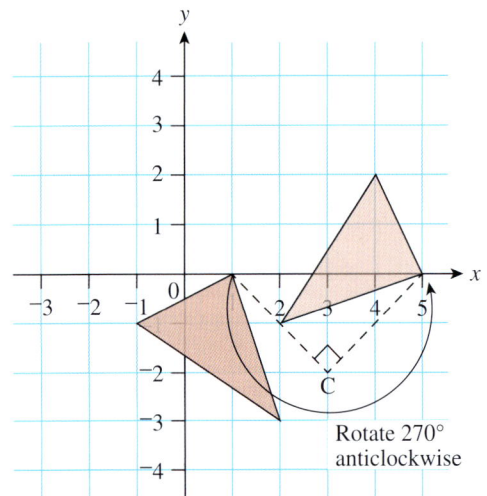

Rotate 270° anticlockwise

b It is equivalent to rotation of the triangle 90° clockwise about centre (3, ‾2).

Example 4

Find the image of the quadrilateral A (0, 3), B (1, 1), C (3, 0), D (4, 4) under an enlargement centred at the origin with a scale factor 2.

....

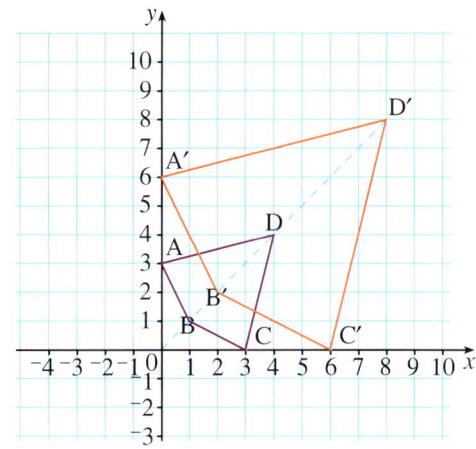

The coordinates of the enlarged quadrilateral are:

A′ (0, 6), B′ (2, 2), C′ (6, 0), D′ (8, 8)

Notice that the distance from the origin to C′ is twice the distance from the origin to C.

Exercise 13

1 The vertices of a triangle ABC are A (2, 1), B (4, 2) and C (4, 5).

a Find the vertices A′, B′ and C′ after a translation $\mathbf{T} = \begin{pmatrix} ^-5 \\ ^-6 \end{pmatrix}$.

b Find the vertices of image A″ B″ C″ of triangle A′B′C′ after a translation $\mathbf{U} = \begin{pmatrix} 0 \\ 6 \end{pmatrix}$.

c Find the vertices of the image of A″ B″ C″ after a translation $\mathbf{V} = \begin{pmatrix} 5 \\ 0 \end{pmatrix}$.

2 **a** Draw the trapezium with vertices A (1, 2), B (4, 2), C (3, 4) and D (2, 4).

b Draw the image of ABCD after a rotation of 90° anticlockwise about the origin.

c Draw the image of ABCD after a rotation of 180° about the origin.

3 S is a rotation of 90° anticlockwise about (‾1, ‾1). Triangle A has vertices at (3, 1), (6, 4), (4, 4).

a Draw, on squared paper, the triangle A.

b Draw S(A), by rotating triangle A.

4 **a** Draw the quadrilateral ABCD, which has vertices A (5, 1), B (7, 2), C (6, 5) and D (4, 4).

b Reflect ABCD in the line:
  i $x = 0$
  ii $y = 0$
  iii $x = 2$

5 On squared paper, draw rectangles ABCD and WXYZ with:

A (2, 1), B (2, 2), C (4, 2), D (4, 1)

W (3, 3), X (3, 5), Y (7, 5), Z (7, 3)

Describe the transformation of ABCD onto WXYZ.

6 a On squared paper, draw a triangle with vertices at (1, 3), (2, 4) and (2, 7).
  b Draw the image of the triangle under an enlargement with centre (0, 0) and scale factor 2.

7 These triangles are all congruent. Write down the sizes of all the unmarked angles and sides.

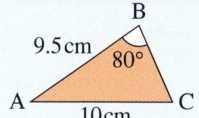
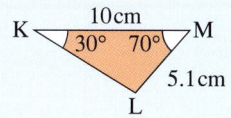
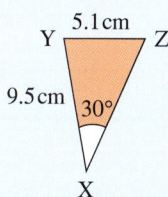

8 A reflection maps these points onto the images given.

$(1, ^-4) \rightarrow (4, ^-1)$

$(^-1, ^-5) \rightarrow (5, 1)$

$(^-1, ^-7) \rightarrow (7, 1)$

What is the equation of the mirror line of the reflection?

9 Write down **a** the order of rotational symmetry and **b** the number of lines of symmetry for these shapes.

i

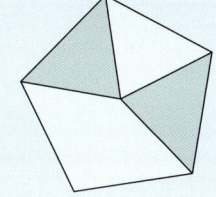

ii

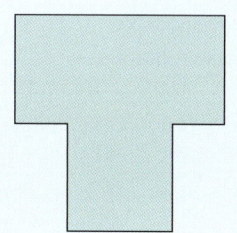

iii

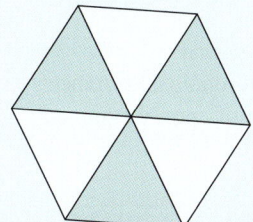

iv

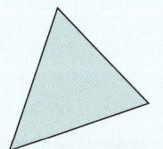

v

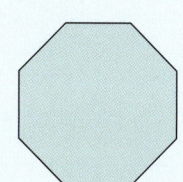

vi

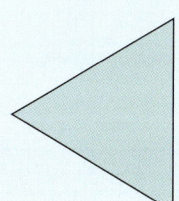

10 ABC and A'B'C' are similar triangles.

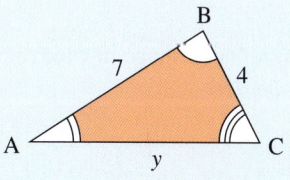
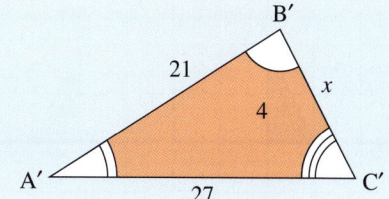

Work out x and y.

11 Find the image of triangle X ($^-$1, 2), Y ($^-$2, 2), Z ($^-$3, 3) under these rotations.

Centre of rotation	Angle of rotation (anticlockwise)
a (0, 0)	90°
b (0, 2)	60°
c ($^-$1, 1)	45°
d (3, 1)	180°

12 For shapes **a** to **d**, find:
  i the order of rotational symmetry
  ii the number of lines of symmetry

  a regular pentagon
  b isosceles triangle
  c square
  d regular octagon

13 The triangle A (2, 3), B (4, 2), C (5, 5) becomes triangle A′ ($^-$4, 3), B′ ($^-$6, 2), C′ ($^-$7, 5) after reflection in a certain line.
  a Draw ABC and A′B′C′ on squared paper.
  b Construct the mirror line.
  c Write down the equation of the mirror line.

Summary

You should know ...

1 How to represent a translation by a column vector.
For example:

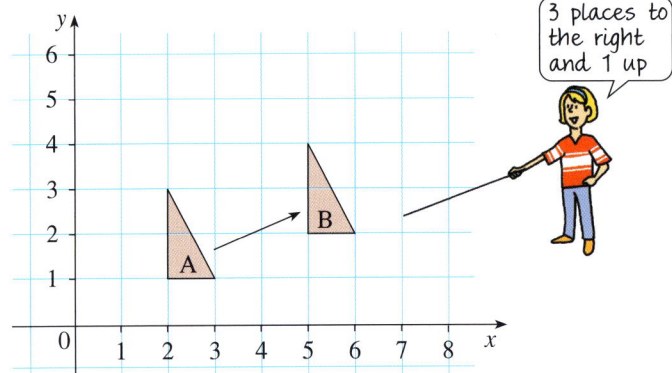

Triangle A is translated to triangle B by the vector $\begin{pmatrix} 3 \\ 1 \end{pmatrix}$.

2 How to reflect a shape in a mirror line.
For example:

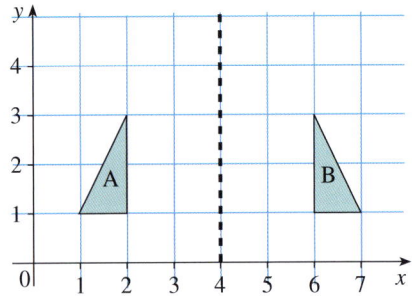

Shape B is the reflection of shape A in the line $x = 4$.

Check out

1 The vertices of a square are at (1, 1), (3, 1), (3, 3) and (1, 3).

Find the coordinates of the vertices after:

  a the translation $\begin{pmatrix} 2 \\ 2 \end{pmatrix}$

  b the translation $\begin{pmatrix} ^-5 \\ ^-7 \end{pmatrix}$

2 Triangle X has vertices at (2, 1), (3, 2) and (4, 0).

Draw, on squared paper, the reflection of triangle X in the line:
  a $y = 0$
  b $x = 0$
  c $x = ^-1$
  d $y = x$

3 A rotation has a centre and an angle.
For example:

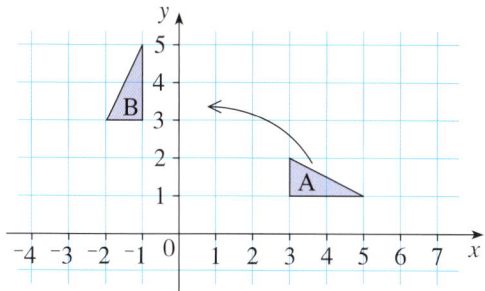

Triangle A is rotated 90° anticlockwise about the origin to triangle B.

3 The triangle with vertices X (⁻2, 4), Y (⁻2, 6) and Z (⁻1, 5) is rotated by 180° about the point (1, 0). Show, on squared paper, triangle XYZ and its image after the rotation.

4 The number of lines of symmetry and the order of rotational symmetry of a regular polygon is equal to the number of sides.
For example:
The order of rotational symmetry of a regular pentagon is 5.

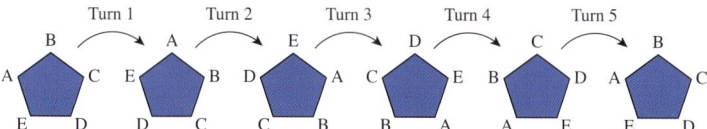

A regular pentagon has 5 lines of symmetry.

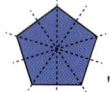

4 For the following shapes, find:
i the order of rotational symmetry
ii the number of lines of symmetry
    a square
    b regular octagon
    c rectangle
    d regular hexagon

5 An enlargement has a scale factor and a centre of enlargement.
For example:

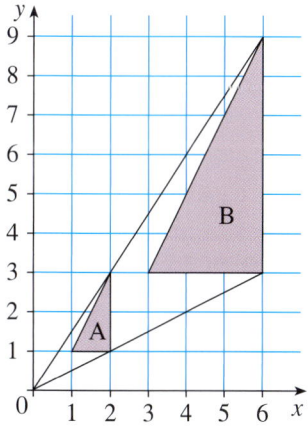

B is the enlargement of A by a scale factor of 3, with centre (0, 0).

5 Triangle XYZ has vertices at (6, 1), (8, 1) and (8, 4) respectively.
a Draw the enlargement of triangle XYZ with scale factor 2, centre (8, 0).
b Draw the enlargement of triangle XYZ with scale factor 3 and centre (6, 0).

Functions and graphs

14

Objectives

In this chapter you will learn about:

- graphs of linear functions
- equations of a line in the form $y = mx + c$
- travel graphs
- functions (including inverses).

What's the point?

Drawing graphs helps you to identify patterns in numbers more easily. From graphs you can make simple predictions and use them to answer such questions as: how much rainfall is expected in September? Will sales increase in March? How much should my baby sister weigh next month?

Before you start

You should know ...

1 How to write down the coordinates of plotted points.

For example:

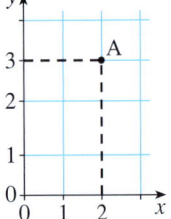

The coordinates of A are (2, 3). Coordinates are also called ordered pairs.

Check in

1

The coordinates are plotted on a graph: P at (1, 4), Q at (3, 0), R at (5, 1).

Write down the coordinates of P, Q and R.

2   How to work with negative numbers.

For example:

$^-3 \times ^-4 = 12$

$^-50 \div 5 = ^-10$

$^-4 - 2 = ^-6$

$7 - ^-1 = 8$

3   How to substitute into an expression.

For example:

Find the value of $2n - 4$ when $n = ^-3$

$2n - 4 = 2 \times ^-3 - 4$

$\qquad = ^-6 - 4$

$\qquad = ^-10$

2   Work out:
 a $^-3 \times 8$
 b $^-30 \div ^-5$
 c $^-3 - 5$
 d $5 - ^-2$

3   Find the value of these expressions when $n = ^-5$

 a $3n + 20$
 b $4n - 2$
 c $60 - 3n$

14.1   Functions

This is a function machine.

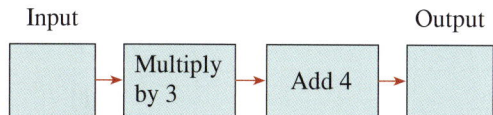

For example:

If 7 is input, then $7 \times 3 + 4 = 25$ is output.

A number entered into the function machine is called the **input**. When a function machine uses the **function** (in this case multiply by 3 then add 4) it produces an **output**. Here, if the input is 2 then the output is 10 because $2 \times 3 + 4 = 10$. If the input is 10 the output is 34, and so on. If the input is n the output is $3n + 4$. We can use a **mapping diagram** like this one to show what outputs go with particular inputs.

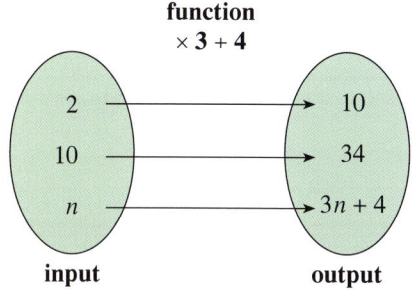

Or inputs and their outputs can be shown in a table.

Input	Output
2	10
10	34
n	$3n + 4$

Example 1

Use this function machine to complete the table below.

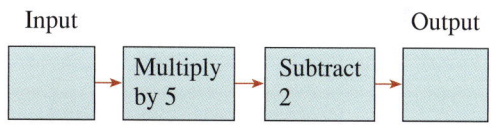

Input	Output
1	
2	
3	
4	
n	

Input	Output
1	3
2	8
3	13
4	18
n	$5n - 2$

Notice when the input is n the output $5n - 2$ is the nth term of the sequence 3, 8, 13, 18, ...

Exercise 14A

1 Use this function machine to complete the table below.

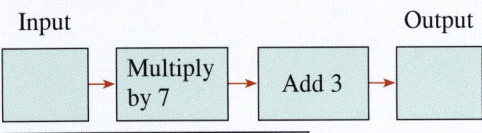

Input	Output
1	
2	
3	
4	
n	

2 Copy and complete this mapping diagram using the function shown.

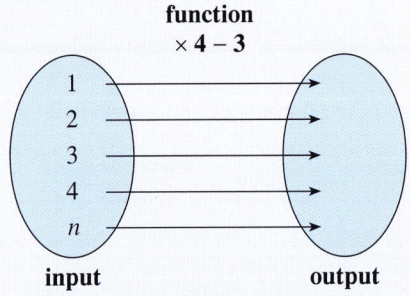

3 For each mapping diagram:
i draw a function machine
ii write the output when the input is *n*

a

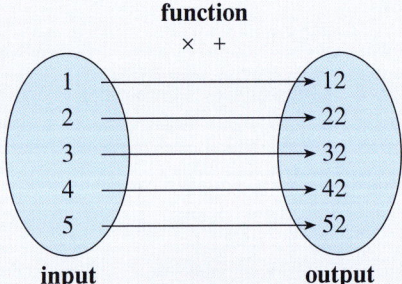

b

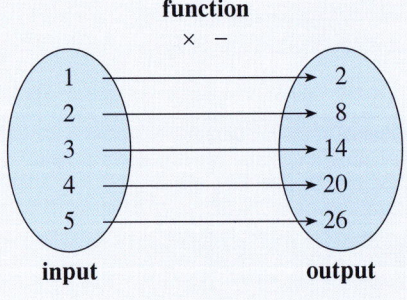

c

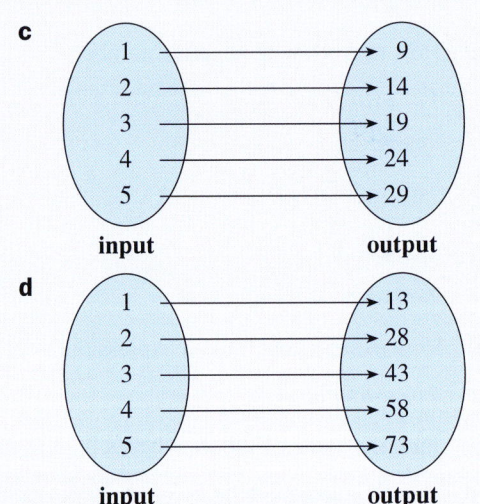

d

Mapping diagrams don't have to have their inputs in the sequence 1, 2, 3, 4, ... Be careful with the next questions.

4 For these mapping diagrams:
i draw a function machine
ii write the output when the input is *n*

a

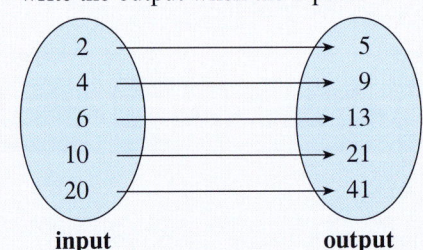

b

5

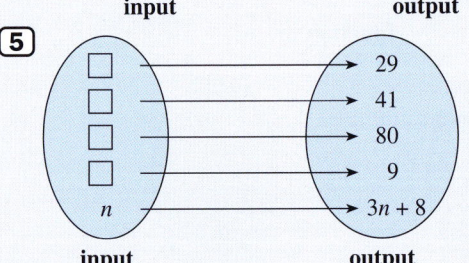

Using the mapping diagram,
Leila wrote the equation $3n + 8 = 29$
then she solved it:
$3n = 21$
$n = 7$

Then Leila wrote:
'The input is 7 for the output of 29.'
Write and solve equations like this to find the missing inputs.

In question **5**, instead of writing an equation, you could use inverse functions to find the solution, so:

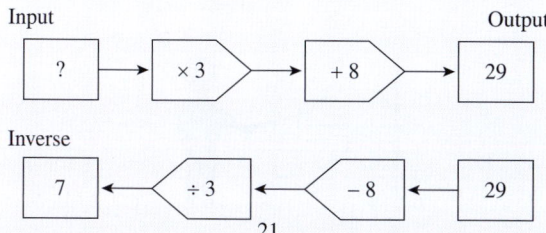

Use this method in question **6a** and whichever method you prefer in question **6b**.

6 Copy and complete these mapping diagrams.

a
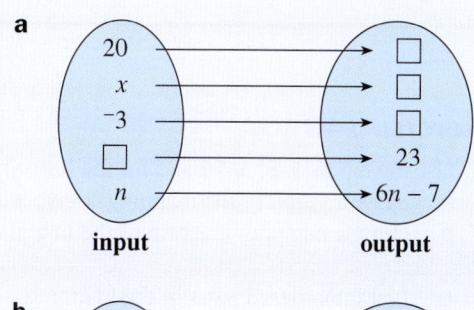

b
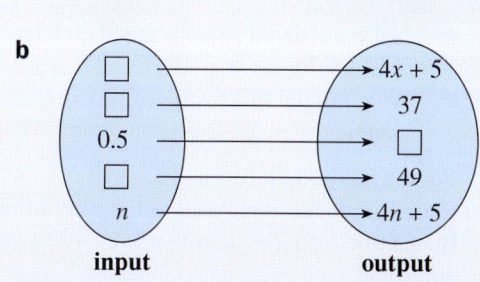

7 A function $y = 4x - 1$ can be represented by this mapping.

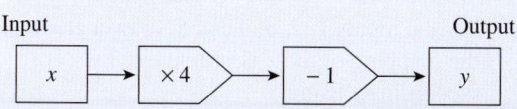

a Complete this inverse function machine to work out the input x when the value of y is 1.

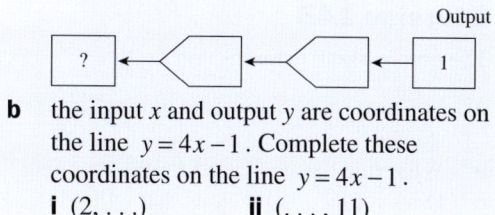

b the input x and output y are coordinates on the line $y = 4x - 1$. Complete these coordinates on the line $y = 4x - 1$.
i $(2, \dots)$ **ii** $(\dots, 11)$

14.2   Linear graphs

Coordinates

On a graph the horizontal axis is often called the x-axis and the vertical axis the y-axis. The **ordered pairs** that describe a mapping on the axes are called **coordinates**.

In the graph below, the black point shows that 5 maps to 8. The coordinates of the point are $(5, 8)$.

5 is the x-coordinate.

8 is the y-coordinate.

The x-coordinate is *always* written first.

- The coordinates (a, b) of a point on a graph represent the x- and y-coordinates respectively.

Exercise 14B

The graphs show mappings on axes.

a For each graph list the set of coordinates shown.

b What rule has been used for the mapping?

c Draw a function machine that will produce this mapping.

1

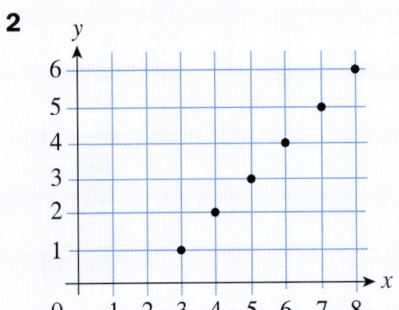

2

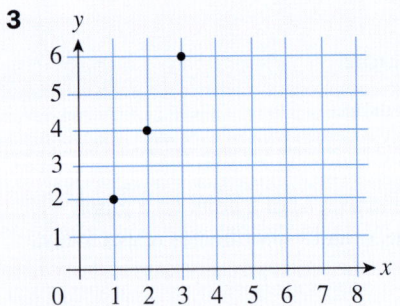

3

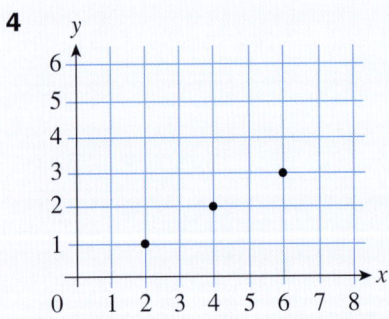

4

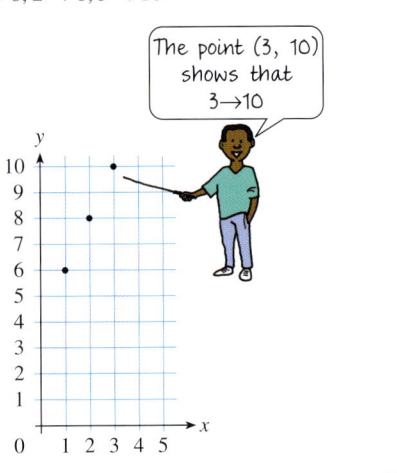

Example 2 illustrates how the rule for a mapping can be shown on a graph.

Draw a graph of the mapping
$$x \rightarrow 2x + 4$$

The rule is *multiply by 2 then add 4* so

$$1 \rightarrow 6, 2 \rightarrow 8, 3 \rightarrow 10$$

> The point (3, 10) shows that $3 \rightarrow 10$

Exercise 14C

1 Use the inputs 1, 2, 3, 4 and 5 for the following.

   a Draw a graph of the mapping $x \rightarrow x + 2$.

   b Draw a mapping diagram of the mapping $x \rightarrow x + 2$.

   c List the ordered pairs of the mapping $x \rightarrow x + 2$.

   d Which of the three ways of showing the mapping $x \rightarrow x + 2$ do you prefer? Why?

2 For the inputs 1, 2, 3, 4 and 5:
the x-coordinate is mapped to the y-coordinate by the rule $x \rightarrow 2x - 1$ so $1 \rightarrow 1, 2 \rightarrow 3, 3 \rightarrow 5$, and so on.

   a List the set of coordinates of the mapping.

   b Draw a graph of the mapping.

3 For the inputs 1, 2, 3, 4 and 5, draw a graph of the mapping of the x-coordinate to the y-coordinate using:

   a $x \rightarrow 3x - 2$

   b $x \rightarrow 2x + 3$

   c $x \rightarrow x$

4 The function machine produces coordinates (input, output).

If 3 is the input, the output is 10. The coordinates are $(3, 10)$.

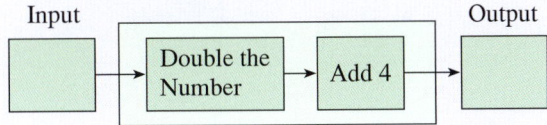

Input → Double the Number → Add 4 → Output

a List the set of coordinates produced when the inputs are 1, 2, 3, 4, 5 and 6.

b Draw a graph to show the mapping. Label the axes.

5 Using the inputs ⁻2, ⁻1, 0, 1, 2, draw, on the same axes, graphs of the mapping of the x-coordinate to the y-coordinate using:

a $x \rightarrow 2x + 3$

b $x \rightarrow 2x + 1$

c $x \rightarrow 2x - 3$

In each case, join the five points with a straight line.

What do you notice?

6 Using the inputs ⁻2, ⁻1, 0, 1, 2, draw, on the same axes, graphs of the mapping of the x-coordinate to the y-coordinate using:

a $x \rightarrow 2x + 2$

b $x \rightarrow x + 2$

c $x \rightarrow 3x + 2$

In each case, join the five points with a straight line.

What do you notice?

A very simple relation is given by the rule *add 3*. This rule is shown in the mapping diagram.

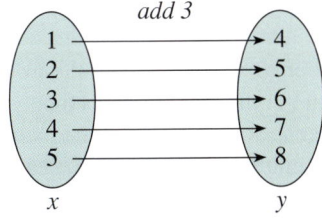

add 3

As a number machine this is:

input → +3 → output

x y

or as a mapping:

$x \rightarrow x + 3$

Alternatively, the mapping $x \rightarrow x + 3$ can be shown as ordered pairs on a graph.

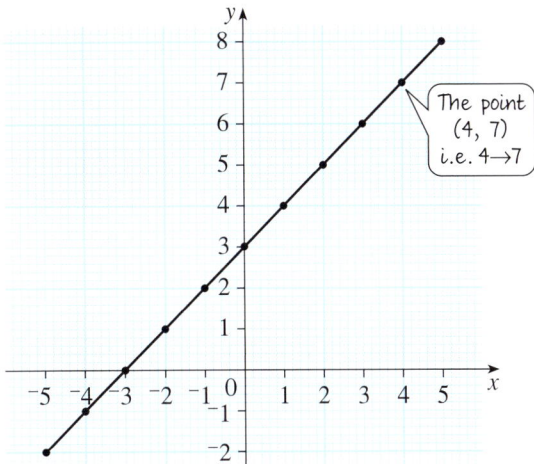

The point (4, 7) i.e. 4→7

Instead of drawing a mapping diagram it is more usual to complete a table for the mapping.

x	1	2	3	4	5
y	4	5	6	7	8

The rule for the mapping $x \rightarrow x + 3$ is rewritten as the equation $y = x + 3$.

The points on the graph can be joined with a straight line.

This is because all of the points in between $x = 1, 2, 3,$ 4 and 5 also follow the same rule. For example, if you had $x = 1.5$ as your input, the output would be $y = 1.5 + 3 = 4.5$, which also lies on the line. The same applies to inputs before 1 and after 5: the line has been extended back to $x = ⁻5$, and in fact carries on forever in both directions. The line represents all the mappings of the linear function.

We say the line is represented by the equation

$$y = x + 3$$

The equation $y = x + 3$ means that to find y-coordinates you apply the function *add 3* to the x-coordinates.

Example 3

a If $y = 2x + 1$, copy and complete this table for the mapping.

x	⁻1	0	1	2	3	4	5
y							

b Make a graph of this relation.

a The equation $y = 2x + 1$ can be written as the mapping

$$x \rightarrow 2x + 1$$

or as the function machine

$$x \rightarrow \boxed{\times 2} \rightarrow \boxed{+ 1} \rightarrow y$$

so when $x = 5$, $y = 2 \times 5 + 1 = 11$

when $x = 4$, $y = 2 \times 4 + 1 = 9$

so the table becomes:

x	⁻1	0	1	2	3	4	5
y	⁻1	1	3	5	7	9	11

(⁻1, ⁻1)   (1, 3)   (3, 7)   (5, 11)
   (0, 1)   (2, 5)   (4, 9)

b These ordered pairs can be plotted on a graph.

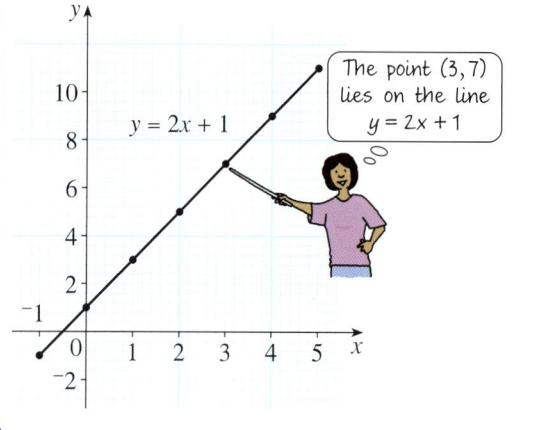

The point (3, 7) lies on the line $y = 2x + 1$

$y = 2x + 1$

Exercise 14D

1 a Copy and complete this mapping diagram for the mapping

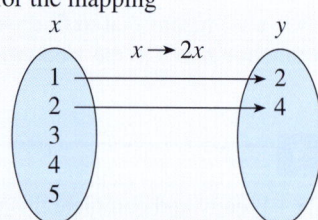

$x \rightarrow 2x$

b Write down the coordinates for the mapping.
c Show the mapping on a coordinate graph.
d Join the points on the graph. What is the equation of the line?

2 Repeat question **1** for each of these mappings.

a
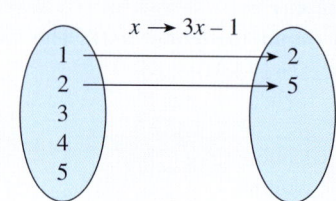
$x \rightarrow 3x - 1$

b
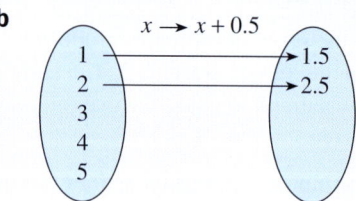
$x \rightarrow x + 0.5$

c
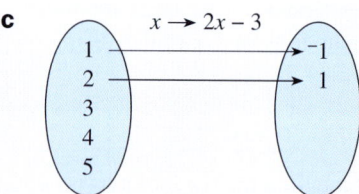
$x \rightarrow 2x - 3$

d
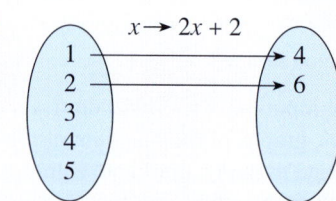
$x \rightarrow 2x + 2$

3 Copy and complete the sentences using the cards (cards can be used more than once).

multiply	add	2	⁻1	subtract	halve

a When $y = 5x$, to find the y-coordinates you . . . the x-coordinates by 5.
b When $y = x - 7$, to find the y-coordinates you . . . 7 from the x-coordinates.
c When $y = 2x - 3$, to find the y-coordinates you . . . the x-coordinates by . . . then you . . . 3.
d When $y = \frac{1}{2}x + 8$, to find the y-coordinates you . . . the x-coordinates then you . . . 8.
e When $y = 6 - x$, to find the y-coordinates you multiply the x-coordinates by . . . then . . . 6 or you can . . . the x-coordinate from 6.

4   The function machine for the equation
$y = 2x + 2$ is

$x \rightarrow \boxed{\times 2} \rightarrow \boxed{+2} \rightarrow 2x + 2 = y$

Write down function machines for the following equations.

a $y = 3x$ **b** $y = x + 4$

c $y = x - 6$ **d** $y = 2x + 4$

e $y = 3x - 9$ **f** $y = \frac{x}{4} + 7$

g $y = \frac{x+2}{5}$ **h** $y = \frac{3x+2}{4}$

5   To draw a straight line you only need to plot 2 points. Why is it a good idea to actually work out 3 or more points?

6   **a**   If $y = 2x + 3$, copy and complete the table for the mapping.

x	-2	-1	0	1	2	3
y			5			9

   b   Plot the points on a graph and join them with a straight line.

7   For each equation, copy and complete the table and then draw its graph.

x	-2	-1	0	1	2	3
y						

a $y = x + 4$ **b** $y = 3x + 4$

c $y = 3x - 2$ **d** $y = 2x - 5$

e $y = 5x - 6$ **f** $y = \frac{1}{2}x + 2$

8   For each equation, copy and complete this table.

x	-5	-4	-3	-2	-1	0	1	2	3	4	5
y											

a $y = 5x + 5$ **b** $y = 3x + 7$

c $y = 2x + 3$

9   Copy and complete these coordinate pairs for the equation $y = 10x + 7$ by calculating the missing y-coordinate using the given x-coordinate.

a $(7, \square)$ **b** $(5, \square)$

c $(-2, \square)$ **d** $(-1, \square)$

10   One of these points doesn't lie on the line $y = 2x + 7$. Which one is it?

$(1, 9)$ $(4, 15)$ $(3, 12)$ $(0, 7)$

11   Which of these points lie on the line $y = 7 - x$?

$(3, 4)$ $(0, 7)$ $(-1, 6)$ $(-2, 9)$

12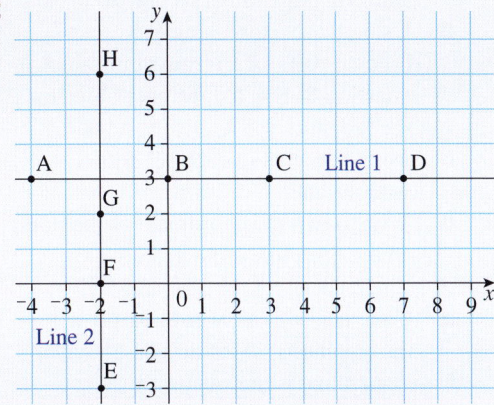

   a   Complete the coordinate pairs for line 1 and line 2 drawn on this graph.

   Line 1:   A $(-4, \square)$ B $(0, \square)$

   C $(3, \square)$ D $(7, \square)$

   Line 2:   E $(\square, -3)$ F $(\square, 0)$

   G $(\square, 2)$ H $(\square, 6)$

   b   What are the equations of these lines?

13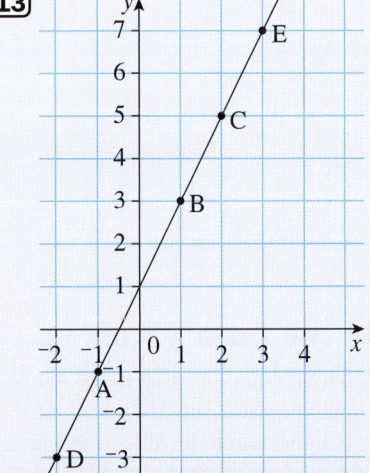

   a   Use the graph to complete the coordinate pairs.

   A $(-1, \square)$ B $(1, \square)$ C $(2, \square)$

   D $(-2, \square)$ E $(3, \square)$

   b   How is the y-coordinate connected to the x-coordinate?

   c   What is the equation of the line?

a On centimetre-square paper, draw a pair of rectangular axes. On the x-axis, show all the numbers from $^-5$ to 5. On the y-axis show all the numbers from $^-20$ to 30.

b Using these axes, show each set of ordered pairs (x, y) from your tables in question **8**. Use a different colour for each set of points.

c Join the points in each set with a straight line.

d Write down the coordinate of the point where each line crosses the y-axis (intercept).

e Compare your answer to part **d** with the equations. What do you notice?

f Which line is the steepest?

 a For each equation, copy and complete this table.

x	$^-2$	$^-1$	0	1	2	3
y						

  i $y = x$
  ii $y = 2x$
  iii $y = 3x$
  iv $y = 4x$

b On suitable axes, plot each of these four graphs.

c Which graph is the steepest?

d Repeat parts **a**, **b** and **c** for:
  i $y = {^-x}$
  ii $y = {^-2x}$
  iii $y = {^-3x}$
  iv $y = {^-4x}$

e What do you notice?

Equations in the form $y = mx + c$

There are many different equations that make many different graphs. Some produce straight lines, some produce curves. It is important to be able to tell just from looking at an equation whether or not it will produce a straight line.

In Student Book 7 you learned that the equation $y = 4$ means that the y-coordinate is always 4, no matter what the x-coordinate is. Any equations of the form $y =$ 'a number' are horizontal lines. Vertical lines are described by $x =$ 'a number'. For example, $x = 2$ means that the x-coordinate is always 2, no matter what the y-coordinate is.

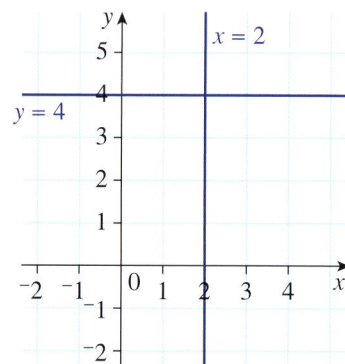

Exercise 14D involved mostly diagonal lines. Diagonal lines contain an x and a y in the equation. There are no powers of x other than x^1 in the equation (for example there is no x^2, $\frac{1}{x}$, x^3). We usually write the equation for a diagonal line in the form $y = mx + c$, where m and c stand for numbers. If the value of c is 0 the line will still be diagonal. If the value of m is 0 the line will be horizontal rather than diagonal. The values m and c can be positive or negative. m is the **gradient** of the line and tells you how steep the line is. c is the **y-intercept** of the line and tells you where the line crosses the y-axis. Diagonal straight lines can be described in other ways – you will learn about this in Student Book 9.

In Exercise 14C question **5**, you drew the lines $y = 2x + 3, y = 2x + 1, y = 2x - 3$
All three lines have the same gradient, 2.
You should have found these lines were all the same steepness and so were parallel.

In Exercise 14C question **6**, you drew the lines $y = 2x + 2, y = x + 2, y = 3x + 2$
All three lines have the same y-intercept, 2.
You should have found these lines all crossed the y-axis at the same point (0, 2).

In Exercise 14D question **15**, part **a**, you drew the lines $y = x, y = 2x, y = 3x, y = 4x$

In each case the gradient is increasing and therefore the steepness of the line is increasing.

In Exercise 14D question **15**, part **d**, you drew the lines $y = {^-x}, y = {^-2x}, y = {^-3x}, y = {^-4x}$

In each case the gradient is negative and the lines slope down rather than up.

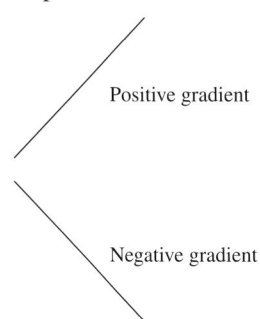

Positive gradient

Negative gradient

Exercise 14E

1 Kanika says that in the line $y = 8 - 3x$, $m = 8$ and $c = {}^-3$. She has made a mistake. What mistake has she made?

2 For these equations of straight lines in the form $y = mx + c$, write down the value of:
  i the gradient, m **ii** the y-intercept, c
  a $y = 5x + 8$
  b $y = {}^-x + 2$
  c $y = 3x$
  d $y = 7 - x$
  e $y = x + 6$
  f $y = 9 - 2x$
  g $y = 4 + x$
  h $y = {}^-3x - 7$

3 Sort these equations of lines into one of the three straight line categories shown.

$y = x$ $y = 3$ $y = {}^-3x$ $x = 7$

$y = 3x - 4$ $y = {}^-x + 3$ $y = 2x$

Categories:

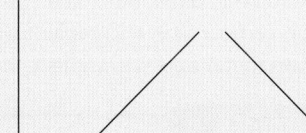

Horizontal line Vertical line   Diagonal line

4 Which of these are straight lines?

$y = 4x - 7$ $y = x^2$ $x = 7$ $y = \frac{1}{x}$

$y = 8 - x$ $y = x^2 - 3x + 4$ $y = 7 - x^3$

$y = {}^-3$

5 Find the number or letter that correctly completes these sentences.
  a To find where a graph crosses the y-axis substitute $x = \ldots$ into the equation $y = mx + c$
  b To find where a graph crosses the x-axis substitute $\ldots = 0$ into the equation $y = mx + c$
  c For the line $y = 4x + 8$, the y-intercept is $\ldots$.
  d For the line $y = x + 3$, the x-intercept is $\ldots$.
  e The line $y = \ldots$ passes through the points $(1, 1)$, $(0, 0)$, $(3, 3)$, $({}^-9, {}^-9)$.
  f In the graph $f = 7 - 4t$, f is plotted on the vertical axis and t is plotted on the horizontal axis. The gradient of this line is $\ldots$.

6 Sort these lines into the two groups shown below.

$y = x$ $y = {}^-4x$ $y = 6x - 2$ $y = 7x$

$y = {}^-x + 9$ $y = 5x + 5$ $y = 11 - 6x$

$y = {}^-x$

Groups:

Line sloping   Line sloping
  upwards   downwards

7 Sort these points into groups of:
  a those that are on the line $y = x$
  b those that are above the line $y = x$
  c those that are below the line $y = x$

  $(1, 2)$ $(7, 7)$ $({}^-3, {}^-4)$ $(3, 2)$ $({}^-8, {}^-8)$

  $({}^-50, {}^-49)$ $(a, a+1)$ $(n \times 3, n+n+n)$

  $(c, c-3)$

8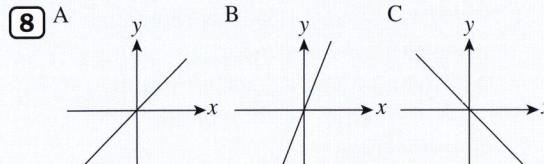

Explain why each of these statements is incorrect.

a Graph A and graph C could be the graph of $y = x$

b Graph B cannot be the graph of $y = x$ because it is not at 45° to the axes.

c Graphs A and B have a positive y-intercept but graph C has a negative y-intercept.

9 For each of these straight lines write down five points on the line and work out the equation of the line.

a

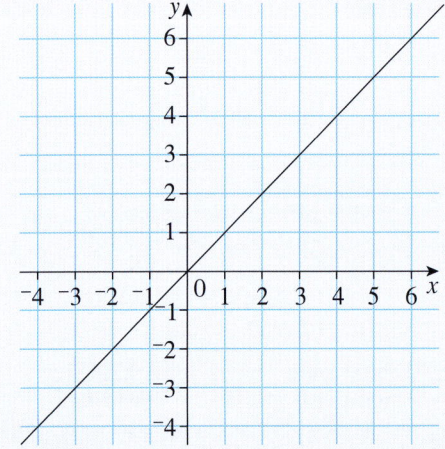

b

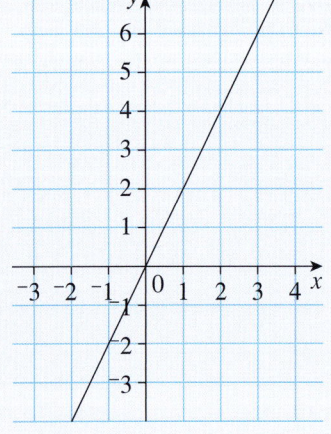

c

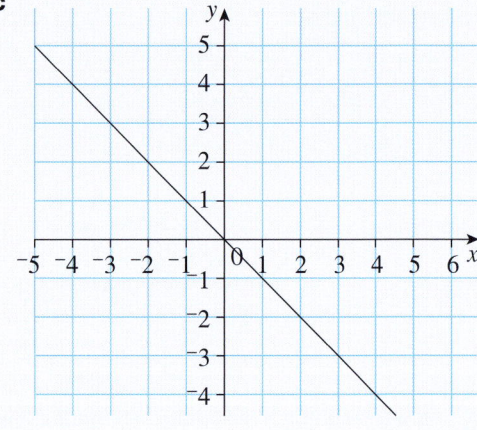

d

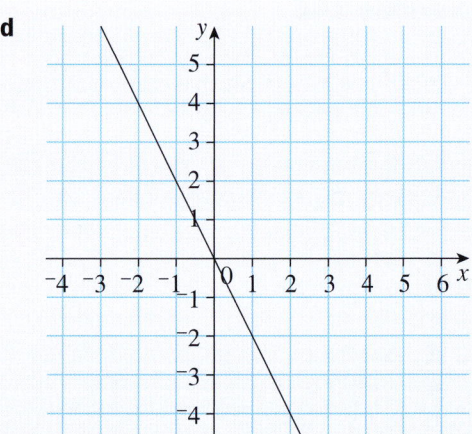

10 Line A has gradient 3 and y-intercept 4.
Line B is parallel to line A and passes through the point $(0, {}^-11)$.
Find the equations of line A and line B.

14.3   Real-life graphs

Labels on graphs don't have to be x and y, they can be any letters. If replacing x and y in $y = mx + c$, the letter on the vertical axis will replace the y and the letter on the horizontal axis will replace the x.

Example 4

A plumber charges a call out fee of $25, then $20 for each hour that he works.

a Draw the graph that shows this.

b Find the equation of the line.

....

a Use c for cost on the vertical axis and h for hours on the horizontal axis.
Create a table to find some points to plot.

h, hours	0	1	2	3
c, cost($)	25	45	65	85

Draw the graph.

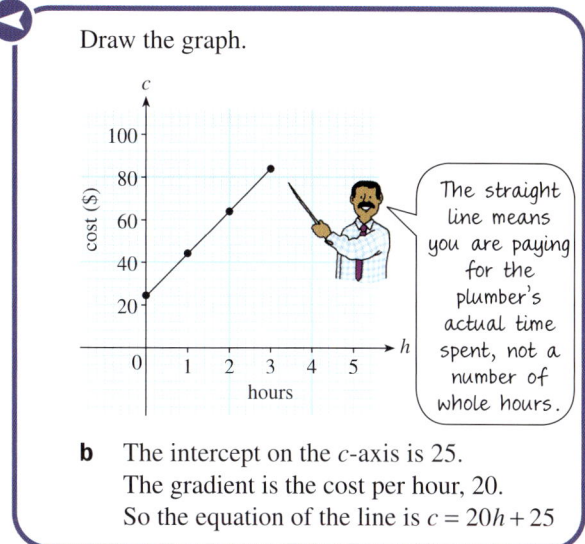

The straight line means you are paying for the plumber's actual time spent, not a number of whole hours.

b The intercept on the *c*-axis is 25.
The gradient is the cost per hour, 20.
So the equation of the line is $c = 20h + 25$

Exercise 14F

1 The cost per unit of electricity is shown in this graph.

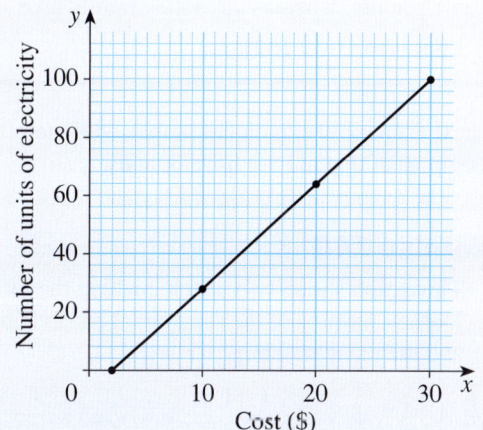

a What is the cost of:
i 20 units **ii** 28 units
iii 76 units?
b How many units of electricity are used if the total cost is:
i $5 **ii** $9.50 **iii** $29?
c What is the standing charge, that is, the charge applied even if no electricity is used?

2 The graph shows the number of US dollars that can be obtained for a given number of Emirati dirhams (AED).

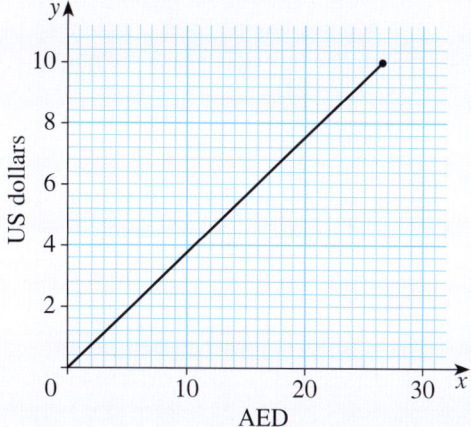

a How many US$ can be obtained for:
i AED15 **ii** AED20
iii AED8?
b How many AED can be obtained for:
i US$6 **ii** US$4.40
iii US$3.60?

3 The volume of water that flows into a water tank over a period of 20 minutes is given in this table.

Time (min)	0	1	3	11	13.5	17	20
Water volume (ℓ)	0	13	39	143	175.5	221	260

a Draw a graph to show this information. Use a scale of 2 cm to represent 4 minutes on the horizontal axis and 2 cm to represent 50 litres on the vertical axis.
b From your graph, find how much water is in the tank after:
i 10 min **ii** 16.4 min
c From your graph, find when the volume of water in the tank is:
i 20 litres **ii** 155 litres

4 A piece of meat is taken out of a freezer. Its temperature rises steadily, as shown in this table.

Time (min)	0	10	20	30	40	50
Temp (°C)	⁻5	0	5	10	15	20

Draw a graph, using a scale of 1 cm to represent 5 minutes on the horizontal axis, and 2 cm to represent 5°C on the vertical axis.

Label the horizontal axis *t* and the vertical axis *d*.

a   Use the graph to estimate:
  i   the temperature of the meat after 17 minutes
  ii the time taken for the meat to reach a temperature of 12°C
b   Complete this equation of the graph
    $d = 0.5t - \square$
c   Use the equation to check your answers to part **a**.

5   The graph shows the cost, c ($) of a taxi for a distance of d km.

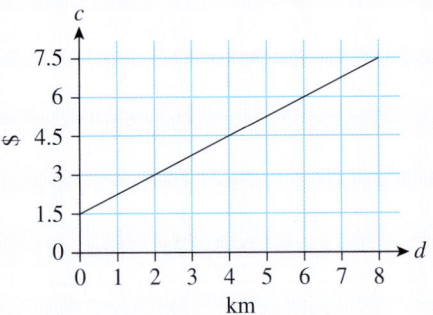

a   Find the intercept on the c-axis. What does this mean?
b   The gradient of the graph is 0.75. What does this mean?
c   What is the equation of the line?

Travel graphs

In Student Book 7 you learned how to draw and interpret distance–time graphs. Here, this is extended to showing more than one person's journey on the same graph.

Example 5

The graph shows the car journeys made by the Astley family (A) and the Brown family (B).

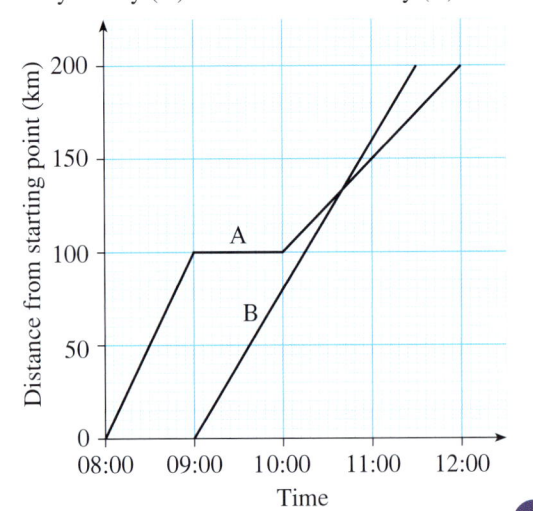

a   What time does each family set off?
    The Astley family left home at 08:00 and the Brown family left home at 09:00.
b   Which family stops for a break in their journey?
    The Astley family stops for a break between 09:00 and 10:00 after driving 100 km, shown by the horizontal line in their graph.
c   At what time have both families travelled the same distance?
    10:40 (the point where both lines intersect, or cross each other).
d   Which family is driving faster at the start of their journey?
    The Astley family's line is steeper and they travel 100 km in the first hour of their journey (or at a speed of 100 km/hr), whereas in the first hour of their journey the Brown family travel 80 km (or at a speed of 80 km/hr).
e   Which family is driving faster for the last 100 km of their journey?
    The Brown family has a steeper line for the last 100 km of the journey. They travel 100 km in $1\frac{1}{4}$ hours, whereas the Astley family take 2 hours to travel 100 km.

This next exercise also includes graphs with more than one component that are not travel graphs.

Exercise 14G

1   This graph shows two people making the same journey, starting at the same time, one cycling and the other walking.

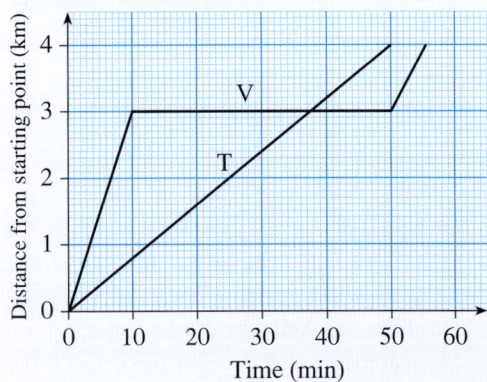

a   Does graph V or T show the cyclist? Give a reason for your answer.
b   For how long did person V stop?
c   Describe what happened 37.5 minutes into the journey.

d What time did person V reach the destination?

e Who arrived at the destination first?

f What speed, in km/hr, was person V travelling
   i before they stopped
   ii after they stopped?

g What speed, in km/hr, was person T travelling?

2 Neema left the park and rode her bike home. This graph shows her journey.

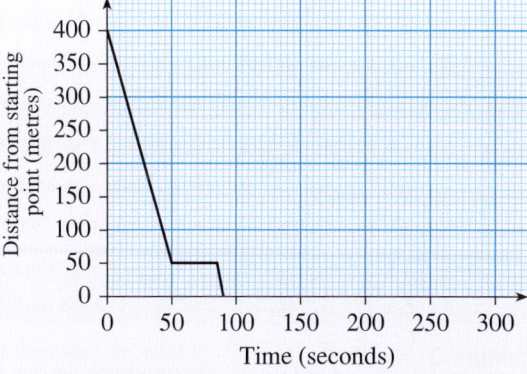

Starting at the same time as Neema, Neema's sister Samira walked from home to the park. This table describes her journey.

Time (seconds)	Distance from home (metres)
0	0
50	50
80	50
300	400

a Copy the graph and draw Samira's journey on it.

b How far is the park from Neema and Samira's house?

c The two sisters met each other.
   i After how many seconds into their journey did they meet?
   ii Having met, how long did they stop for?

d Describe how the graph shows who is travelling faster in the first 50 seconds of the journeys.

e In m/s, how fast was Neema cycling for the first part of her journey?

f In m/s, how fast was Samira walking during the first part of her journey?

3 The graph below shows two friends, Devaj and Farhad, going for a run and coming home again.

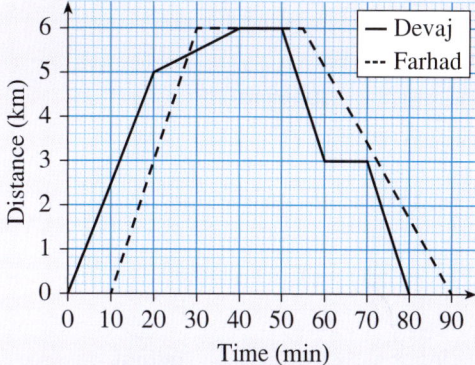

Are these statements true or false?

a Farhad stopped for half an hour.

b The slowest running was by Devaj, in the last kilometre he ran.

c After 70 minutes Devaj was nearer home than Farhad.

d Farhad began running at 15 km/hr.

e Devaj ran the last part of his journey at 18 km/hr.

f Devaj and Farhad were away from home for the same length of time.

g The friends met each other twice on their run.

h On the way home Devaj stopped for a 10-second break.

4 Azhar buys a phone for $70 and the call rate is $0.10 per minute.
Wasim buys a phone for $60 and the call rate is $0.15 per minute.
They each make 400 minutes of calls.

a Draw a graph to show this information with total cost in dollars, c, on the vertical axis and time in minutes, t, on the horizontal axis. Start your vertical axis at 50 rather than 0.

b How does the graph show who paid least?

c If Azhar had made 300 minutes of calls what would the total cost be?

d If Wasim had paid a total of $75 how many minutes of calls is this?

e What is the number of minutes of calls where the total cost is the same for Azhar and Wasim?

Consolidation

Example 1

a Complete this mapping diagram using the function shown.

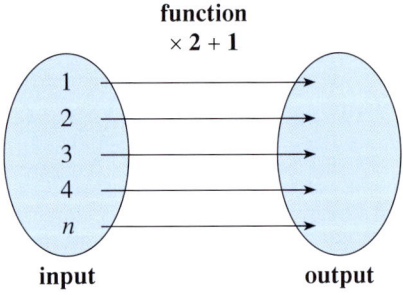

function
× **2** + **1**

1
2
3
4
n

input   output

b If the inputs are the x-coordinates and the outputs are the y-coordinates write the equation of the line.

c Copy and complete this table using the equation you found in part **b**.

x	$^-2$	$^-1$	0	1	2
y					

d Write down the coordinates of the five points you found from the table.

e Draw the line of the equation you found in part **b**.

....

a

function
× **2** + **1**

1 → 3
2 → 5
3 → 7
4 → 9
n → 2n + 1

input   output

b Using $x \rightarrow 2x + 1$ gives us the equation $y = 2x + 1$. Notice this equation is of the form $y = mx + c$ so it will be a straight line graph.

c

x	$^-2$	$^-1$	0	1	2
y	$^-3$	$^-1$	1	3	5

d The table gives us the coordinates:
$(^-2, ^-3)$ $(^-1, ^-1)$ $(0, 1)$ $(1, 3)$ $(2, 5)$

e Plotting these points and joining with a straight line gives the following.

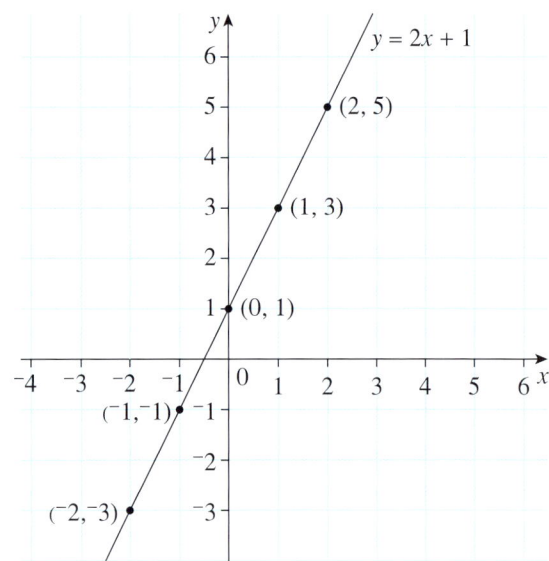

Example 2

The graph below shows the journeys of Yasmina and Jadee as they went out for a run.

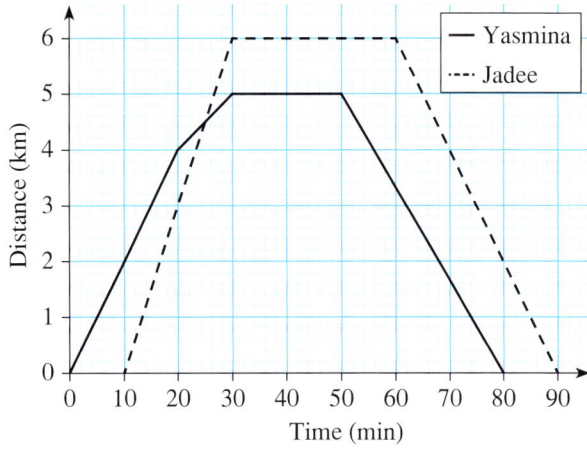

a Who travelled furthest?

b How long did each runner stop for?

c When did Jadee pass Yasmina?

....

a Jadee travelled furthest, as she ran 12 km in total while Yasmina only ran 10 km.

b Yasmina stopped for 20 minutes, Jadee stopped for $\frac{1}{2}$ hour.

c Jadee passed Yasmina after 25 minutes.

Exercise 14

1 a Copy and complete this mapping diagram using the function shown.

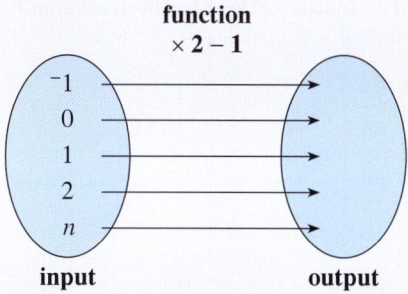

function
× 2 − 1

input   output

b If the inputs are the x-coordinates and the outputs are the y-coordinates write the equation of the line.

c Copy and complete this table using the equation you found in part **b**.

x	⁻1	0	1	2
y				

d Write down the coordinates of the four points you found from the table.

e Draw the line of the equation you found in part **b**.

2 What can you say about your answers to question **1**, parts **a**, **c** and **d**?

3 a Using the inputs ⁻2, ⁻1, 0, 1, 2, draw mapping diagrams for the functions:
 i × 2 + 4   **ii** × 3 − 5

b Using your answers to part **a**, draw the lines:
 i $y = 2x + 4$ **ii** $y = 3x − 5$

4 What are the missing numbers in these sequences?
 a 32, 38, 44, □, 56, 62, □, 74, …
 b □, 46, 42, 38, 34, □, 26, …

5 Find the nth term for the multiples of 8.

6 One of these points doesn't lie on the line $y = 8x − 3$. Which is it?
 (1, 5)   (4, 29)   (0, ⁻3)   (7, 59)

7 Copy and complete the table below for the equation $y = 2x − 3$.
 Use it to draw the graph of $y = 2x − 3$.

x	⁻2	⁻1	0	1	2
y					

8 Write down the first five terms of the sequence with nth term $11n + 4$.

9 Using the inputs ⁻2, ⁻1, 0, 1, 2, draw graphs of the mapping of the x-coordinate to the y-coordinate using:
 a $x \rightarrow x − 1$ **b** $x \rightarrow 3x + 1$
 c $x \rightarrow 2x + 4$ **d** $x \rightarrow 5 − 2x$

10 Using the equation $y = 9x + 10$, copy and complete these coordinate pairs by calculating the missing y-coordinate using the given x-coordinate.
 a (6, □)   **b** (3, □)
 c (⁻4, □)   **d** (⁻5, □)

11 Write the equation that goes with each sentence.
 a To find the y-coordinate multiply the x-coordinate by 4 then subtract 3.
 b To find the y-coordinate subtract the x-coordinate from 2.
 c To find the y-coordinate halve the x-coordinate then add 8.
 d To find the y-coordinate add 3 to the x-coordinate then halve that result.

12 Which of these points lie on the line $y = 8 − 2x$?
 (3, 4)   (0, 8)   (⁻1, 10)   (⁻2, 4)

13 This table shows the conversion from metres per second to kilometres per hour.

Speed (m/s)	0	10	20	30	40	50	60
Speed (km/h)	0	36	72	108	144	180	216

a Draw a graph to show this information. Use a scale of 2 cm to 10 m/s on the horizontal axis and 1 cm to 10 km/h on the vertical axis.

b Use your graph to convert these speeds to km/h.
 i 15 m/s   **ii** 45 m/s

c Use your graph to convert these speeds to m/s.
 i 90 km/h   **ii** 150 km/h

14 The masses of different volumes of kerosene are shown in this graph.

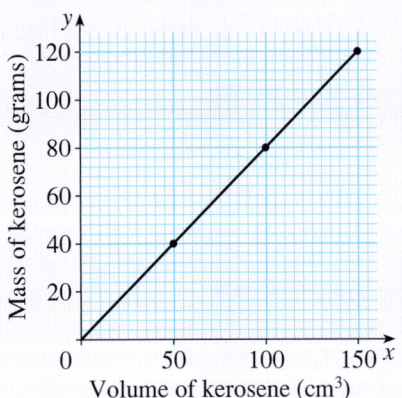

a What mass of kerosene has a volume of:
    i 50 cm³   **ii** 70 cm³
    iii 106 cm³?
b What volume of kerosene has a mass of:
    i 20 g   **ii** 32 g   **iii** 105 g?
c What is the mass of 1 cm³ of kerosene?
    (Note: this is the density of kerosene.)

15 One car travels from town A to town B. Starting at the same time, another car travels from town B to town A.

The journey of one car is shown on this graph.

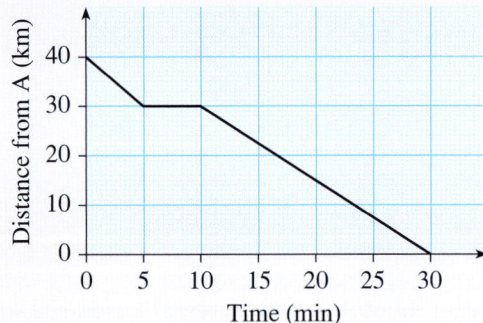

The other car started at town A, drove 30 km in 15 minutes, stopped at the garage for 5 minutes for fuel, then drove the rest of the way at a speed of 120 km/h.
a Describe the journey of the car starting at town B.
b Copy the graph above and draw on it the journey of the car starting at town A.
c After how long did the cars pass each other? How far from town A were they?

Summary

You should know ...

1 Graphs with equations in the form $y = mx + c$ are straight lines.
For example:
$y = 12x - 4$ and $y = 8 - x$ are straight lines.
For $y = 12x - 4$, $m = 12$ and $c = {}^-4$
For $y = 8 - x$, $m = {}^-1$ and $c = 8$

Check out

1 For these lines with equations in the form $y = mx + c$ write down the value of:
    i m
    ii c
a $y = 2x + 7$
b $y = x - 4$
c $y = 8x$
d $y = 10 - x$

2 The function $x \rightarrow 2x + 3$ represents the equation $y = 2x + 3$.
To find the y-coordinates you need to double the
x-coordinates and add 3.
Completing a table gives you:

x	$^-1$	0	1	2
y	1	3	5	7

This is the graph of $y = 2x + 3$.

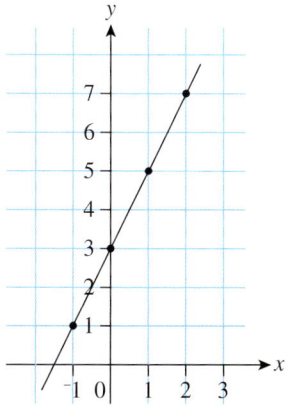

3 How to draw and interpret graphs with more than one
component.
For example:
This graph shows a journey to and from the town centre
made by bus and by car, starting at the same time.

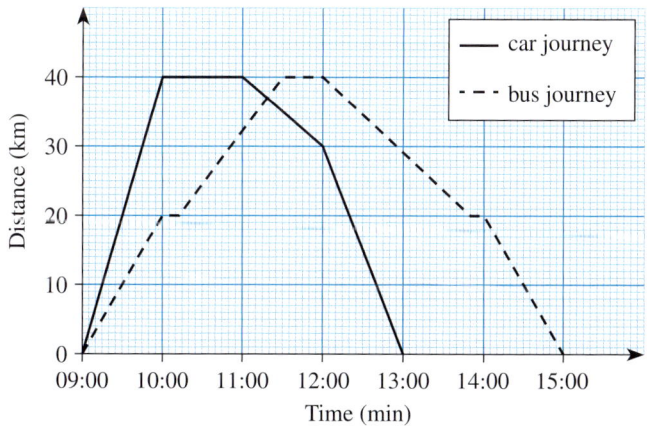

How many stops did the bus make on the way to the
bus station (which is in the town centre)?

There was one stop at 10 am.

2 For each equation, complete this
table and draw the graph.

x	$^-2$	$^-1$	0	1	2
y					

a $y = 2x + 3$
b $y = 4 - x$

3 Answer these questions using the
graph on the left.
a How long did the car driver
stop in the town centre?
b How long did the bus driver
stop for at the bus station?
c The car driver was on the way
home while the bus was still on
its way into town. At what time
did they pass each other?
d Between which times was the
car driver's journey:
i fastest
ii slowest?

15 Probability

Objectives

In this chapter you will learn about:
- experimental and theoretical probability
- the probability of an event not happening
- using lists and diagrams to identify combined events
- finding the probability of combined events.

What's the point?

How likely is it that your team will win a football match? What are the chances that it will rain today? Probability gives answers to such questions and many more.

Before you start

You should know ...

1 How to simplify a fraction by dividing the numerator and denominator by the highest common factor.
   For example:

$$\frac{8}{12} = \frac{2}{3}$$
   $\div 4$ (top), $\div 4$ (bottom)

   > 4 is the HCF of 8 and 12.

2 How to add and subtract decimals.
   For example:
   $0.2 + 0.35 = 0.55$
   $1 - 0.6 = 0.4$

Check in

1 Write these fractions in their simplest form.

   a $\frac{9}{18}$ b $\frac{4}{16}$

   c $\frac{12}{30}$ d $\frac{15}{35}$

2 Work out:
   a $0.4 + 0.2$
   b $0.5 + 0.25$
   c $1 - 0.8$
   d $1 - 0.35$

3 How to add and subtract fractions using a common denominator.

For example:

$$\frac{1}{2} + \frac{1}{4} = \frac{2}{4} + \frac{1}{4} = \frac{3}{4} \quad \text{or} \quad 0.75$$

$$1 - \frac{2}{5} = \frac{5}{5} - \frac{2}{5} = \frac{3}{5} \quad \text{or} \quad 0.6$$

3 Work out:

a $\frac{1}{4} + \frac{2}{5}$ **b** $\frac{3}{8} + \frac{4}{7}$

c $1 - \frac{3}{5}$ **d** $1 - \frac{7}{8}$

15.1 The idea of probability

Some events are more likely to happen than others. In mathematics, the word **probability** is used to describe this situation.

There is a **high probability** that you will put on shoes tomorrow.

There is a **low probability** that you will stay awake for 24 hours.

In mathematics, the probability of an event is given as a number from 0 to 1.

The diagram shows the meaning of different probabilities:

```
1   ┬ certain
0.9 ┤ nearly certain
0.8 ┤
0.7 ┤
0.6 ┤
0.5 ┤ just as likely to happen as not (even chance)
0.4 ┤
0.3 ┤
0.2 ┤
0.1 ┤ very unlikely
0   ┴ impossible
```

You can see that the probability of a **certain** event is **1**; the probability of an **impossible** event is **0**; if an event is **as likely to happen as not**, its probability is 0.5.

Probability can be written as a fraction as well as a decimal. Probability can also be written as percentages. When using percentages the probability scale goes from 0% to 100%. Most of the work in this chapter will be using decimals and fractions.

Never write probability in words or using ratios. For example, '3 : 4' or '2 out of 5' are both commonly used but incorrect ways of writing probability.

Exercise 15A

1 *I will live to be a hundred.*

How likely is this event?

2 Three students answered question **1**. Their answers are shown on the diagram. Try to write their answers in words.

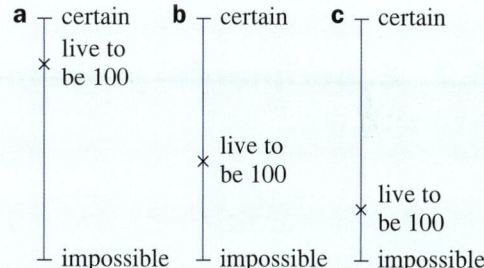

3 Draw a diagram like those in question **2**.

Mark and label a cross on it to show whether you think each event is nearly certain, nearly impossible, or as likely as not.
a I will get married.
b If I buy a new pen it will write.
c It will rain tomorrow.
d I will get to school on time tomorrow.
e I will take 20 catches in a cricket match.

4 Are some of the events in question **3** more likely to happen than others? Pick out the most likely event.

5 Write down three events that you think are:
a impossible
b certain
c just as likely to happen as not.

6 Try to decide for yourself how certain or impossible an event is, if its probability is described as:
a 0   **b** 1   **c** $\frac{1}{2}$

d $\frac{9}{10}$ **e** $\frac{2}{10}$

7 Look again at question **3**. Give each event a probability between 0 and 1.

8 Assign a probability of 0 or 1 to each event.
   a One day I will die.
   b Tomorrow it will snow.
   c I will get 100% in my maths exam.
   d If I put ice outside in the sunshine it will melt.
   e I shall grow 2 cm taller today.

9 Choose a fraction between 0 and 1 that you think describes the probability of each event:
   a The next baby born at the local hospital will be a boy.
   b A throw of a dice will give a 5.
   c A number chosen from 1 to 10 will begin with F when spelled out in words.

15.2 Experimental probability

You will need a coin.

You can define the probability that an event occurs, that is, a 'successful outcome' P(S), as:

● $\mathbf{P(S)} = \dfrac{\text{number of successful outcomes}}{\text{total number of outcomes}}$

For example, if you throw a coin 100 times the probability of getting heads is:

$P(\text{heads}) = \dfrac{\text{number of times heads occurred}}{\text{total number of throws}}$

$= \dfrac{\text{about } 50}{100}$ or about $\dfrac{1}{2}$ or 0.5

Many probabilities can be found either by experiment or survey.

Example 1

Fifty people were asked their favourite sport.

The results are shown in the table.

Sport	Cricket	Basketball	Netball	Football
No. of people	17	8	20	5

Calculate the probability that:
a a person's favourite sport is football
b a person's favourite sport is *not* netball

....

a P(favourite sport is football)

$= \dfrac{\text{number of people liking football}}{\text{total number of people}}$

$= \dfrac{5}{50} = \dfrac{1}{10}$

b P(favourite sport is not netball)

$= \dfrac{\text{number of people who like sports other than netball}}{\text{total number of people}}$

$= \dfrac{17+8+5}{50} = \dfrac{30}{50} = \dfrac{3}{5}$

Exercise 15B

1 Call the face of a one-cent coin with a 1 written on it *tails*. Call the other face *heads*. Toss the coin 100 times. Keep a record of which face turns up each time in a table like this:

Tails	Heads
⊬⊬ ‖	⊬⊬ ⊬⊬

For what fraction of the total number of throws was the outcome:
a tails   **b** heads?

2 Now collect the results from nine students in your class and record them in a table like this:

	Tails	**Heads**
My result	48	52
Student 1		
2		
⋮		
9		
Total		

a Compare your results with other students' results.

b For what fraction of the total number of throws was the outcome:
    i tails   **ii** heads?

3 Look at your results for questions **1** and **2**. When you throw a coin are you:
  a more likely to get tails
  b more likely to get heads
  c equally likely to get either?

4 Packets of flower seeds are checked for purity, in case any weed seeds have been included. Here is a frequency table showing the results of checking 100 packets.

Number of weed seeds	0	1	2	3	4	5	6	7	8
Number of packets	5	16	26	19	13	12	5	2	2

What is the probability that a packet contains:
  a 6 weed seeds   **b** 1 weed seed
  c more than 6 weed seeds?

5 A scientist weighs kittens at birth, with the following results.

Mass (g)	190	200	210	220	230	240	250
Frequency	13	34	57	50	29	12	5

  a How many kittens did the scientist weigh?
  b Find the probability that a kitten has a mass of:
    i 250 g   **ii** less than 250 g.

6 A test report on 32 cars gives the maximum speeds shown in this table.

Maximum speed (km/h)	Number of cars
100 119	3
120–139	9
140–159	5
160–179	5
180–199	5
200–219	3
220–239	1
240–259	1

Find the probability that a car picked at random has a maximum speed of:
  a 120–139 km/h
  b less than 120 km/h

7 These are the marks of 50 people who took a skills test. The maximum mark was 240.

Marks	Number of candidates
0–29	2
30–59	5
60–89	9
90–119	16
120–149	8
150–179	7
180–209	2
210 or over	1

What is the probability that a person chosen at random scored:
  a 120–149   **b** less than 30
  c 120–179   **d** 180 or over?

15.3 Theoretical probability

Often it is impractical or too time-consuming to carry out an experiment. In these cases you can still calculate the theoretical probability of an event. It is defined in the same way as experimental probability, that is:

$$P(\text{success}) = \frac{\text{number of successful outcomes}}{\text{total number of outcomes}}$$

Example 2

One letter is chosen at random from the word STATISTICS. What is the probability that it is
  a S   **b** a vowel   **c** not a vowel?

....

  a $P(S) = \dfrac{\text{number of Ss}}{\text{total number of letters}} = \dfrac{3}{10}$

  b $P(\text{vowel}) = \dfrac{\text{number of vowels}}{\text{total number of letters}} = \dfrac{3}{10}$

  c $P(\text{not a vowel}) = \dfrac{\text{number of consonants}}{\text{total number of letters}} = \dfrac{7}{10}$

You will see from the answers to Example 2, parts **b** and **c** that the probability of getting a vowel and the probability of not getting a vowel add up to 1:

$$\frac{3}{10} + \frac{7}{10} = 1$$

So there is a quick way of working out the answer to part **c**:

$$1 - \frac{3}{10} = \frac{7}{10}$$

• If the probability of an event occurring is p, then the probability of it not occurring is $1 - p$.

The event 'A' and the event 'not A' are said to be complementary. We use this notation:

$$P(A) + P(A') = 1$$

Exercise 15C

1 A dice is thrown.
   a How many different ways could it land?
   b What is the probability that a 3 is thrown?
   c What is the probability of not throwing a 3?

2 What is the probability of picking a 6 at random from 10 cards labelled 1 to 10?

3 A bag contains 2 white beads and 3 black beads. One bead is chosen at random.

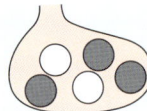

What is the probability that the bead is
   a white   **b** black?

4 A letter is chosen at random from the word MATHEMATICS. What is the probability that the letter is
   a A   **b** M   **c** not A   **d** not I?

5 25 yellow cards numbered 1 to 25 and 25 blue cards numbered 1 to 25 are mixed together.

Here are *some* of the cards.

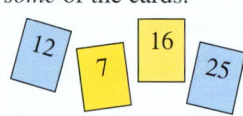

What is the probability that a card picked at random is
   a a 3
   b blue
   c not a 7
   d not blue
   e a number that is a multiple of 3
   f a blue 24?

6 **a** One month of the year is to be chosen. List the possible outcomes.
   b How many of the possible outcomes in part **a** begin with the letter J?

c What is the probability that the chosen month will begin with the letter:
   i J   **ii** M   **iii** D   **iv** P?
d What is the probability that the month will not begin with J?

7 A class has 15 boys and 30 girls. The teacher chooses the class monitor as follows.

The students' names are written on slips of paper, which are folded and put in a bag; the bag is shaken and one name is taken out. This person will be the monitor.

What is the probability that the monitor will be:
   a a boy   **b** a girl?

8 Look at your answers for question **7**.
   a Is it more likely that the monitor will be a boy or a girl?
   b What can you say about the sum of the two probabilities?
   c What does a probability of 1 mean?

9 A bag contains 3 blue beads, 5 green beads and 2 red beads. One bead is picked from the bag. Which colour is:
   a most likely to be picked
   b least likely to be picked?

10 A box of buttons is made up as follows: 5 blue, 6 green, 8 red, 4 yellow, 7 white. One button is picked from the box.
   a How many buttons are there altogther?
   b Are the following true or false?

   i $P(R) = \frac{4}{15}$ **ii** $P(B) = \frac{1}{5}$
   iii $P(G) = 20\%$ **iv** $P(\text{not Y}) = 0.8$

11 A bag contains the following currency notes: ten \$1, twenty \$5, forty \$10, thirty \$20. The notes are shaken up and one is chosen.
   a How many notes are there altogether?
   b Calculate:
      i $P(\$1)$ **ii** $P(\$5)$
      iii $P(\$10)$ **iv** $P(\$20)$

12 A box contains 20 pens. Three of them are faulty. One pen is chosen from the box. What is the probability that it is:
   a faulty   **b** not faulty?

13 A dice is rolled once. Calculate:
  a P(5)
  b P(6)
  c P(even number)
  d P(odd number)
  e P(number < 5)
  f P(number > 2)
  g P(multiple of 3)
  h P(number not divisible by 3)

14 The probability it will rain tomorrow is 0.1. Find the probability it will not rain tomorrow.

15 The probability that Anna will win her tennis match is 0.45. Find the probability Anna will not win her tennis match.

16 If P(C) = 25% find P(C′).

17

A bag contains 3 black balls and 2 white balls. A ball is picked from the bag at random. Each time the ball is picked it is put back in the bag.

Some extra balls are added to the original bag or balls are removed from the original bag. Decide whether the probability of picking a **white** ball stays the same, increases or decreases for each of the following.

  a 6 more black balls and 4 more white balls are added to the original bag.
  b 2 grey balls are added to the original bag.
  c 5 more black balls and 4 more white balls are added to the original bag.
  d 2 black balls and 1 white ball are removed from the original bag.
  e 1 black ball and 1 white ball are removed from the original bag.

18 a What set of whole numbers should be written in each of the 10 sections of this spinner so that all of the following statements are true?

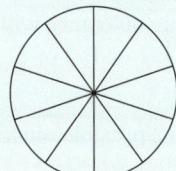

  i The probability of the spinner landing on a number less than 17 is 1.
  ii The probability of the spinner landing on a number less than 4 is 0.
  iii The probability of any two numbers on the spinner being the same is 0.
  iv The probability of the spinner landing on an odd number is 0.6.
  v The probability of the spinner landing on a square number is 30%.
  vi The probability of the spinner landing on a multiple of 4 is $\frac{2}{5}$.

  b Make up your own similar puzzle. Try to make it so that there is only one possible set of numbers. You can use spinners a different number of sections. Ask your friend to solve your puzzle.

Comparing experimental probabilities with theoretical probabilities

Experiments can be used to see if something is **fair**.

For something that is **biased** (not fair) the experimental probabilities and the theoretical probabilities will be very different.

In Year 7 you learned that the more times you repeat an experiment, the closer the experimental probabilities will be to the theoretical probabilities. You will have to be cautious about conclusions if the experiment was only repeated a few times.

Exercise 15D

1 A fair coin is one that has two sides, heads and tails, each of which it is equally likely to land on when flipped. Sort these coins into one of these categories:
  i coin likely to be biased
  ii coin likely to be fair
  iii too few trials to tell
  a Coin A gets 45 heads when it is flipped 100 times.
  b Coin B gets 3 heads when it is flipped 8 times.
  c Coin C gets 350 heads when it is flipped 1000 times.
  d Coin D gets 22 heads when it is flipped 40 times.

2 A fair 6-sided dice has a probability of $\frac{1}{6}$ of showing a score of 5 when it is rolled.

   a How many 5s would you expect Anya to get if she rolled the dice:
      i 60 times
      ii 6000 times
      iii 300 times
      iv 2400 times?

   b Anya rolls four dice 480 times each. Only one of them is a fair dice. She records the number of times she gets a 5. Which of A, B, C or D is most likely to be the fair dice?

Dice	A	B	C	D
Number of 5s	98	71	84	62

3 Sachin says, 'My science test only has true or false questions. It's certain I'll get at least half of the questions correct.'
Do you agree with Sachin? Explain why.

4 Here are three spinners.

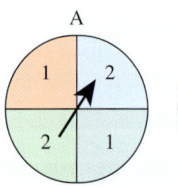 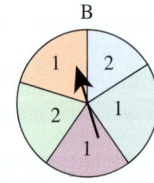 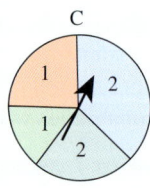

   a Dilvender spins one of these spinners 100 times. 42 of these spins are a 2. Which of these spinners do you think she used?

   b **i** How many 1s would you expect Dilvender to get if she spins spinner C 200 times?

      ii Explain why it is not certain she will get exactly this number.

5 Emma writes the four letters of her name on separate cards.

She picks a card at random and notes which letter it is. She puts it back and then repeats this until she has picked 150 times in total. Are the following statements true or false?

   a Picking M 70 times is likely.
   b Picking E 60 times is unlikely.
   c Picking a vowel 80 times is unlikely.
   d Picking A 35 times is likely.

15.4 Combined events

So far you have just been looking at single events such as rolling one dice.

Combined events are when you look at two or more events together. For example, rolling two dice, or rolling one dice and spinning a spinner at the same time.

Mutually exclusive outcomes means 'outcomes that cannot happen at the same time'. For example, if you roll a dice, this is an event with lots of possible outcomes.

Two possible outcomes are rolling a 1 and rolling a 6. These two outcomes are mutually exclusive because they cannot both happen at the same time.

Two other possible outcomes are rolling a 1 and rolling an odd number. These are not mutually exclusive because if you roll a 1 both events have happened.

In Exercise 15C, question **6a** you were asked to list all the possible outcomes. When there is only a single event, listing all possible mutually exclusive outcomes is easy.

When there are two or more events you need to make sure that you follow a logical method when listing outcomes, to make sure that you do not miss any out. For example, if you were writing the possible outcomes for throwing two dice, it does not make sense to list them randomly, e.g. 1, 4 or 6, 2 or 2, 5 . . . . You can see that it would be very easy to miss out some outcomes. The next example shows a more logical way of listing.

Example 3

A dice is rolled and a coin is thrown.
a Write down all the possible outcomes.
b What is the probability of throwing a heads and rolling a 5?

a First, list all the possible outcomes with a heads thrown with the coin: H, 1 or H, 2 or H, 3 or H, 4 or H, 5 or H, 6.

Then list all the possible outcomes with tails thrown with the coin: T, 1 or T, 2 or T, 3 or T, 4 or T, 5 or T, 6.

b You can see there are 12 possible outcomes. Only one outcome is H, 5, so $P(H,5) = \frac{1}{12}$.

If there are more outcomes, you can use a two-way table to list them. This is shown in the next example.

$P(\text{score same on both dice}) = \frac{4}{16} = \frac{1}{4}$

If there are not many outcomes for each event you can use a **tree diagram**.

Example 4

a Two tetrahedral (4-sided) dice are rolled. Each dice is numbered 1 to 4. Draw a two-way table to show all the possible outcomes.

b What is the probability of the score on the second dice being higher than the score on the first dice?

a

		Dice 1			
		1	**2**	**3**	**4**
Dice 2	**1**	1, 1	2, 1	3, 1	4, 1
	2	1, 2	2, 2	3, 2	4, 2
	3	1, 3	2, 3	3, 3	4, 3
	4	1, 4	2, 4	3, 4	4, 4

b In the table, those outcomes for which the score on the second dice is higher than the score on the first dice are shown in blue: there are 6. There are 16 possible outcomes, so:

$P(\text{score on the second dice is higher}) = \frac{6}{16} = \frac{3}{8}$

Example 5

Harry flips a coin and spins this spinner.

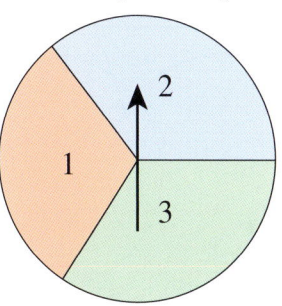

Use a tree diagram to find the probability of getting tails and an odd number.

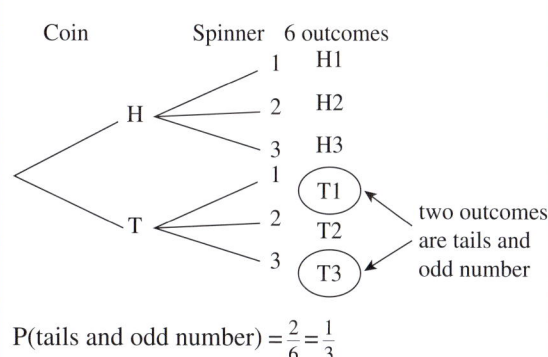

two outcomes are tails and odd number

$P(\text{tails and odd number}) = \frac{2}{6} = \frac{1}{3}$

You can also find probabilities from **Venn diagrams** (see Student Book 7 Chapter 16 for more on Venn diagrams).

You don't always have to put numbers (or letters) in a table, as in Example 4. You can use a **possibility diagram**. For example:

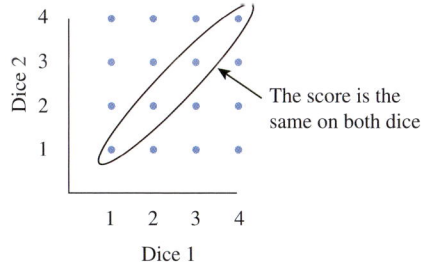

The score is the same on both dice

Example 6

Sourav picks a number at random from the numbers 1 to 15. Find the probability that it is odd and a multiple of 5.

You can sort the numbers in a Venn diagram.

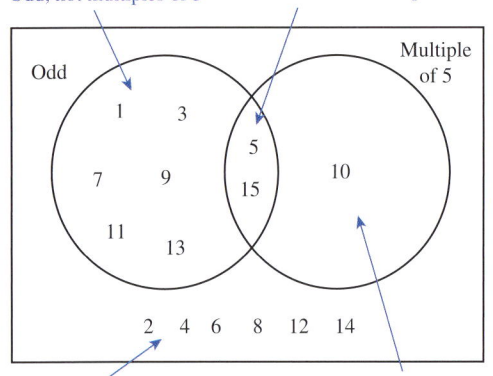

Odd, not multiples of 5   Both odd and a multiple of 5

Even, not multiples of 5   Even multiples of 5

$P(\text{odd number and a multiple of 5}) = \dfrac{2}{15}$

Exercise 15E

1 Two coins are thrown. This is Sarah's working.

The possible outcomes are HH, HT, TT so $P(TT) = \frac{1}{3}$

a Sarah has made a mistake. What mistake has she made?
b What should the probability of two tails be?

2 Two dice are rolled. Each dice is numbered 1 to 6.

a Copy and complete this two-way table to show all the possible outcomes.

		Dice 1					
		1	2	3	4	5	6
Dice 2	1		2, 1				
	2						
	3						
	4					5, 4	
	5						
	6						

b What is the probability of rolling:
  i a 3 on the first dice and a 6 on the second dice
  ii the same score on both dice
  iii a higher score on the first dice than the second?

3 Two children are born in a family. List all possible outcomes and work out the probability that they are both boys.

4 On this restaurant set menu you can choose a starter and a main course.

a Copy and complete this list of possible outcomes (using just the first letter of each dish):
M, L   M, T   etc.
b What is the total number of possible combinations of meals?
c Assuming a starter and main course are chosen at random, what is the probability that at least one of those dishes will contain fish (sardines or trout)?
d If there were 6 starters and 8 main courses, what would the total number of possible combinations of meals be?

5 Alex, Budi, Carl and Dinesh all run a race.
a Write down all the possible outcomes for who wins and who comes second.
b What is the probability that Carl wins and Dinesh comes second (assuming that the runners all have roughly the same ability)?

6 Shashi flips a coin and spins this spinner.

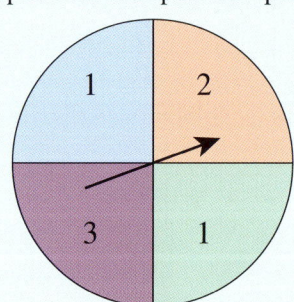

Here is her working to find the probability of getting heads and an even number.

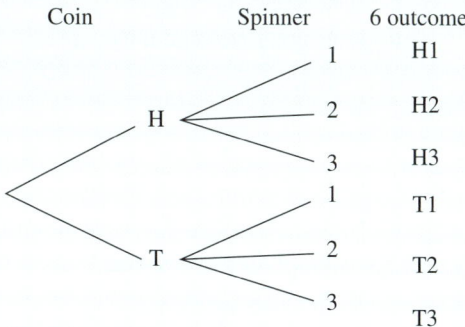

Coin   Spinner   6 outcomes

P(head and even number) = $\frac{1}{6}$

Explain Shashi's mistake.

7 Draw a possibility diagram to find the probability of getting:

a the same number on two 6-sided dice

b a higher number on dice 1 than dice 2

8 Emma flips a coin three times. Complete the tree diagram to find the probability that two of the outcomes are heads and one is tails.

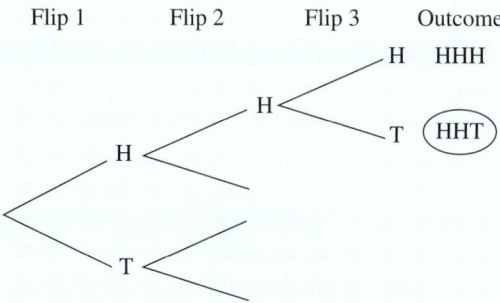

Flip 1   Flip 2   Flip 3   Outcome

9 This Venn diagram shows information about the holiday destinations of 100 people in a particular year.

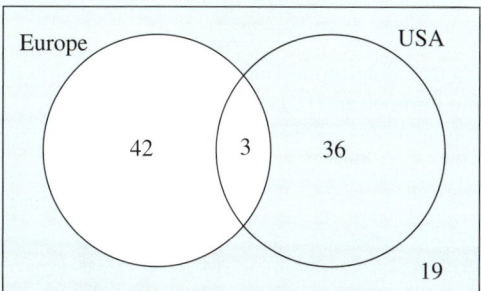

One of these people is picked at random. Find the probability that this person:

a went to both Europe and the USA

b went to Europe but not the USA

c went to Europe

d did not go to Europe or the USA

Give your answers as percentages.

10 a Write down all the different three-digit numbers you can make using the digits 2, 4 and 5. You can use each digit only once.

b What is the probability that the number is greater than 250?

11 Jane is having pizza. She is allowed two toppings on her pizza. (If she really likes something she is allowed two lots of the same topping.)

a If Jane chooses her toppings from pineapple, meat, sweetcorn, mushroom and olives, list all the possible outcomes for her two toppings.

b Assuming Jane chooses her two toppings at random, what is the probability that her pizza has no meat?

12 Jamil wants to phone a friend. He knows the phone number contains six digits. He knows the first four digits of the phone number are 3027 but he cannot remember the last two digits.

a Jamil decides to phone all the possible numbers until he gets the right one. Calls cost $0.25 each. How much would it cost if he had to try all the numbers?

b What is the probability of him calling the correct number first time?

13 Danushka has three bags full of coloured sweets. Each bag has an equal number of red, yellow, green and purple sweets. Without looking, Danushka picks a sweet from each of the three bags at random.

a List all the possible outcomes for the colours of his three sweets.

b What is the probability that at least two of the sweets are red?

c What is the probability that at most one of the sweets is yellow?

14 This table shows how 400 students travel to school.

	Car	Walk	Bus	Train	Total
Boys	26	68	34	56	184
Girls	48	52	40	76	216
Total	74	120	74	132	400

a Find the probability that one student picked at random:
  i is a boy
  ii is a girl who gets the train to school
  iii is a boy who walks to school

b One of the boys is picked at random. Find the probability that he:
  i gets the bus to school
  ii does not go to school by car

Activity

This is a game for two people, player A and player B. You will need two dice and some paper.

One player rolls both dice. The score is the sum of the two rolls.

Player A gets a point if the score is 2, 3, 4, 10, 11 or 12.
Player B gets a point if the score is 5, 6, 7, 8 or 9.

Jane said: 'This game is not fair because player A gets a point if 6 different scores are rolled but player B only gets a point for 5 different scores.'

Play the game for 10 rolls. Who won?
Play the game for 100 rolls. Who won?

Do you think this game is fair? Why?

Change the scoring system of the game to make it fair.

Make up a game of chance of your own.
Write the rules so that each player has an equal chance of winning.
Now change the rules so one player has a better chance of winning.

Consolidation

Example 1

The probability that Emil picks a blue pen from a box is 0.3. What is the probability he does not pick a blue pen?

....

If the probability of an event occurring is p, then the probability of it not occurring is $1 - p$.

$1 - 0.3 = 0.7$

Example 2

James flips a coin 300 times. He gets 142 tails. Is the coin likely to be fair?

....

The theoretical probability of a tail on a fair coin is $\frac{1}{2}$ so you would expect around 150 tails. As 142 is close to this, the coin is likely to be fair.

Example 3

Jenny flips a coin and spins this spinner.

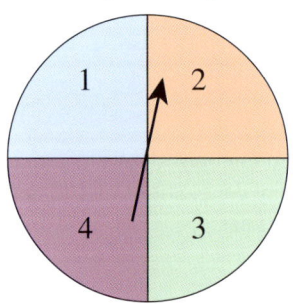

Use a tree diagram to find the probability of getting a prime number and tails.

....

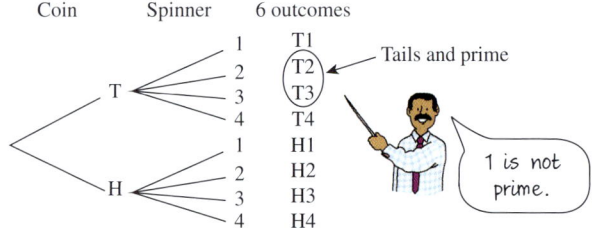

$\text{P(prime number and tails)} = \frac{2}{8} = \frac{1}{4}$

Exercise 15

1 What is the probability of:
  a rolling a dice and getting an even number
  b picking a single digit at random from the digits 1 to 9 and it being less than 3
  c picking a yellow ball from a bag containing 3 blue, 4 red and 8 yellow balls
  d not picking a green ball from a bag containing 7 blue, 2 red, 3 yellow and 8 green balls
  e picking a letter from the alphabet at random that is a vowel (a, e, i, o or u)
  f picking a letter from the alphabet at random that is not a vowel?

2 Jason rolls two dice. Here are his results.

Dice A

Score	1	2	3	4	5	6
Frequency	61	69	58	56	65	64

Dice B

Score	1	2	3	4	5	6
Frequency	103	81	113	90	68	75

  a How many times did he roll:
    i dice A   **ii** dice B?
  b What is the probability of getting a 1 on:
    i dice A   **ii** dice B?
  c Can you tell if the dice are fair? Explain.

3 In the last year, 129 days were recorded as 'wet' days in Cambridge, UK.

  What is the probability that it will be dry in Cambridge today?

4 Mr Masood decided to give his class double homework if both the spinners below were the same colour when spun.

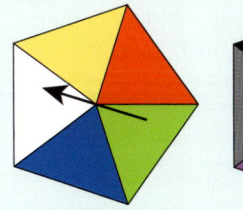 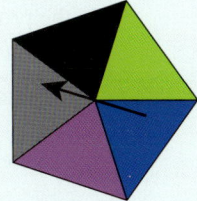

List all possible outcomes.
What is the probability of Mr Masood's class getting double homework?

5 A spinner with numbers 1, 2, 3 and 4 is spun. Sort these spinners into the best category from these three:
    i spinner likely to be biased
    ii spinner likely to be fair
    iii too few trials to tell
  a Spinner A gets 45 even numbers when it is spun 100 times.
  b Spinner B gets 2 number 4s when it is spun 12 times.
  c Spinner C gets 250 number 1s when it is spun 2000 times.
  d Spinner D gets 22 number 3s when it is spun 80 times.

6 The probability that Amie picks a red sweet from a bag of sweets is 0.15. What is the probability she does not pick a red sweet?

7 a A spinner with the numbers 1, 2 and 3 is spun and a coin is flipped. Draw a tree diagram to show all the possible outcomes.
  b What is the probability of an even number and a tail?

8 For lunch Sarah has:

  ● a sandwich that is either tuna (T) or cheese (C)
  ● a drink that is either water (W) or juice (J)
  ● fruit that is either a banana (B) or a mango (M).

  a List all the possible outcomes for Sarah's lunch. The first is done for you.
    TWB, . . .
  b What is the probability that Sarah has a cheese sandwich, water and a banana?

9 The probability that Mahendra has pizza for dinner is 38%. What is the probability that he does not have pizza for dinner?

Summary

You should know …

1 How to find the probability of an event by performing an experiment.

The occurrence of a particular event is called a successful outcome.

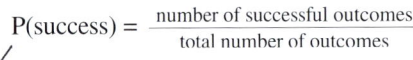 P(success) = $\dfrac{\text{number of successful outcomes}}{\text{total number of outcomes}}$

This means the probability of a success.

For example:
A coin was thrown 100 times and 52 heads occurred, so

P(heads) = $\dfrac{52}{100} = \dfrac{13}{25}$

is the experimental probability.

2 How to find the probability of simple events without performing experiments.

For example:
If a coin is tossed, the theoretical probability of getting heads is

P(heads) = $\dfrac{\text{number of successful outcomes}}{\text{total number of outcomes}}$

= $\dfrac{1}{2}$

3 If the probability of an event occurring is p, then the probability of it not occurring is $1 - p$.

For example:
The probability of picking a red sweet is $\dfrac{3}{8}$ then the

probability of not picking a red sweet is $1 - \dfrac{3}{8} = \dfrac{5}{8}$.

Check out

1 In a survey of 40 people, 6 said they wore small t-shirts, 15 wore medium t-shirts and 19 wore large t-shirts.
  a What is the probability that a person chosen at random from the sample wears a small t-shirt?
  b What is the probability that a person wears a medium or large t-shirt?

2 From the letters in the word ADDING, one letter is chosen. What is the probability that it is:
  a a D
  b not a D
  c a vowel?

3 **a** If the probability of picking a yellow sweet is $\dfrac{2}{5}$, what is the probability of not picking a yellow sweet?
  b If the probability of it raining tomorrow is 0.7, what is the probability of it not raining tomorrow?
  c If the probability of winning a football match is 45%, what is the probability of not winning the football match?

4 When comparing estimated, experimental probabilities with theoretical probabilities, it is important to remember that:
- when experiments are repeated different outcomes may result
- increasing the number of times an experiment is repeated generally leads to better estimates of probability.

4 Three groups of students conducted an experiment using the same biased coin. Here are their results.

Group 1 results

Outcome	Heads	Tails
Frequency	1	9

Group 2 results

Outcome	Heads	Tails
Frequency	31	69

Group 3 results

Outcome	Heads	Tails
Frequency	256	744

a For each of the three groups, work out the experimental probability for getting heads.

b Which of the three groups' results will give the best estimate of probability? Why?

c The theoretical probability for getting heads with this biased coin is actually 0.25. Why do you think that none of the groups' results, from part **a**, were 0.25?

5 How to list systematically all possible mutually exclusive outcomes for two successive events.

For example:
A 6-sided dice and a tetrahedral dice are both rolled.

The possible outcomes can be listed or shown in a two-way table like this.

		Tetrahedral dice			
		1	**2**	**3**	**4**
6-sided dice	**1**	1, 1	2, 1	3, 1	4, 1
	2	1, 2	2, 2	3, 2	4, 2
	3	1, 3	2, 3	3, 3	4, 3
	4	1, 4	2, 4	3, 4	4, 4
	5	1, 5	2, 5	3, 5	4, 5
	6	1, 6	2, 6	3, 6	4, 6

5 List all the possible outcomes when a letter from A, B or C is chosen at random and a 6-sided dice is rolled. The list has been started below for you.

A1, A2, . . .

Review C

1 The probability that Francine picks a white ball from a bag of balls is 0.65. What is the probability that she does not pick a white ball?

2 For the following shapes, find:
  i the order of rotational symmetry
  ii the number of lines.
  a equilateral triangle
  b regular pentagon
  c isosceles trapezium
  d regular heptagon

3 Copy this diagram.

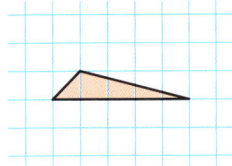

Translate the triangle through these vectors.

a $\begin{pmatrix} 6 \\ 3 \end{pmatrix}$ **b** $\begin{pmatrix} 0 \\ 4 \end{pmatrix}$

c $\begin{pmatrix} {}^-5 \\ 0 \end{pmatrix}$ **d** $\begin{pmatrix} 4 \\ {}^-8 \end{pmatrix}$

4 For the mapping below:
  a draw the function machine
  b write down the ordered pairs
  c draw an X, Y graph on a grid

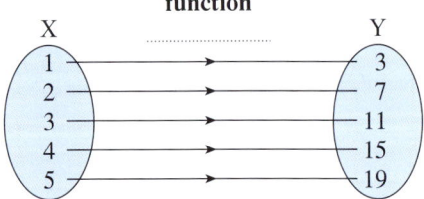

5 Use a ratio to compare the quantities:
  a 1 m; 20 cm
  b 20 mm; 1 cm
  c 3 minutes; 30 seconds
  d 45 minutes; 2 hours
  e 25 g; 0.75 kg
  f 3.6 kg; 90 g

6 Two people start out on the same journey at the same time. One person walks and the other cycles.

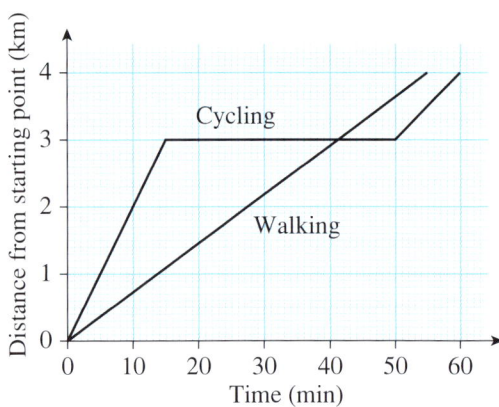

  a For how long did the cyclist stop?
  b Describe what happened just after 41 minutes.
  c After how long did each person reach the destination?
  d What speed, in km/h, was the cyclist travelling
    i before they stopped
    ii after they stopped?

7 a Three coins are flipped. Draw a tree diagram to show all the possible outcomes.
  b What is the probability that all three coins show the same result?

8 Copy this diagram. The quadrilateral on the left is a mirror image of the quadrilateral on the right.

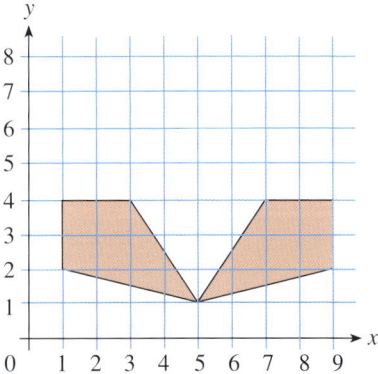

  a On your diagram, draw in the mirror line.
  b Write down the coordinates of three points on that line.
  c Write down the equation of the mirror line.

9 a Write the equation that goes with these sentences.
    i To find the y-coordinate multiply the x-coordinate by 2 then add 5.

ii The x-coordinate is always ⁻4.

iii To find the y-coordinate subtract the x-coordinate from 15.

iv The y-coordinate is always 3.

b Draw the graphs of all the lines from part **a**.

10 Look at this diagram.

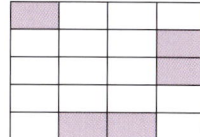

How many more rectangles need to be shaded so that the ratio of shaded to unshaded is 2 : 3?

11

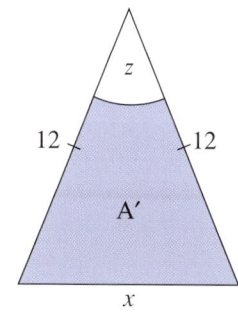

Triangle A is enlarged to make triangle A′.

a What is the scale factor of the enlargement?

b Find the side length x.

c What can you say about angles y and z?

12 Jim flips a coin twice then spins a fair spinner with numbers 1, 2 and 3 on it.

a List all the possible outcomes.

b What is the probability that he gets one heads, one tails and an odd number?

13 The cost of petrol for a given number of litres is shown in this table.

No. of litres	2	6	16	22	38	46
Cost ($)	3.70	11.10	29.60	40.70	70.30	85.10

a Draw a graph of this information. Use a scale of 1 cm to represent 5 litres on the horizontal axis and 1 cm to represent $10 on the vertical axis.

b From your graph, find the cost of the following quantities of petrol.

i 5 litres   **ii** 33 litres

c From your graph, find how much petrol can be bought for:

i $20   **ii** $55

14 Find the image of triangle A (1, 1), B (2, 2), C (2, 4) under an enlargement with:

a scale factor 2 and centre (0, 0)

b scale factor 2 and centre (1, 0)

c scale factor 3 and centre (1, 1).

15 It takes Mr Speedy 32 minutes to cycle to work, or 1 hour 48 minutes to walk to work. Write as a ratio the cycling time to the walking time. Simplify your ratio.

16 Using the numbers 1, 2, 3 and 4 for x, draw a mapping diagram to show:

a $x \rightarrow x + 5$ **b** $x \rightarrow 2x + 1$

17 A kite ABCD has vertices A (1, ⁻4), B (3, ⁻3), C (1, 5) and D (⁻1, ⁻3)

Find the image of the kite under an anticlockwise rotation of:

a 180°   **b** 270° about the origin

18 Write each of these as a ratio in its simplest whole number form.

a $\frac{1}{2} : 3$ **b** $3.5 : 2.8 : \frac{7}{10}$

c $1.4 : 8.2 : 6.3$ **d** $\frac{2}{25} : 6.4$

19 Find the first 5 terms of these sequences.

a The position-to-term rule is *multiply by 3*.

b The position-to-term rule is *multiply by 4 then subtract 1*.

20 Draw the rotation of the shaded triangle 90° clockwise about centre (2, ⁻1).

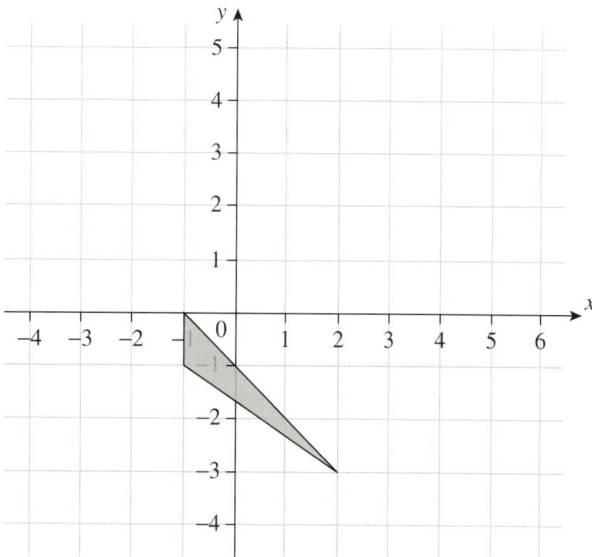

21 A farm has hens, goats and sheep in the proportion 1 : 2 : 5.
  a Find how many goats and sheep there are if there are 8 hens.
  b Find how many hens and goats there are if there are 35 sheep.
  c Find how many hens and sheep there are if there are 12 goats.
  d Find how many of each type of animal there are, if there are altogether:
    i 16 animals   **ii** 48 animals
    iii 72 animals   **iv** 120 animals.

22 Draw a possibility diagram to find the probability of getting an odd number on dice 1 and an even number on dice 2 when two 4-sided dice are rolled.

23 Divide:
  a 540 kg in the ratio 3 : 2 : 4
  b 0.35 m in the ratio 2 : 3
  c 0.119 ℓ in the ratio 5 : 2
  d 0.91 kg in the ratio 7 : 5 : 1
  e 1 hour in the ratio 1 : 9 : 10

24 Look at these diagrams.

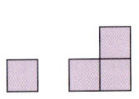

Each time new squares are added around the outside of the previous shape.
  a Draw the next few shapes.
  b Copy and complete this table.

Shape	Number of squares
1	1
2	3
3	
4	
5	
	85
13	
n	

  c Looking at the diagrams, explain why the nth term works.

25 a Write down all the different three-digit numbers you can make using the digits 1, 3 and 6. You can use each digit only once.
  b What is the probability that a number you make is even?

26 Which of these points lie on the line $y = 8 - 2x$?
  (3, 18)   (0, 8)   (⁻1, 6)   (⁻2, 12)

27 The triangle P (2, 1), Q (2, 3), R (1, 3) is mapped to P′ (4, 1), Q′ (4, 5), R′ (2, 5) by an enlargement.
  a Draw PQR and its image on squared paper.
  b By joining PP′, QQ′ and RR′ find the centre of enlargement. Write down its coordinates.
  c What do you notice about the sides and angles of triangle PQR and its image?

28 The probability that a school's netball team wins the next match is $\frac{1}{5}$. What is the probability that it will not win?

29 Describe the transformations that map these triangles to their images.
  a A (2, 0), B (2, 2), C (0, 0)
    → A′ (3, 1), B′ (3, 3), C′ (1, 1)
  b A (⁻1, 2), B (1, ⁻1), C (⁻2, ⁻2)
    → A′ (1, 3), B′ (3, 0), C′ (0, ⁻1)
  c A (3, ⁻1), B (⁻2, 2), C (⁻1, 4)
    → A′ (1, ⁻2), B′ (⁻4, 1), C′ (⁻3, 3)

30 This table shows the number of customers and skate rental at a rollerskating rink during one week in summer.

Day	Customers	Pairs of skates rented
Monday	192	130
Tuesday	328	212
Wednesday	296	222
Thursday	325	195
Friday	456	292

  a What is the probability that a customer will rent skates on Wednesday?
  b What is the probability that a customer will rent skates on Thursday?
  c What is the probability that a customer has their own skates (and doesn't rent any) on Thursday?
  d During all five days, what is the probability that a customer will rent skates?

31 Find the image of triangle A (⁻3, 1), B (⁻1, ⁻1), C (2, 2) under the translation:
  a $\begin{pmatrix} 1 \\ 0 \end{pmatrix}$ **b** $\begin{pmatrix} 0 \\ 1 \end{pmatrix}$ **c** $\begin{pmatrix} 2 \\ 3 \end{pmatrix}$

  d $\begin{pmatrix} ⁻2 \\ 3 \end{pmatrix}$ **e** $\begin{pmatrix} 2 \\ ⁻3 \end{pmatrix}$ **f** $\begin{pmatrix} ⁻2 \\ ⁻3 \end{pmatrix}$

32 Compare these quantities using ratio.
  a 300 mm and 35 cm
  b 7.6 ℓ and 800 ml
  c 6.24 kg and 460 g
  d 23 cm² and 6900 mm²

33 Write down the first five terms of the sequences described.
  a The first term is 80, the term-to-term rule is *subtract 11*.
  b The first term is 0.5, the term-to-term rule is *multiply by 2*.
  c The first term is 11, the term-to-term rule is *add 8*.
  d The *third* term is 200, the term-to-term rule is *divide by 10*.

34 A box contains a set of tennis balls. Six are red, four are white and five are green.
They are mixed and one is chosen at random.
Find the probability of each of these events.
  a A red ball is chosen.
  b A white ball is chosen.
  c A red or green ball is chosen.
  d A red ball is not chosen.

35 a Draw a quadrilateral Q with vertices (0, 1), (1, 3), (-2, 5) and (3, 4).
  b Reflect Q in the line $x = 4$
  c Reflect Q in the line $y = {}^-x$

36

x	-2	-1	0	1	2
y					

Copy and complete the table for each of these equations. Use the table to draw a graph.
  a $y = 3x + 4$ **b** $y = 2x - 1$
  c $y = 7 - x$

37 Natalia has pens and pencils in the ratio 9 : 16.
Venla has pens and pencils in the ratio 4 : 9.
Who has the greatest proportion of pencils?
Show your working.

38 The Venn diagram shows the number of students studying these subjects in a language school with 200 students.

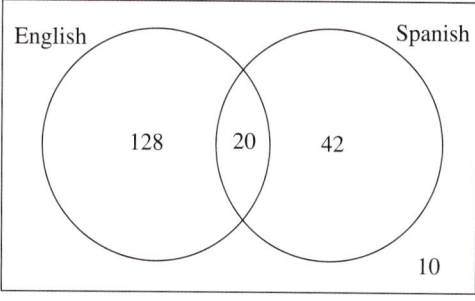

A student is selected at random. What is the probability that this student studies:
  a both English and Spanish
  b English but not Spanish
  c English
  d neither English nor Spanish.
Give your answers as percentages.

39 a Draw a quadrilateral Q with vertices (-3, 1), (-1, 6), (1, 6) and (2, 4).
  b Reflect Q in the line $y = x$
  c Reflect Q in the line $y = 2$

40 Write down the first five terms of the sequence with nth term $3n - 2$

41 A letter from the word STATISTICS is picked at random. Find the probability that the letter is:
  a a vowel
  b an S
  c a consonant
  d an I or a C
  e also a letter from the word PROBABILITY

42 Triangle P with vertices (7, 0), (7, 2), (9, 3) is mapped to triangle Q with vertices (11, 0), (11, 4), (15, 6) by a transformation.
  a Draw the two triangles.
  b Describe the transformation.

43 A plumber charges a call out fee of $30 then $15 for each hour that he works.
  a Draw the graph that shows this information. The maximum number of hours he works in one day is 8 hours.
  b Find the equation of the line if T is the total cost and h is the number of hours worked.

44 Copy and complete:

	Transformation			
	Rotation	**Enlargement**	**Translation**	**Reflection**
Object and image are	Congruent	Similar		
Side lengths preserved?			Yes	
Angles preserved?	Yes			

(**Note:** 'preserved' means 'stay the same'.)

45 Describe fully these transformations.

a

b

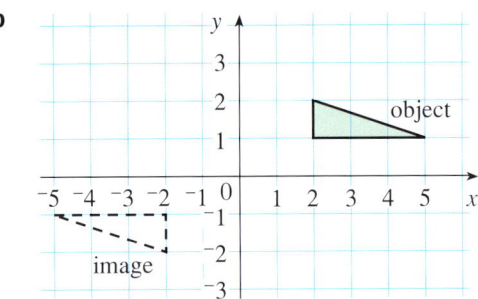

46 Sort these lines into one of the three straight line categories shown below.

$y = 2x$ $x = {}^-4$
$y = {}^-x$ $y = 7$
$y = 3x + 0.5$ $y = \frac{1}{2}x$
$y = 6x - 2$

Categories:

_____ | /\

Horizontal line   Vertical line   Diagonal line

Vectors

Objectives

In this chapter you will learn about:
- column vectors
- adding and subtracting vectors
- using vectors in geometry
- position vectors.

The work in Chapter 16 is not in the Cambridge Lower Secondary Mathematics curriculum. It is in the Cambridge IGCSE® curriculum and is included to stretch and challenge high-attaining students.

What's the point?

We use vectors simply for getting from A to B. They are a great way of describing the space around us and even within us. Vectors are quantities that have both magnitude and direction. They have many applications. A sailboat captain, for example, has to take account of both the wind and the current in order to steer a course correctly. Both of these are vectors.

Before you start

You should know ...

How to find the midpoint of a line segment AB, given the coordinates of points A and B.

For example:

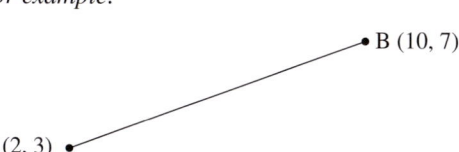

A (2, 3)   B (10, 7)

The midpoint, M, is at $\left(\dfrac{10+2}{2}, \dfrac{7+3}{2} \right)$

M (6, 5)

Check in

Find the midpoint of the line segment AB if:
a   A is at (4, 3) and B is at (10, 5)
b   A is at (17, 1) and B is at (11, 7)

16.1 Introduction to vectors

You will need squared paper.

A **vector** is a quantity that has both **size** (magnitude) and **direction**.

A vector is distinguished from a **scalar** quantity, which possesses size only.

For example:

Scalar	Vector
Mass	Weight
Speed	Velocity
Distance	Displacement

Note: the **velocity** of a car is its **speed** in a particular direction.

Vectors are represented geometrically by arrows. The length of the arrow represents the size of the vector and the direction of the arrow gives the direction of the vector. Such vectors are usually described in terms of their components as **column vectors**.

Note there are many different ways to describe a vector. In Example 1 the vectors can be described as:

- $\overrightarrow{PQ}$ using the letters at the start and end of the line (in part **a**) with an arrow indicating the direction you move between the two points P and Q.

- Using the column vector $\begin{pmatrix} 3 \\ 2 \end{pmatrix}$ the top number being positive means move 3 in the positive x direction (right), the bottom number being positive means move 2 in the positive y direction (up).

- Using the column vector $\begin{pmatrix} ^-2 \\ ^-3 \end{pmatrix}$ the top number being negative means move 2 in the negative x direction (left), the bottom number being negative means move 2 in the negative y direction (down).

- **x** using bold font (for part **b**).

Example 1

Write the vectors as column vectors.

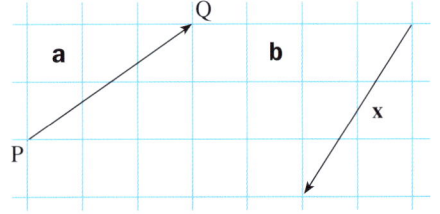

a $\quad \overrightarrow{PQ} = \begin{pmatrix} 3 \\ 2 \end{pmatrix}$ **b** $\quad \mathbf{x} = \begin{pmatrix} ^-2 \\ ^-3 \end{pmatrix}$

Parallel vectors have the same or opposite directions but may differ in magnitude.

Example 2

Look at the vectors below.

Explain how **b** and **c** are related to **a**.

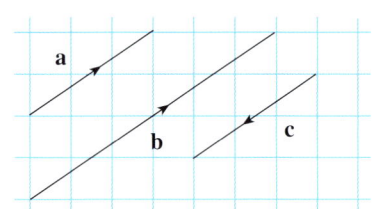

$\mathbf{a} = \begin{pmatrix} 3 \\ 2 \end{pmatrix}, \quad \mathbf{b} = \begin{pmatrix} 6 \\ 4 \end{pmatrix}, \quad \mathbf{c} = \begin{pmatrix} ^-3 \\ ^-2 \end{pmatrix}$

$\mathbf{b} = \begin{pmatrix} 6 \\ 4 \end{pmatrix} = 2 \times \begin{pmatrix} 3 \\ 2 \end{pmatrix} = 2\mathbf{a}$

b is parallel to **a** and twice its length.

$\mathbf{c} = \begin{pmatrix} ^-3 \\ ^-2 \end{pmatrix} = -\begin{pmatrix} 3 \\ 2 \end{pmatrix} = ^-\mathbf{a}$

c is parallel to **a** and has the same length but goes in the opposite direction.

Exercise 16A

1 Write the vectors in the diagram in the form
$$\overrightarrow{AB} = \begin{pmatrix} a \\ b \end{pmatrix}$$

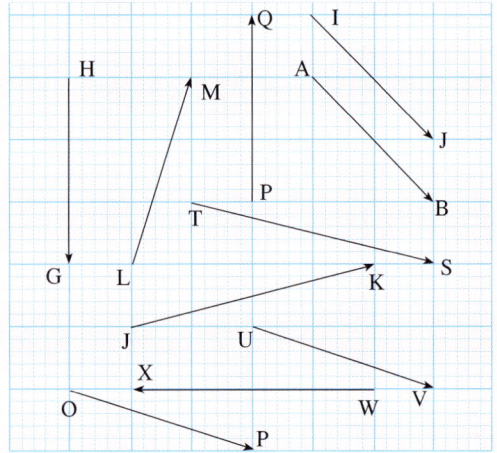

2 Draw these column vectors.

a $\mathbf{a} = \begin{pmatrix} 1 \\ 2 \end{pmatrix}$ **b** $\mathbf{b} = \begin{pmatrix} 1 \\ -2 \end{pmatrix}$

c $\mathbf{c} = \begin{pmatrix} -1 \\ 2 \end{pmatrix}$ **d** $\mathbf{d} = \begin{pmatrix} 2 \\ -4 \end{pmatrix}$

e $\mathbf{e} = \begin{pmatrix} 2 \\ 4 \end{pmatrix}$ **f** $\mathbf{f} = \begin{pmatrix} -1 \\ -2 \end{pmatrix}$

g $\mathbf{g} = \begin{pmatrix} 3 \\ -6 \end{pmatrix}$ **h** $\mathbf{h} = \begin{pmatrix} 3 \\ 0 \end{pmatrix}$

i $\mathbf{i} = \begin{pmatrix} 0 \\ -2 \end{pmatrix}$ **j** $\mathbf{j} = \begin{pmatrix} 0 \\ 1 \end{pmatrix}$

3 a In question **2**, which pairs of vectors are parallel?
b Of the parallel pairs, $\mathbf{e} = 2\mathbf{a}$. What other relationships can you find?

4 Using the vectors **a** and **b** given below, show on squared paper the vectors
$\mathbf{a} + \mathbf{b}, \mathbf{a} + 2\mathbf{b}, \mathbf{a} - \mathbf{b}, 2\mathbf{a} + \mathbf{b}, 2\mathbf{a} + 2\mathbf{b}$.

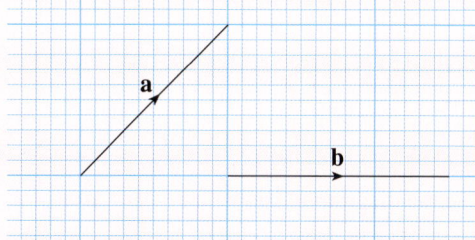

5 Write these vectors in terms of **a**.

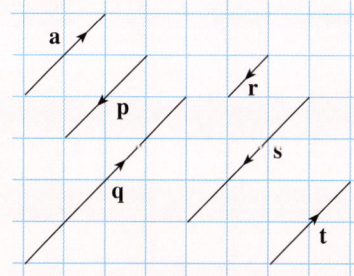

6 If $\mathbf{a} = \begin{pmatrix} 1 \\ 2 \end{pmatrix}$ draw the vectors:
a $2\mathbf{a}$ **b** $3\mathbf{a}$
c $^-\mathbf{a}$ **d** $^-4\mathbf{a}$

7 Draw $\frac{1}{2}\mathbf{a}, \frac{5}{4}\mathbf{a}, 3\mathbf{a}, \frac{-2}{3}\mathbf{a}, ^-3\mathbf{a}$ for each of these vectors.

a 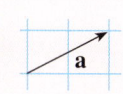 **b**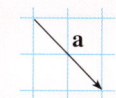

c **d**

Adding vectors

In geometrical applications of vectors, it is useful to think of vectors as representing translations.

The addition of two vectors can then be thought of as the single translation that is equivalent to the two successive translations.

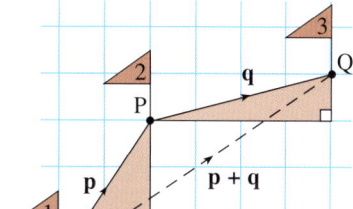

$\mathbf{p} = \begin{pmatrix} 2 \\ 3 \end{pmatrix}$

$\mathbf{q} = \begin{pmatrix} 4 \\ 1 \end{pmatrix}$

In the diagram, the translation of flag 1 to flag 2 is shown by the arrow from O to P.

The translation of flag 2 to flag 3 is shown by the arrow from P to Q.

The translation of flag 1 to flag 3 could be shown by an arrow from O to Q. It is equivalent to translation **p** followed by translation **q**, that is:

$\begin{pmatrix} 2 \\ 3 \end{pmatrix} + \begin{pmatrix} 4 \\ 1 \end{pmatrix} = \begin{pmatrix} 6 \\ 4 \end{pmatrix}$

Vectors are added by placing the arrows that represent them tip to tail:

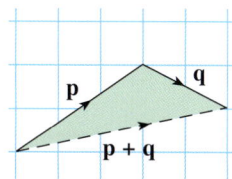

$\begin{pmatrix} 3 \\ 2 \end{pmatrix} + \begin{pmatrix} 2 \\ -1 \end{pmatrix} = \begin{pmatrix} 5 \\ 1 \end{pmatrix}$

Example 3

If $\mathbf{a} = \begin{pmatrix} ^-3 \\ ^-2 \end{pmatrix}$ and $\mathbf{b} = \begin{pmatrix} 2 \\ ^-1 \end{pmatrix}$ draw a diagram to show $\mathbf{a} + \mathbf{b}$

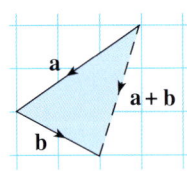

$\mathbf{a} + \mathbf{b} = \begin{pmatrix} ^-3 \\ ^-2 \end{pmatrix} + \begin{pmatrix} 2 \\ ^-1 \end{pmatrix} = \begin{pmatrix} ^-1 \\ ^-3 \end{pmatrix}$

You should notice that you can add vectors by adding their components:

$$\begin{pmatrix} x_1 \\ y_1 \end{pmatrix} + \begin{pmatrix} x_2 \\ y_2 \end{pmatrix} = \begin{pmatrix} x_1 + x_2 \\ y_1 + y_2 \end{pmatrix}$$

You can subtract vectors in the same way.

For example:

if $\mathbf{a} = \begin{pmatrix} 2 \\ ^-1 \end{pmatrix}$ and $\mathbf{c} = \begin{pmatrix} ^-1 \\ 4 \end{pmatrix}$ then

$\mathbf{a} - \mathbf{c} = \begin{pmatrix} 2 \\ ^-1 \end{pmatrix} - \begin{pmatrix} ^-1 \\ 4 \end{pmatrix}$

$\quad\quad = \begin{pmatrix} 2 \\ ^-1 \end{pmatrix} + \begin{pmatrix} 1 \\ ^-4 \end{pmatrix} = \begin{pmatrix} 3 \\ ^-5 \end{pmatrix}$

Pictorially, this is

$\mathbf{a} - \mathbf{c} = \mathbf{a} + (^-\mathbf{c})$

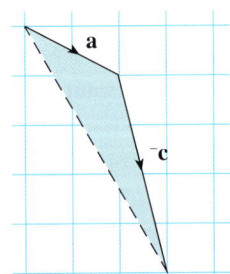

$\mathbf{c} = \begin{pmatrix} ^-1 \\ 4 \end{pmatrix}$

$^-\mathbf{c} = \begin{pmatrix} 1 \\ ^-4 \end{pmatrix}$

Exercise 16B

1 If $\mathbf{a} = \begin{pmatrix} 2 \\ ^-1 \end{pmatrix}$, $\mathbf{b} = \begin{pmatrix} ^-3 \\ 2 \end{pmatrix}$, $\mathbf{c} = \begin{pmatrix} ^-1 \\ 4 \end{pmatrix}$

and $\mathbf{d} = \begin{pmatrix} ^-2 \\ ^-3 \end{pmatrix}$ draw diagrams to find:

a $\mathbf{a} + \mathbf{b}$	**b** $\mathbf{b} + \mathbf{a}$
c $\mathbf{a} + \mathbf{c}$	**d** $\mathbf{c} + \mathbf{a}$
e $\mathbf{b} + \mathbf{d}$	**f** $\mathbf{c} + \mathbf{d}$
g $\mathbf{a} + \mathbf{d}$	**h** $\mathbf{a} - \mathbf{c}$

2 Draw diagrams to show $\mathbf{p} - \mathbf{q}$ when:

a $\mathbf{p} = \begin{pmatrix} 4 \\ 3 \end{pmatrix}$, $\mathbf{q} = \begin{pmatrix} 1 \\ 2 \end{pmatrix}$

b $\mathbf{p} = \begin{pmatrix} 1 \\ 3 \end{pmatrix}$, $\mathbf{q} = \begin{pmatrix} 2 \\ 1 \end{pmatrix}$

Check your answers by subtracting the column vectors.

3 Draw diagrams to illustrate these vector additions.

a $\begin{pmatrix} 1 \\ 3 \end{pmatrix} + \begin{pmatrix} 2 \\ 4 \end{pmatrix} = \begin{pmatrix} 3 \\ 7 \end{pmatrix}$

b $\begin{pmatrix} 1 \\ 3 \end{pmatrix} + \begin{pmatrix} ^-2 \\ ^-4 \end{pmatrix} = \begin{pmatrix} ^-1 \\ ^-1 \end{pmatrix}$

c $\begin{pmatrix} 2 \\ 0 \end{pmatrix} + \begin{pmatrix} 0 \\ 1 \end{pmatrix} + \begin{pmatrix} ^-2 \\ ^-1 \end{pmatrix} = \begin{pmatrix} 0 \\ 0 \end{pmatrix}$

4 Use the diagram to show that $\mathbf{p} + \mathbf{q} = \mathbf{q} + \mathbf{p}$

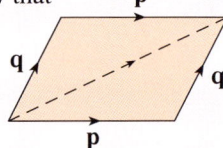

5 Use the diagram to show that:
a $\mathbf{p} - \mathbf{q} = \mathbf{p} + (^-\mathbf{q})$
b $\mathbf{p} + \mathbf{q} + (^-\mathbf{q}) = \mathbf{p}$

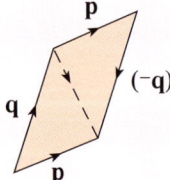

6 By drawing a sketch find:

a $\begin{pmatrix} 3 \\ 8 \end{pmatrix} + \begin{pmatrix} 1 \\ ^-2 \end{pmatrix}$ **b** $\begin{pmatrix} ^-7 \\ 2 \end{pmatrix} + \begin{pmatrix} 2 \\ ^-5 \end{pmatrix}$

c $\begin{pmatrix} 6 \\ 5 \end{pmatrix} + \begin{pmatrix} 6 \\ 7 \end{pmatrix}$ **d** $\begin{pmatrix} 5 \\ ^-5 \end{pmatrix} + \begin{pmatrix} 6 \\ ^-2 \end{pmatrix}$

e $\begin{pmatrix} 7 \\ 4 \end{pmatrix} + \begin{pmatrix} 2 \\ ^-3 \end{pmatrix}$ **f** $\begin{pmatrix} ^-3 \\ 0 \end{pmatrix} + \begin{pmatrix} 4 \\ 1 \end{pmatrix}$

Check your answers by vector addition.

7 Calculate:

a $\begin{pmatrix} 1 \\ 3 \end{pmatrix} + \begin{pmatrix} ^-2 \\ 1 \end{pmatrix} + \begin{pmatrix} 4 \\ 2 \end{pmatrix}$

b $\begin{pmatrix} 3 \\ 4 \end{pmatrix} + \begin{pmatrix} ^-4 \\ 2 \end{pmatrix} + \begin{pmatrix} ^-6 \\ ^-1 \end{pmatrix}$

c $\begin{pmatrix} 13 \\ 15 \end{pmatrix} - \begin{pmatrix} 20 \\ 25 \end{pmatrix} + \begin{pmatrix} 3 \\ 8 \end{pmatrix}$

d $\begin{pmatrix} 0 \\ ^-1 \end{pmatrix} + \begin{pmatrix} 1 \\ 2 \end{pmatrix} - \begin{pmatrix} ^-1 \\ ^-1 \end{pmatrix}$

Use sketches to check your answers.

Vector algebra

All vectors have the following properties.

1 Two vectors **a** and **b** can be added.

For example, if $\mathbf{a} = \begin{pmatrix} 2 \\ 3 \end{pmatrix}$ and $\mathbf{b} = \begin{pmatrix} 4 \\ 1 \end{pmatrix}$

then $\mathbf{a} + \mathbf{b} = \begin{pmatrix} 2+4 \\ 3+1 \end{pmatrix} = \begin{pmatrix} 6 \\ 4 \end{pmatrix}$

2 Vector addition is commutative.

That is, $\mathbf{a} + \mathbf{b} = \mathbf{b} + \mathbf{a}$

3 Vector **a** can be subtracted from vector **b** to give vector **b** − **a**

For example, if $\mathbf{b} = \begin{pmatrix} 4 \\ 1 \end{pmatrix}$ and $\mathbf{a} = \begin{pmatrix} 2 \\ 3 \end{pmatrix}$

then $\mathbf{b} - \mathbf{a} = \begin{pmatrix} 4-2 \\ 1-3 \end{pmatrix} = \begin{pmatrix} 2 \\ ^-2 \end{pmatrix}$

4 A vector **a** can be multiplied by a number k. The result is another vector, $k\mathbf{a}$. When the vector is written as a column vector, each component is multiplied by k.

For example, if $\mathbf{a} = \begin{pmatrix} 2 \\ 3 \end{pmatrix}$

then $5\mathbf{a} = \begin{pmatrix} 5\times2 \\ 5\times3 \end{pmatrix} = \begin{pmatrix} 10 \\ 15 \end{pmatrix}$

5 For each vector **a** there is an opposite or **inverse vector** ($^-\mathbf{a}$), and $\mathbf{a} + (^-\mathbf{a}) = 0$

The result **0** is called the **zero vector** or **null vector**.

For example, if $\mathbf{a} = \begin{pmatrix} 3 \\ ^-2 \end{pmatrix}$ then $^-\mathbf{a} = \begin{pmatrix} ^-3 \\ 2 \end{pmatrix}$

and $\mathbf{a} + (^-\mathbf{a}) = \begin{pmatrix} 0 \\ 0 \end{pmatrix}$

Exercise 16C

1 $\mathbf{r} = \begin{pmatrix} ^-1 \\ 5 \end{pmatrix}$, $\mathbf{s} = \begin{pmatrix} 4 \\ ^-3 \end{pmatrix}$, $\mathbf{t} = \begin{pmatrix} 3 \\ ^-15 \end{pmatrix}$

a Write down:

i $^-\mathbf{r}$ **ii** $^-\mathbf{s}$ **iii** $^-\mathbf{t}$

b Show that:
i $\mathbf{r} + \mathbf{s} = \mathbf{s} + \mathbf{r}$ **ii** $\mathbf{t} + (^-\mathbf{t}) = 0$
iii $\mathbf{t} = ^-3\mathbf{r}$ **iv** $\mathbf{t} + \mathbf{r} = ^-2\mathbf{r}$

c Find:
i $\mathbf{r} + \mathbf{s} + \mathbf{t}$ **ii** $3\mathbf{s} + 2\mathbf{r} + \mathbf{t}$
iii $\frac{1}{3}\mathbf{t}$ **iv** $\frac{2}{3}\mathbf{t}$
v $3\mathbf{t} - 2\mathbf{s}$

2 If $\mathbf{r} = \begin{pmatrix} 5 \\ 6 \end{pmatrix}$ and $\mathbf{s} = \begin{pmatrix} 3 \\ 4 \end{pmatrix}$, find:

a $\mathbf{r} - \mathbf{s}$ **b** $\mathbf{s} - \mathbf{r}$
c $2\mathbf{r} - \mathbf{s}$ **d** $3\mathbf{r} - 2\mathbf{s}$

Show also that $3(\mathbf{r} + \mathbf{s}) = 3\mathbf{r} + 3\mathbf{s}$

3 Find the values of x and y if:

a $\begin{pmatrix} x \\ y \end{pmatrix} + \begin{pmatrix} 2 \\ 3 \end{pmatrix} = \begin{pmatrix} 4 \\ 5 \end{pmatrix}$

b $\begin{pmatrix} 3 \\ x \end{pmatrix} + \begin{pmatrix} y \\ 2 \end{pmatrix} = \begin{pmatrix} 6 \\ ^-3 \end{pmatrix}$

c $3\begin{pmatrix} x \\ 2 \end{pmatrix} - 2\begin{pmatrix} 1 \\ y \end{pmatrix} = \begin{pmatrix} 7 \\ 4 \end{pmatrix}$

4 Given $\mathbf{a} = \begin{pmatrix} 4 \\ ^-3 \end{pmatrix}$, $\mathbf{b} = \begin{pmatrix} ^-6 \\ ^-5 \end{pmatrix}$ and $\mathbf{c} = \begin{pmatrix} ^-8 \\ ^-4 \end{pmatrix}$ find:

a $3\mathbf{a} + 2\mathbf{c}$
b $^-3\mathbf{c} + 4\mathbf{b}$
c $2\mathbf{a} - 2\mathbf{b} - \mathbf{c}$

5 If $\mathbf{a} = \begin{pmatrix} 4 \\ 3 \end{pmatrix}$, $\mathbf{b} = \begin{pmatrix} ^-4 \\ 2 \end{pmatrix}$,

$\mathbf{c} = \begin{pmatrix} ^-3 \\ ^-2 \end{pmatrix}$ and $\mathbf{d} = \begin{pmatrix} ^-6 \\ ^-5 \end{pmatrix}$

find:

a $\mathbf{a} - \mathbf{b}$ **b** $2\mathbf{a} + \mathbf{b}$
c $3\mathbf{a} + 2\mathbf{b}$ **d** $\mathbf{a} + 5\mathbf{b}$
e $^-2\mathbf{c} - 3\mathbf{d}$ **f** $2\mathbf{c} - 5\mathbf{d}$
g $\mathbf{a} - 2\mathbf{b} + \mathbf{c}$ **h** $\mathbf{a} - \mathbf{b} - 3\mathbf{c}$
i $^-\mathbf{a} - 4\mathbf{d} - 2\mathbf{c}$ **j** $\mathbf{a} - 2\mathbf{b} - 3\mathbf{c}$
k $^-2\mathbf{c} - \mathbf{d} - \mathbf{a}$ **l** $^-3\mathbf{d} - \mathbf{c} - \mathbf{b}$

16.2 Using vectors in geometry

You can use vectors to prove simple geometrical properties of shapes.

There are two basic ideas.

1

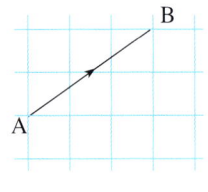

$$\overrightarrow{AB} = \overrightarrow{-BA}$$

2

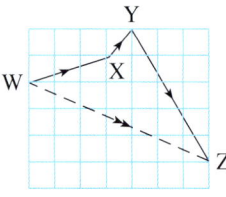

$$\overrightarrow{WX} + \overrightarrow{XY} + \overrightarrow{YZ} = \overrightarrow{WZ}$$

Example 4

Two lines, AC and DB, intersect at M. M is the midpoint of both AC and DB.

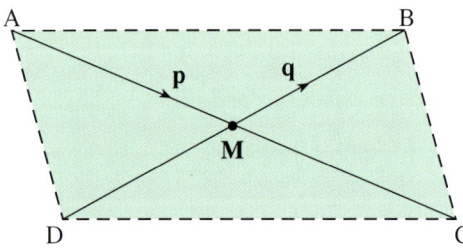

If $\overrightarrow{AM} = \mathbf{p}$ and $\overrightarrow{MB} = \mathbf{q}$ show that AB = DC and AB is parallel to DC.

....

$\overrightarrow{AB} = \overrightarrow{AM} + \overrightarrow{MB} = \mathbf{p} + \mathbf{q}$

As M is the midpoint of AC and DB,

$\overrightarrow{DM} = \overrightarrow{MB} = \mathbf{q}$ and $\overrightarrow{MC} = \overrightarrow{AM} = \mathbf{p}$

So, $\overrightarrow{DC} = \overrightarrow{DM} + \overrightarrow{MC} = \mathbf{q} + \mathbf{p}$

Hence $\overrightarrow{AB} = \overrightarrow{DC}$

Since the vectors $\overrightarrow{AB}$ and $\overrightarrow{DC}$ are equal, they have the same length (AB = DC) and the same direction, so they are parallel.

Exercise 16D

1

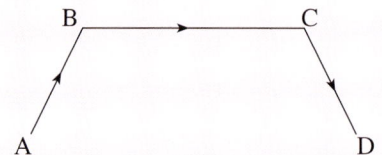

Copy and complete:

a $\overrightarrow{AB} + \overrightarrow{BC} = \square$

b $\overrightarrow{BC} + \overrightarrow{CD} = \square$

c $\overrightarrow{AB} + \overrightarrow{BC} + \overrightarrow{CD} = \square$

2

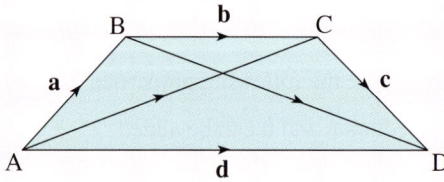

Copy and complete:

a $\overrightarrow{AB} + \overrightarrow{BD} = \square$

b $\overrightarrow{AC} + \overrightarrow{CD} = \square$

c $\overrightarrow{BA} + \overrightarrow{AD} + \overrightarrow{DC} = \square$

3 X and Y are the midpoints of AB and DC, which are sides of the parallelogram ABCD.

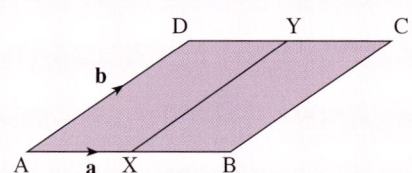

If $\overrightarrow{AX} = \mathbf{a}$ and $\overrightarrow{AD} = \mathbf{b}$ write in terms of $\mathbf{a}$ and $\mathbf{b}$:

a $\overrightarrow{AB}$ **b** $\overrightarrow{DY}$ **c** $\overrightarrow{AY}$

d $\overrightarrow{AC}$ **e** $\overrightarrow{YB}$ **f** $\overrightarrow{CA}$

4 Use the figure to complete:

a $\overrightarrow{AB} + \overrightarrow{BC} = \square$

b $\overrightarrow{CD} + \overrightarrow{DE} = \square$

c $\overrightarrow{AC} + \overrightarrow{CD} = \square$

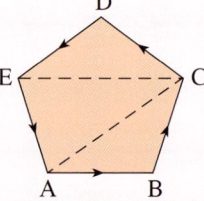

Now explain why

$\overrightarrow{AB} + \overrightarrow{BC} + \overrightarrow{CD} + \overrightarrow{DE} + \overrightarrow{EA}$ is the zero vector.

5 Draw any hexagon ABCDEF. Copy and complete:

a $\overrightarrow{AB} + \overrightarrow{BC} = \square$ **b** $\overrightarrow{AC} + \overrightarrow{CD} = \square$

c $\overrightarrow{AD} + \overrightarrow{DE} = \square$ **d** $\overrightarrow{AE} + \overrightarrow{EF} = \square$

Use your results to show that:

$\overrightarrow{AB} + \overrightarrow{BC} + \overrightarrow{CD} + \overrightarrow{DE} + \overrightarrow{EF} + \overrightarrow{FA} = \mathbf{0}$

6 Look at your answers for questions **4** and **5**. Explain why the sum of the vectors representing the sides of any closed polygon, taken in order, is always the zero vector.

7 $\overrightarrow{OA} = \mathbf{a}$ and $\overrightarrow{OB} = \mathbf{b}$
Write in terms of **a** and **b**:

a $\overrightarrow{AO}$

b $\overrightarrow{AB}$

c $\overrightarrow{BA}$

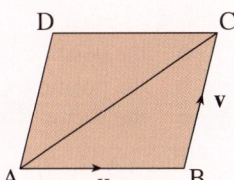

8 In the parallelogram ABCD, $\overrightarrow{AB} = \mathbf{u}$ and $\overrightarrow{BC} = \mathbf{v}$. Write in terms of **u** and **v**:

a $\overrightarrow{AD}$

b $\overrightarrow{CD}$

c $\overrightarrow{AC}$

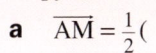

9 In the diagram, M and N are the midpoints of AB and AC respectively. Copy and complete:

a $\overrightarrow{AM} = \frac{1}{2}(\ \ )$

b $\overrightarrow{AN} = \frac{1}{2}(\ \ )$

c $\overrightarrow{AB} + \square = \overrightarrow{AC}$

d $\overrightarrow{AM} + \square = \overrightarrow{AN}$

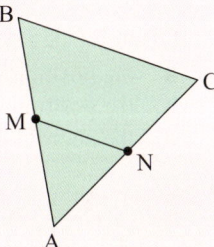

10 In the diagram in question **9**, if $\overrightarrow{AB} = \mathbf{p}$ and $\overrightarrow{AC} = \mathbf{q}$ write expressions in terms of **p** and **q** for:

a $\overrightarrow{AM}$ **b** $\overrightarrow{AN}$ **c** $\overrightarrow{BC}$ **d** $\overrightarrow{MN}$

Use your answers to show that $\overrightarrow{MN} = \frac{1}{2}\overrightarrow{BC}$

11 In question **9**, the vector representing MN is half the vector representing BC. Explain why this shows that MN is parallel to BC and why the length of MN is half that of BC.

12 OACB is a square with $\overrightarrow{OA} = 2\mathbf{a}$ and $OB = 3\mathbf{b}$. M is the midpoint of BC and N is a point one third of the way along AC.

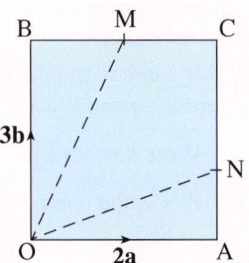

Write down in terms of **a** and **b** the vectors for:

a $\overrightarrow{BM}$ **b** $\overrightarrow{MC}$ **c** $\overrightarrow{AN}$ **d** $\overrightarrow{CN}$

Use your results to write down the vectors for:

e $\overrightarrow{OM}$ **f** $\overrightarrow{ON}$ **g** $\overrightarrow{BN}$ **h** $\overrightarrow{MN}$

13 In the triangle OAB, C is the midpoint of AB, $\overrightarrow{OA} = \mathbf{a}$ and $\overrightarrow{OB} = \mathbf{b}$

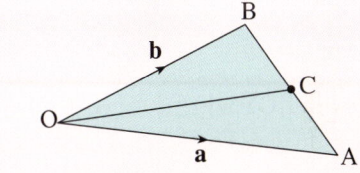

Write in terms of **a** and **b**:

a $\overrightarrow{AO}$ **b** $\overrightarrow{AB}$

c $\overrightarrow{AC}$ **d** $\overrightarrow{OC}$

Position vectors

Instead of using coordinates to describe the position of a point on a graph, you could use a **vector**.

Any point P in the plane defines the vector $\overrightarrow{OP}$ which joins the origin O to P. $\overrightarrow{OP}$ is called the **position vector** of P. The position vector for another point Q is $\overrightarrow{OQ}$

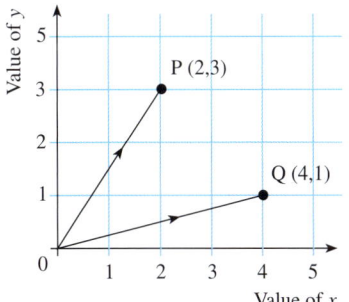

$$\overrightarrow{OP} = \begin{pmatrix} 2 \\ 3 \end{pmatrix}$$

$$\overrightarrow{OQ} = \begin{pmatrix} 4 \\ 1 \end{pmatrix}$$

The coordinates of P are (2, 3).

The position vector of $\mathbf{p} = \overrightarrow{OP} = \begin{pmatrix} 2 \\ 3 \end{pmatrix}$

A column vector is used for $\overrightarrow{OP}$ to make sure you don't confuse it with the coordinates of P.

Example 5

XY is a line with midpoint M, where X is the point (2, 3) and Y is the point (4, $^-$2)

a Write down the position vectors of X, Y and M.

b If N is the point (6, 4), find the components of $\overrightarrow{MN}$

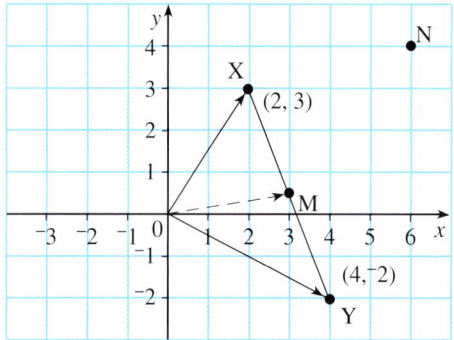

a $\overrightarrow{OX} = \begin{pmatrix} 2 \\ 3 \end{pmatrix}, \overrightarrow{OY} = \begin{pmatrix} 4 \\ -2 \end{pmatrix}$

M has coordinates $\left(\dfrac{2+4}{2}, \dfrac{3+(^-2)}{2} \right)$

$= (3, 0.5)$

So $\overrightarrow{OM} = \begin{pmatrix} 3 \\ 0.5 \end{pmatrix}$

b $\overrightarrow{ON} = \begin{pmatrix} 6 \\ 4 \end{pmatrix}$

$\overrightarrow{MN} = \overrightarrow{MO} + \overrightarrow{ON} = ^-\overrightarrow{OM} + \overrightarrow{ON}$

$= -\begin{pmatrix} 3 \\ 0.5 \end{pmatrix} + \begin{pmatrix} 6 \\ 4 \end{pmatrix} = \begin{pmatrix} 3 \\ 3.5 \end{pmatrix}$

The next example shows how position vectors can assist you when a line is divided in a ratio.

Example 6

Find the coordinates of the point K which divides the line PQ in the ratio 2:3

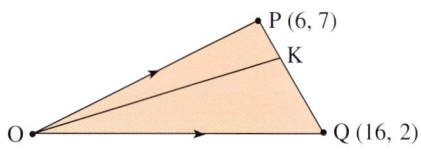

In the diagram PK : KQ = 2:3
So if PQ is divided into five parts, PK is two parts and KQ is three parts.
In other words PK = $\frac{2}{5}$ PQ

$\overrightarrow{OP} = \begin{pmatrix} 6 \\ 7 \end{pmatrix}, \overrightarrow{OQ} = \begin{pmatrix} 16 \\ 2 \end{pmatrix}$

$\overrightarrow{PQ} = \overrightarrow{PO} + \overrightarrow{OQ} = ^-\overrightarrow{OP} + \overrightarrow{OQ}$

$= ^-\begin{pmatrix} 6 \\ 7 \end{pmatrix} + \begin{pmatrix} 16 \\ 2 \end{pmatrix} = \begin{pmatrix} 10 \\ -5 \end{pmatrix}$

$\overrightarrow{OK} = \overrightarrow{OP} + \overrightarrow{PK} = \overrightarrow{OP} + \frac{2}{5}\overrightarrow{PQ}$

So $\overrightarrow{OK} = \begin{pmatrix} 6 \\ 7 \end{pmatrix} + \frac{2}{5}\begin{pmatrix} 10 \\ -5 \end{pmatrix}$

$= \begin{pmatrix} 6 \\ 7 \end{pmatrix} + \begin{pmatrix} 4 \\ -2 \end{pmatrix}$

$= \begin{pmatrix} 10 \\ 5 \end{pmatrix}$

The coordinates of K are therefore (10, 5).

Exercise 16E

1 A has coordinates (1, 3) and B has coordinates ($^-$1, 4) What is:
   a the position vector of A
   b the position vector of B
   c the vector $\overrightarrow{AB}$?

2 Given A (3, $^-$2) and $\overrightarrow{AB} = \begin{pmatrix} 2 \\ 4 \end{pmatrix}$ find:
   a the position vector of A
   b the position vector of B
   c the coordinates of B

3 Given B ($^-$3, 5) and $\overrightarrow{AB} = \begin{pmatrix} 7 \\ 3 \end{pmatrix}$ find:
   a the position vector of A
   b the position vector of B
   c the coordinates of A.

4
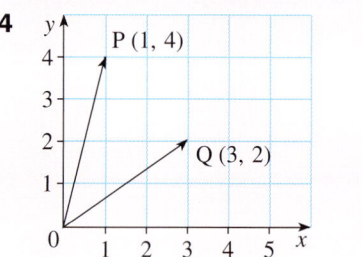

$\overrightarrow{OP} = \begin{pmatrix} 1 \\ 4 \end{pmatrix}$

$\overrightarrow{OQ} = \begin{pmatrix} 3 \\ 2 \end{pmatrix}$

   a For the diagram above, complete the statement: $\overrightarrow{PQ} = \overrightarrow{PO} + \square = {}^-\overrightarrow{OP} + \square$
   b Find the column vector for PQ.
   c If M is the midpoint of PQ, write down the column vector for $\overrightarrow{PM}$
   d Complete the statement $\overrightarrow{OM} = \overrightarrow{OP} + \square$.
   e Write down the position vector of M.

5 Compare the column vectors for $\overrightarrow{OP}$, $\overrightarrow{OQ}$ and $\overrightarrow{OM}$ in question **4**.
Can you find a quick way of obtaining $\overrightarrow{OM}$ from $\overrightarrow{OP}$ and $\overrightarrow{OQ}$?

6 In the diagram, O is the origin. R and S are points with position vectors **r** and **s**.
M is the midpoint of RS.

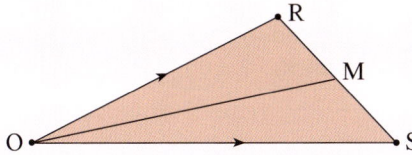

a Write $\overrightarrow{RS}$ in terms of **r** and **s**.
b By first finding $\overrightarrow{RM}$ write the position vector of M in terms of **r** and **s**.
c Use this result to check your answer for question **5**.

Consolidation

Example 1

Draw the column vectors:

a $\begin{pmatrix} 3 \\ -2 \end{pmatrix}$ **b** $\begin{pmatrix} -2 \\ -5 \end{pmatrix}$

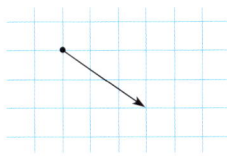

(3 across, 2 down)   (2 left across, 5 down)

Example 2

If $\mathbf{A} = \begin{pmatrix} 3 \\ -2 \end{pmatrix}$, $\mathbf{B} = \begin{pmatrix} 2 \\ 1 \end{pmatrix}$ and $\mathbf{C} = \begin{pmatrix} 0 \\ 2 \end{pmatrix}$ find:

a $\mathbf{A} + \mathbf{B} + \mathbf{C}$ **b** $3\mathbf{A} - 2\mathbf{B}$

a $\mathbf{A} + \mathbf{B} + \mathbf{C} = \begin{pmatrix} 3 \\ -2 \end{pmatrix} + \begin{pmatrix} 2 \\ 1 \end{pmatrix} + \begin{pmatrix} 0 \\ 2 \end{pmatrix}$

$= \begin{pmatrix} 5 \\ 1 \end{pmatrix}$

b $3\mathbf{A} - 2\mathbf{B} = 3\begin{pmatrix} 3 \\ -2 \end{pmatrix} - 2\begin{pmatrix} 2 \\ 1 \end{pmatrix}$

$= \begin{pmatrix} 9 \\ -6 \end{pmatrix} - \begin{pmatrix} 4 \\ 2 \end{pmatrix} = \begin{pmatrix} 5 \\ -8 \end{pmatrix}$

Example 3

In the parallelogram PQRS, $\overrightarrow{PQ} = A$ and $\overrightarrow{PS} = B$

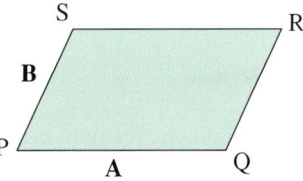

Write **a** $\overrightarrow{PR}$ and **b** $\overrightarrow{QS}$ in terms of **A** and **B**.

a $\overrightarrow{PR}$

$\overrightarrow{PR} = \overrightarrow{PQ} + \overrightarrow{QR}$

$= \mathbf{A} + \mathbf{B}$

b $\overrightarrow{QS}$

$\overrightarrow{QS} = \overrightarrow{QP} + \overrightarrow{PS}$

$= -\mathbf{A} + \mathbf{B}$

$= \mathbf{B} - \mathbf{A}$

Exercise 16

1 Draw these column vectors.

a $\begin{pmatrix} 2 \\ 3 \end{pmatrix}$ **b** $\begin{pmatrix} 3 \\ 5 \end{pmatrix}$ **c** $\begin{pmatrix} 2 \\ 0 \end{pmatrix}$

d $\begin{pmatrix} 0 \\ -2 \end{pmatrix}$ **e** $\begin{pmatrix} -2 \\ 1 \end{pmatrix}$ **f** $\begin{pmatrix} -3 \\ -1 \end{pmatrix}$

2 Given $\mathbf{A} = \begin{pmatrix} 1 \\ 3 \end{pmatrix}$, $\mathbf{B} = \begin{pmatrix} 2 \\ -1 \end{pmatrix}$ and $\mathbf{C} = \begin{pmatrix} -3 \\ -5 \end{pmatrix}$ find:

a $\mathbf{A} + \mathbf{B}$ **b** $3\mathbf{A}$
c $2\mathbf{B}$ **d** $\mathbf{A} - \mathbf{C}$
e $3\mathbf{A} + \mathbf{C}$ **f** $3\mathbf{A} + 2\mathbf{B}$
g $\mathbf{A} + \mathbf{B} - \mathbf{C}$ **h** $4\mathbf{C} - \mathbf{A}$
i $5\mathbf{C} + 2\mathbf{B}$

3 In triangle OAB, the midpoints of OA, OB and AB are W, X and Y respectively.

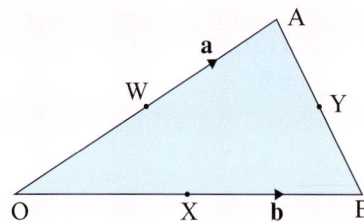

If $\overrightarrow{OA} = \mathbf{a}$ and $\overrightarrow{OB} = \mathbf{b}$

a Write down expressions for:

   i $\overrightarrow{OY}$ **ii** $\overrightarrow{AX}$ **iii** $\overrightarrow{BW}$

   in terms of **a** and **b**

b Show that $\overrightarrow{OY} + \overrightarrow{AX} + \overrightarrow{BW} = 0$

4 A ship leaves port and sails 5 km south and 4 km west. It then changes course and sails 2 km north and 3 km west.

a Using a scale of 1 cm = 1 km, draw the ship's course on a grid.

b Write down the ship's journey as the sum of two vectors.

c How far is the ship from port?

d The ship is now returning directly to port. Write down a vector that represents its track.

Summary

You should know ...

1 How to use a column vector to describe a translation.

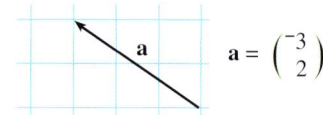

$$\mathbf{a} = \begin{pmatrix} ^-3 \\ 2 \end{pmatrix}$$

2 How to add or subtract vectors.
For example:

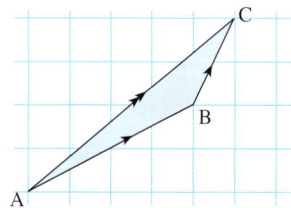

$$\overrightarrow{AB} + \overrightarrow{BC} = \begin{pmatrix} 4 \\ 2 \end{pmatrix} + \begin{pmatrix} 1 \\ 2 \end{pmatrix} = \begin{pmatrix} 5 \\ 4 \end{pmatrix} = \overrightarrow{AC}$$

Check out

1 Describe these translations as column vectors.

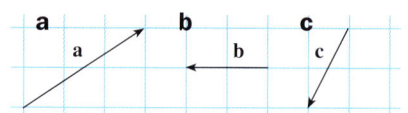

2 If $\mathbf{a} = \begin{pmatrix} 4 \\ 3 \end{pmatrix}$, $\mathbf{b} = \begin{pmatrix} ^-2 \\ 4 \end{pmatrix}$, $\mathbf{c} = \begin{pmatrix} ^-3 \\ ^-2 \end{pmatrix}$ find the value of:

a $\mathbf{a} + \mathbf{b}$

b $\mathbf{a} - \mathbf{b}$

c $2\mathbf{a} - 3\mathbf{c}$

d $3\mathbf{a} + \mathbf{b} - \mathbf{c}$

3 How vectors can be used to solve geometric problems.
For example:

In the diagram $\overrightarrow{AD}$ can be written in terms of **a** and **b**.

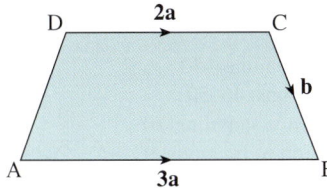

$$\overrightarrow{AD} = \overrightarrow{AB} + \overrightarrow{BC} + \overrightarrow{CD}$$
$$= 3\mathbf{a} - \mathbf{b} - 2\mathbf{a}$$
$$= \mathbf{a} - \mathbf{b}$$

4 Any point can be represented by a position vector.
For example:

the point P (a, b) has position vector $\overrightarrow{OP} = \begin{pmatrix} a \\ b \end{pmatrix}$

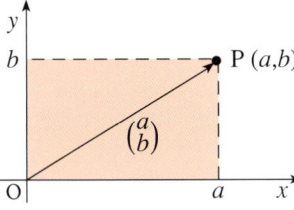

3

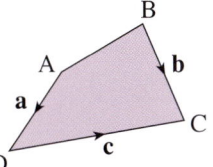

ABCD is a quadrilateral with
$\overrightarrow{AD} = \mathbf{a}$, $\overrightarrow{BC} = \mathbf{b}$ and $\overrightarrow{DC} = \mathbf{c}$

a Find in terms of **a**, **b** and **c**:
  i $\overrightarrow{AC}$
  ii $\overrightarrow{AB}$

b If M is the midpoint of DC,
find $\overrightarrow{AM}$.

4 a Write down the position vectors of
the points A (2, 3) and B (6, 1).

b Hence find the position vector of the
midpoint, M, of AB.

Index